Dynamics and Nonlinear Control of Integrated Process Systems

Presenting a systematic model reduction and hierarchical controller design framework for broad classes of integrated process systems encountered in practice, this book first studies process systems with large material recycle and/or with small purge streams, followed by systems with energy integration. Step-by-step model reduction procedures are developed to derive nonlinear reduced models of the dynamics in each time scale. Hierarchical control architectures, consisting of coordinated levels of control action in different time scales, are proposed for each class of process systems considered in order to enforce stability, tracking performance, and disturbance rejection. Numerous process applications are discussed in detail to illustrate the application of the methods and their potential to improve process operations. Matlab codes are also presented to guide further application of the methods developed and facilitate practical implementations.

Michael Baldea is Assistant Professor in the Department of Chemical Engineering at The University of Texas at Austin. Prior to joining The University of Texas, he held industrial research positions with Praxair Technology Center in Tonawanda, NY and GE Global Research in Niskayuna, NY. He has received several research and service awards, including the Model Based Innovation Prize from Process Systems Enterprise and the Best Referee Award from the *Journal of Process Control*, and has co-authored over 60 papers and presentations.

Prodromos Daoutidis is Professor in the Department of Chemical Engineering and Materials Science at the University of Minnesota. He also held a position as Professor at the Aristotle University of Thessaloniki in 2004–2006. He is the recipient of several research and teaching awards and recognitions, including the NSF CAREER Award, the Model Based Innovation Prize from Process Systems Enterprise, the Ted Peterson Award of CAST, the George Taylor Career Development Award, the Mc'Knight Land Grant Professorship, the Ray D. Johnson/ Mayon Plastics Professorship, and the Shell Chair at the University of Minnesota. He has also been a Humphrey Institute Policy Fellow. He has co-authored two books and 175 refereed papers.

Cambridge Series in Chemical Engineering

Books in Series

Baldea and Daoutidis, *Dynamics and Nonlinear Control of Integrated Process Systems*
Chau, *Process Control: A First Course with MATLAB*
Cussler, *Diffusion: Mass Transfer in Fluid Systems, Third Edition*
Cussler and Moggridge, *Chemical Product Design, Second Edition*
Denn, *Chemical Engineering: An Introduction*
Denn, *Polymer Melt Processing: Foundations in Fluid Mechanics and Heat Transfer*
Duncan and Reimer, *Chemical Engineering Design and Analysis: An Introduction*
Fan and Zhu, *Principles of Gas–Solid Flows*
Fox, *Computational Models for Turbulent Reacting Flows*
Leal, *Advanced Transport Phenomena: Fluid Mechanics and Convective Transport*
Mewis and Wagner, *Colloidal Suspension Rheology*
Morbidelli, Gavriilidis and Varma, *Catalyst Design: Optimal Distribution of Catalyst in Pellets, Reactors, and Membranes*
Noble and Terry, *Principles of Chemical Separations with Environmental Applications*
Orbey and Sandler, *Modeling Vapor–Liquid Equilibria: Cubic Equations of State and their Mixing Rules*
Petlyuk, *Distillation Theory and its Applications to Optimal Design of Separation Units*
Rao and Nott, *An Introduction to Granular Flow*
Russell, Robinson and Wagner, *Mass and Heat Transfer: Analysis of Mass Contactors and Heat Exchangers*
Slattery, *Advanced Transport Phenomena*
Varma, Morbidelli and Wu, *Parametric Sensitivity in Chemical Systems*

Dynamics and Nonlinear Control of Integrated Process Systems

MICHAEL BALDEA
The University of Texas at Austin

PRODROMOS DAOUTIDIS
University of Minnesota

CAMBRIDGE
UNIVERSITY PRESS

CAMBRIDGE UNIVERSITY PRESS
Cambridge, New York, Melbourne, Madrid, Cape Town, Singapore,
São Paulo, Delhi, Mexico City

Cambridge University Press
The Edinburgh Building, Cambridge CB2 8RU, UK

Published in the United States of America by Cambridge University Press, New York

www.cambridge.org
Information on this title: www.cambridge.org/9780521191708

First published 2012

Printed and Bound in the United Kingdom by the MPG Books Group

A catalogue record for this publication is available from the British Library

Library of Congress Cataloguing in Publication data

Baldea, Michael, author.
Dynamics and nonlinear control of integrated process systems / Michael Baldea, Prodromos Daoutidis.
pages cm. – (Cambridge series in chemical engineering)
ISBN 978-0-521-19170-8 (Hardback)
1. Chemical process control. 2. Systems engineering. 3. Nonlinear control theory. I. Daoutidis, Prodromos, author. II. Title.
TP155.75.B35 2012
515′.724–dc23

2012018949

ISBN 978-0-521-19170-8 Hardback

To our families

Contents

Preface

The chemical process industry is an intensely competitive environment, where cost reduction represents a critical factor towards increasing profit margins. Over the last few decades, an ever growing need to lower utility costs and energy consumption, and to improve raw material use, has spurred the development and implementation of increasingly integrated process designs that make extensive use of material recycling and energy recovery.

The significant reduction in capital and operating costs associated with process integration does, however, come at the price of additional operational and control challenges. Research on the control of interconnected process systems and entire chemical plants has been driven both by developments in control and optimization theory, and by shifts in market demands and industry needs. Initial efforts focused on decentralized multi-loop control structures and on including plant-wide considerations in the tuning of PID controllers. The associated benefits dwindled, however, with the rise of modern, tightly integrated processes with strong dynamic coupling between the different process units. More recently, control systems developed within the linear model predictive control (MPC) paradigm have allowed centralized decision making and accounting for economic optimality under operating constraints. In the (petro)chemical industry, MPC remains the established means for regulatory control and plant operation around a given steady state.

The current economic environment is, however, highly dynamic. Economically optimal plant operations thus entail frequent switching among different operating conditions (i.e., different steady states), having different product grades and production rates. Adopting or adapting the existing fully centralized or completely decentralized control designs for enforcing such transitions is neither practical nor effective in the context of integrated processes, where the interactions between the process units become significant and unique dynamic features emerge.

Developed around an extensive body of recent research by the authors, this book provides a new paradigm for the effective control of tightly integrated process systems, by

- documenting rigorously the dynamic behavior that emerges at the plant level when tight integration through material recycling and energy recovery is employed
- presenting the means for deriving explicit and physically meaningful low-dimensional models of the dominant plant dynamics
- describing a hierarchical controller design framework that discerns and coordinates between regulatory control at the unit level and supervisory, plant-wide control, and enables the design of nonlinear controllers for enforcing plant-wide transitions
- illustrating the application of the theoretical concepts to several integrated processes found in the chemical and energy industries

The chapters strive to balance rigor and practicality. The systematic analysis of generic, prototypical processes that exemplify the process integration structures encountered in practice is emphasized together with the unique dynamic features and control challenges that they present. Illustrative examples and extensive case studies on specific problems support the theoretical developments and provide a practical vista. The text adopts a unique and quintessentially chemical engineering perspective by introducing the concept of a process-level dimensionless number to characterize process integration from both a process *design* and a process *control* point of view. We are hopeful that our approach will allow readers to rapidly master the underlying theory and develop extensions to other classes of problems. Implementation details (sample computer codes) are provided in order to further encourage the rapid deployment of practical applications.

The book targets graduate students and researchers interested in dynamics and control, as well as practitioners involved in advanced control in industry. It can serve as a reference text in an advanced process systems engineering or process control course and as a valuable resource for the researcher or practitioner. Written at a basic mathematical level (and largely self-contained from a mathematical point of view), the material assumes some familiarity with process modeling and an elementary background in nonlinear dynamical systems and control.

We are grateful to our colleagues at the Department of Chemical Engineering and Materials Science at Minnesota for maintaining an environment of scientific excellence and collegiality over the years. M.B. is also grateful to the fellow researchers at the Praxair Technology Center in Tonawanda, NY for creating an intellectually stimulating atmosphere. We owe special thanks to Ed Cussler for his advice and encouragement in the initial stages of the writing of this book, the staff at Cambridge University Press for their support and advice, and the National Science Foundation for the support it provided for the research that formed the basis for this book. We also owe a special note of appreciation to Aditya Kumar for his instrumental role in the initial phase of research on this subject, and to Sujit Jogwar, whose recent work further solidified the basic thesis and direction of the book.

This book is dedicated to my parents, with gratitude for their unconditional love and support, and to the memory of my grandparents, who fondly followed my childhood scientific pursuits.

M.B.

I dedicate this book to my wife Aphrodite for her uncompromising pursuit of beauty in all aspects of our life, and to my children Stylianos and Euphrosyne for the immeasurable joy and inspiration they bring.

P.D.

Part I

Preliminaries

1 Introduction

Integrated process systems, such as the one in Figure 1.1, consisting of multiple reaction and separation units, heat integrated and interconnected through material recycle streams, represent the rule rather than the exception in the modern process industries. The dynamics and control of such systems present distinct challenges: in addition to the nonlinear behavior of the individual units, the feedback interactions caused by the recycle connections typically give rise to a more complex, *overall* process dynamics. The use of design modifications, such as surge tanks and unit oversizing, and the choice of mild operating conditions, preventing the propagation of disturbances through the plant, initially allowed the problem of controlling chemical plants with material recycling to be dealt with at the unit level, using the "unit operations" approach (Umeda *et al.* 1978, Stephanopoulos 1983): control loops were designed for each unit, their tuning being subsequently adjusted to improve the operation of the entire plant. However, the shortage of raw materials, rising energy prices, and the need to lower capital costs have, over the past few decades, spurred the process industry's tendency to build "lean,"[1] integrated plants, relying heavily on material recycles and energy recovery.

Owing to dwindling fossil-fuel supplies (and the associated increase in the cost of energy), improving energy efficiency has become particulary important. Energy integration and recovery are key enablers to this end. Fundamentally, energy integration involves identifying the energy sources and sinks *within* a system and establishing the means for energy transfer between them,[2] thereby reducing the use of external energy sources and utility streams. Chemical reactors and distillation columns inherently contain such sources and sinks and clearly constitute prime targets for energy integration. Numerous energy-integrated process configurations have been proposed at the conceptual level: reactor-feed effluent heat exchanger systems, heat exchanger networks, heat-integrated and thermally coupled distillation columns, etc.

The design and optimization of energy integration schemes has been an active research area from the early days of process systems engineering. Initial efforts (Rathore *et al.* 1974, Sophos *et al.* 1978, Nishida *et al.* 1981) focused on the

[1] With little, if any, design margin (Stephanopoulos 1983).
[2] Assuming, of course, that such transfer is thermodynamically feasible.

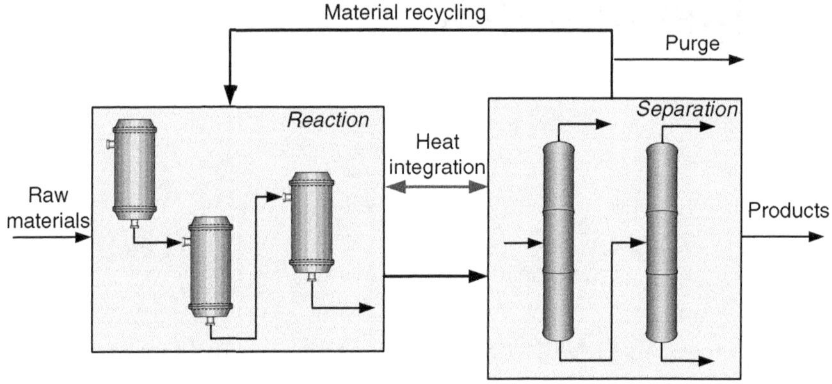

Figure 1.1 An integrated process system.

synthesis of energy-integrated processes using heuristics. Later, pinch analysis (Linnhoff and Hindmarsh 1983, Linnhoff *et al.* 1983) and bounding techniques for utility usage (Morari and Faith III 1980, Andrecovich and Westerberg 1985a, Mészáros and Fonyó 1986) were introduced, and they have since seen numerous successful applications in the synthesis of new energy integration systems as well as in plant retrofits. Mathematically rigorous formulations such as mixed-integer linear/nonlinear programming (Andrecovich and Westerberg 1985b, Floudas and Paules 1988, Yeomans and Grossmann 1999, Wei-Zhong and Xi-Gang 2009) and genetic algorithms (Wang *et al.* 1998, Yu *et al.* 2000, Wang *et al.* 2008) were subsequently developed to ensure the *optimality* of integrated processes. The significant reduction in capital and operating costs resulting from energy integration is now well documented (Muhrer *et al.* 1990, Yee *et al.* 1990, Annakou and Mizsey 1996, Reyes and Luyben 2000b, Westerberg 2004, El-Halwagi 2006, Diez *et al.* 2009).

As integrated process designs continued to gain acceptance owing to their improved economics, the process control community also became aware of the distinct challenges posed by the operation of such plants, and a number of research studies ensued.

An initial theoretical study (Gilliland *et al.* 1964) established that, for a simple plant model consisting of a continuous stirred-tank reactor (CSTR) and a distillation column, the material recycle stream increases the sensitivity to disturbances together with increasing the time constant of the overall plant over those of the individual units. Moreover, it was shown that in certain cases the plant can become unstable even if the reactor itself is stable.

Several papers have since focused on either reaction–separation–recycle processes (Verykios and Luyben 1978, Denn and Lavie 1982, Luyben 1993a, Scali and Ferrari 1999, Lakshminarayanan *et al.* 2004) or individual multi-stage processes (Kapoor *et al.* 1986) and have shown that recycle streams can "slow down" the overall process dynamics (described by a small number of time constants) compared with the dynamics of the individual units, and may even lead to the recycle loop being unstable. An analogy was drawn (Denn and Lavie 1982) between the recycle system and a closed-loop system with positive feedback, thus concluding that the presence of a recycle stream may increase the overall response time of the plant and may increase the steady-state gain by a significant amount. The effect of the recycle on the zero dynamics was studied (Jacobsen 1999), and it was demonstrated that the feedback effect of the recycle stream can induce a non-minimum-phase behavior even for the transfer function of single units. Most of the aforementioned analyses were based on simplified transfer function models and linear analysis tools. More recently, a number of studies (Morud and Skogestad 1994, Mizsey and Kalmar 1996, Bildea *et al.* 2000, Pushpavanam and Kienle 2001, Kiss *et al.* 2002, Larsson *et al.* 2003, Kiss *et al.* 2005, Vasudevan and Rangaiah 2009) have indicated that, even in simple, prototype models of reactor–separator systems, the recycle stream can lead to strongly nonlinear overall dynamics, manifested in the form of multiple steady states, limit cycles or even chaotic behavior (Jacobsen and Berezowski 1998). The above results indicate that recycle streams are responsible for the complex behavior of process systems, and place the control of recycle loops at the heart of the plant-wide control problem.

The necessity to develop systematic procedures for coordinating distributed (i.e., unit-level) and plant-wide control objectives and strategies was thus acknowledged, and several studies have been dedicated to this purpose. Dynamic process control (DPC) (Buckley 1964) was the first control strategy to divide the control actions for a process plant (with or without recycle streams) into two categories: material-balance control (necessary for the management of the plant's operation in the presence of *low-frequency* (slow) changes, such as production flow rate), and product-quality control (for countering the effects of *high-frequency* (fast) disturbances acting at the unit level). Although it was a pioneering effort at the time, DPC is not effective in modern, tightly integrated plants, where the strong coupling induced by mass and energy recycling leads to the propagation of disturbances across the frequency spectrum through multiple process units.

Later on, the complexities introduced by process integration were fully acknowledged by researchers in the field, and motivated a series of studies on the effect of the material recycle streams on the design, controllability, and control structure selection for specific reaction/separation processes.

Luyben (1993a) provided valuable insights into the characteristics of recycle systems and their design, control, and economics, and illustrated the challenges caused by the feedback interactions in such systems, within a multi-loop linear control framework. Also, in the context of steady-state operation, it was shown (Luyben 1994) that the steady-state recycle flow rate is very sensitive to disturbances in feed flow rate and feed composition and that, when certain control configurations are used, the recycle flow rate increases considerably facing feed flow rate disturbances. This behavior was termed "the snowball effect."

The publication of an actual industrial plant-wide control problem, the Tennessee Eastman challenge process (Downs and Vogel 1993) generated several valuable studies on the control of recycle processes, both within a linear control framework (McAvoy and Ye 1994, Banerjee and Arkun 1995, Lyman and Georgakis 1995, Ricker 1996, Wu and Yu 1997, Larsson *et al.* 2001, Wang and McAvoy 2001, Tian and Hoo 2005) and within a nonlinear (Ricker and Lee 1995) control framework.

The control challenges posed by the feedback interactions induced by the recycle were also illustrated in studies carried out on other problems, such as supercritical fluid extraction (Ramchandran *et al.* 1992) and recycle reactors (Kanadibhotla and Riggs 1995, Antoniades and Christofides 2001).

The above results have revealed that process integration severely limits the effectiveness of the traditional, unit-operations approach, with fully decentralized controllers for individual process units, which assumes that the combination of these controllers (possibly with some adjustments) would constitute an effective control scheme for the overall plant. The strong coupling between the control loops in different process units in an integrated process system was thus recognized early on (Foss 1973) as a major issue that must be addressed in a plant-wide control setting, and several generic strategies to this end have been proposed.

Drawing on the ideas of Buckley (1964), Price and Georgakis (1993) provided guidelines for designing inventory-control structures that are consistent with the main mass and energy flows of the process, surmising that the best performance is achieved when some empirically selected control loops are tightly tuned and the others have loose tuning. Banerjee and Arkun (1995) presented a procedure for screening possible control configurations for a plant, using linearized models for assessing the robustness of the control loops, without specifically accounting for the presence of mass or energy recycles. Georgakis (1986) suggested the use of empirically identified extensive fast and slow variables for the synthesis of controllers for a process. In Ng and Stephanopoulos (1996), a hierarchical procedure for plant-wide controller synthesis is proposed, recommending a multiple-time-horizon control structure, with the longest horizon being that of the plant itself. Luyben *et al.* (1997) presented a tiered, heuristic controller design procedure for process systems that addresses both energy management and inventory and product purity control. A multi-step heuristic design procedure was also introduced in Larsson and Skogestad (2000), advocating a top-down plant analysis for identifying control objectives, followed by a bottom-up controller implementation. A set of criteria for designing and assessing the performance of plant-wide controllers has been proposed in Vasudevan and Rangaiah (2010).

In a different vein, Kothare *et al.* (2000) formally defined the concept of partial control on the basis of the practical premise that, in some cases, complex chemical processes can be reasonably well controlled by controlling only a small subset of the process variables, using an equally small number of "dominant" manipulated variables. An analysis method for identifying the dominant variables of a process was proposed in Tyreus (1999).

Using the concept of passivity, Farschman *et al.* (1998), Ydstie (2002), Jillson and Ydstie (2007), Bao and Lee (2007), Rojas *et al.* (2009) introduced a formal framework for stability analysis and stabilization of process systems using decentralized control, subject to thermodynamic and equipment constraints. Within this context, the passivity/dissipativity properties of individual units in a process are established using thermodynamic arguments, and existing results for the interconnections of passive/dissipative systems (*e.g.*, Desoer and Vidyasagar 2009) are used to determine the closed-loop stability properties of the overall process. Within this framework, the stabilization of the process dynamics is achieved via decentralized inventory controllers.

Following the ideas of Morari *et al.* (1980), Skogestad (2000, 2004), and Downs and Skogestad (2009) proposed an algorithm for determining a "self-optimizing" plant-wide control structure, consisting of identifying a set of controlled variables that, when kept at constant setpoints, indirectly lead to near-optimal operation with respect to a given economic objective. The proposed approach relies on steady-state optimization and thus additional simulation steps are needed in order to select the control structure with the best dynamic performance.

A hierarchical decision procedure for formulating control structures on the basis of the minimization of economic penalties, while also accounting for the process dynamics, was also proposed in Zheng et al. (1999), following Douglas's hierarchical method for conceptual process design (Douglas 1988). However, the formulated control structures often require that additional surge capacities be provided/installed in the process in order to achieve reasonable dynamic performance, and may therefore increase the capital cost of the plant.

McAvoy (1999) advanced the use of optimization calculations at the controller design stage, proposing the synthesis of plant-wide control structures that ensure minimal actuator movements. The initial work relying on steady-state models (McAvoy 1999) was recast into a controller synthesis procedure based on linear dynamic plant models (Chen and McAvoy 2003, Chen *et al.* 2004), whereby the performance of the generated plant-wide control structures was evaluated through dynamic simulations.

The plant-wide control techniques referenced above are generally based on the use of linear, multi-loop, decentralized control structures. Model predictive control (MPC) constitutes a different class of control techniques, consisting of determining the manipulated inputs of a process by minimizing an objective function capturing either the deviation between the process states and the corresponding setpoints (Prett and Garcia 1988) or an economic objective (Edgar 2004, Diehl *et al.* 2011), possibly under the physical constraints associated with the plant operation, over a receding time horizon. MPC can be applied to plant-wide control problems, having multivariable control and constraint-handling capabilities. However, calculating the manipulated inputs involves the solution of an often computationally expensive optimization problem (owing to the use of high-dimensional plant models in the problem formulation) at each time step, and, although they are numerous (Qin and Badgwell 2003), successful practical implementations have been confined to the realm of plants with *slow dynamics*, such as oil refineries.

A more recent direction relies on the use of *distributed* model-based control strategies as an alternative to centralized controllers (based on the full plant model) for large, integrated systems. Local controller design has been approached both via MPC techniques (see, e.g., Zhu *et al.* 2000, Zhu and Henson 2002, Venkat *et al.* 2006, 2008, Rawlings and Stewart 2008, Liu *et al.* 2008, 2009, Scattolini 2009, Stewart *et al.* 2010) and as an agent-based problem (e.g., Tatara *et al.* 2007, Tetiker *et al.* 2008). Typically, the analysis and implementation of distributed architectures considers the plant as a set of interconnected subsystems, with each subsystem being assumed to have a controller that exchanges (some of the) subsystem state information with the controllers of all the other subsystems. Within the distributed MPC framework, it has been shown that predictive control applications are possible for large plants with *fast dynamics*, since closed-loop stability is assured at all times by formulating the optimization problem to be feasible at every iteration.

The challenge posed by establishing and maintaining communication between distributed controllers has also stimulated research in the area of networked process control (El-Farra *et al.* 2005, Mhaskar *et al.* 2007, Sun and El-Farra 2008, 2010). The central issue of maintaining closed-loop stability in the presence of bandwidth constraints and limitations in transmitter battery longevity is typically addressed by a judicious distribution of computation and communication burdens between local/distributed control systems and a centralized supervisory controller.

In general, MPC implementations (including those cited above) rely on the use of data-driven linear plant models for computing the optimal plant inputs. However, chemical processes are inherently nonlinear, and these models lose accuracy when economic circumstances call for operating the process under conditions that differ significantly from the operating region in which model identification was carried out. The implementation of MPC to processes with nonlinearities (nonlinear MPC, NMPC) remains one of the most difficult problems associated with plant-wide MPC applications: because NMPC relies on using a nonlinear dynamic model, a nonlinear optimization problem must be solved at each time step in order to calculate the optimal plant inputs, and the computation time scales very unfavorably with the dimension of the plant model. To date, NMPC implementations for integrated processes (e.g., Ricker and Lee 1995, Zhu and Henson 2002) have made extensive use of modeling and controller simplifications in order to reduce computational complexity.

Many of the aforementioned heuristic decentralized control synthesis approaches rely on engineering judgement rather than rigorous analysis. On the other hand, the implementation of advanced, model-based, control strategies for process systems is hindered by the often overwhelming size and complexity of their dynamic models. The results cited above indicate that the design of fully centralized controllers on the basis of entire process models is impractical, such controllers being almost invariably ill-conditioned, difficult to tune, expensive to implement and maintain, and sensitive to measurement errors and noise. Thus, the need to find a rational and transparent paradigm for synthesizing process-wide model-based nonlinear control structures has emerged as (and remains) a key issue in modern process control. This need is also an integral part of the ongoing smart manufacturing initiative of twenty-first-century industry (Christofides *et al.* 2007, Edgar and Davis 2009).

A salient feature of integrated process systems is their *multiple-time-scale behavior*, owing to physical and chemical phenomena that occur at vastly different rates, a feature that translates into their dynamic models being described by stiff systems of differential equations. Stiffness represents in effect one of the main hindrances to the implementation of plant-wide model-based control techniques. It is at the origin of the ill conditioning of linear and nonlinear inversion-based and optimization-based controller designs, and greatly increases the difficulty of obtaining a numerical solution for optimal control problems.[3]

Although repeatedly acknowledged (directly or unwittingly) in plant-wide control studies (Buckley 1964, Georgakis 1986, Price and Georgakis 1993, Ng and Stephanopoulos 1996, Wang and McAvoy 2001, Lakshminarayanan *et al.* 2004), the issue of time-scale multiplicity at the plant level has not been accounted for in a mathematically rigorous way until recently (Kumar and Daoutidis 2002, Baldea and Daoutidis 2007, Jogwar *et al.* 2009). The goal of this text is thus to explain the origin of time-scale multiplicity at the process level, and to elucidate its impact on the development of systematic, hierarchical controller design procedures for the control of integrated process systems featuring material recycling and/or energy recovery. To this end, we will make use of generic, prototype systems that are representative for the design and operation of broad classes of integrated processes. Moreover, we will introduce a novel set of *process-level dimensionless numbers* that capture the salient *steady-state* design features of the processes under consideration, and establish a connection between these design features and process *dynamics and control.* Our goal is therefore to develop fundamental, rather than heuristic, results that are widely applicable in process systems engineering and beyond our discipline. Evidently, we illustrate the use of these results through numerous examples as well as an extensive case study at the end of each chapter.

The book is organized as follows. Chapter 2 provides an introduction to the mathematical description of multiple-time-scale systems and to singular

[3] The term *ill conditioning* refers to the condition number of the linearized model of a plant, defined as $\gamma = \lambda_{\max}/\lambda_{\min}$, with λ being the eigenvalues of the model. For large values of γ, the plant dynamics will span more time scales (its time constants being defined as the reciprocals of the eigenvalues), and the larger γ is, the more ill-conditioned (stiff) the plant is considered to be. By way of consequence, model-based controllers that are designed on the basis of inverting the (linear or nonlinear) plant model will be ill-conditioned as well. Ill-conditioned controllers tend to amplify disturbances and modeling errors, and even induce closed-loop instability.

perturbation theory used in their analysis. Chapter 3 discusses the design, dynamics and control of integrated process systems with significant material recycle streams. Chapter 4 focuses on processes with *small* purge streams (an important and common feature in chemical plants). Chapter 5 provides a modeling and model reduction framework for process systems featuring purge streams *and* large material recycle streams. The impact of energy recovery on process dynamics and control is analyzed in Chapter 6, while Chapter 7 concentrates on the dynamic behavior of process systems with high energy throughput.

2 Singular perturbation theory

2.1 Introduction

The review in the previous chapter pointed out that, while long acknowledged, the multiple-time-scale dynamic behavior of integrated chemical plants has been dealt with mostly empirically, both from an analysis and from a control point of view. In the remainder of the book, we will develop a mathematically rigorous approach for identifying the causes, and for understanding and mitigating the effects of time-scale multiplicity at the process system level.

The present chapter introduces the reader to singular perturbation theory as the framework for modeling and analyzing systems with multiple-time-scale dynamics, which we will make extensive use of throughout the text.

2.2 Properties of ODE systems with small parameters

The analysis of ordinary differential equation (ODE) systems with small parameters ε (with $0 < \varepsilon \ll 1$) is generally referred to as perturbation analysis or perturbation theory. Perturbation theory has been the subject of many fundamental research contributions (Fenichel 1979, Ladde and Siljak 1983), finding applications in many areas, including linear and nonlinear control systems, fluid mechanics, and reaction engineering (see, e.g., Kokotović *et al.* 1986, Kevorkian and Cole 1996, Verhulst 2005). The main concepts of perturbation theory are presented below, following closely the developments in (Kokotović *et al.* 1986).

Let us consider the following system of equations:

$$\begin{aligned} \frac{d\mathbf{x}_1}{dt} &= \mathbf{f}(\mathbf{x}_1, \mathbf{x}_2), \ \mathbf{x}_1(0) = \mathbf{x}_1^0 \\ \frac{d\mathbf{x}_2}{dt} &= \mathbf{g}(\mathbf{x}_1, \mathbf{x}_2), \ \mathbf{x}_2(0) = \mathbf{x}_2^0 \end{aligned} \tag{2.1}$$

where $\mathbf{f}(\mathbf{x}_1, \mathbf{x}_2)$ and $\mathbf{g}(\mathbf{x}_1, \mathbf{x}_2)$ are assumed to be sufficiently many times differentiable with respect to their variables $\mathbf{x}_1$ and $\mathbf{x}_2$. For our purposes, we can assume that $\mathbf{x}_1 \in \mathbb{R}^n$, $\mathbf{x}_2 \in \mathbb{R}^m$ (and hence $\mathbf{x} = [\mathbf{x}_1\ \mathbf{x}_2]^{\mathrm{T}} \in \mathbb{R}^{n+m}$), $\mathbf{f} : \mathbb{R}^n \rightarrow \mathbb{R}^n$, and $\mathbf{g} : \mathbb{R}^m \rightarrow \mathbb{R}^m$.

Equation (2.1) is an ODE system, and, since the values of the variables $\mathbf{x}_1$ and $\mathbf{x}_2$ at $t=0$ are provided, it is an initial value problem. By employing a small perturbation parameter $0 < \varepsilon \ll 1$, (2.1) and implicitly its solution are perturbed. The perturbation can occur in different manners. Consider, for example,

$$\begin{aligned}\frac{d\mathbf{x}_1}{dt} &= \mathbf{f}(\mathbf{x}_1, \mathbf{x}_2) + \varepsilon \mathbf{f}_1(\mathbf{x}_1, \mathbf{x}_2), \ \mathbf{x}_1(0) = \mathbf{x}_1^0 \\ \frac{d\mathbf{x}_2}{dt} &= \mathbf{g}(\mathbf{x}_1, \mathbf{x}_2) + \varepsilon \mathbf{g}_1(\mathbf{x}_1, \mathbf{x}_2), \ \mathbf{x}_2(0) = \mathbf{x}_2^0\end{aligned} \tag{2.2}$$

Equation (2.2) is said to be a *regular perturbation problem*. Notice that in the limiting case, as $\varepsilon \to 0$, the regular perturbation problem reduces to the original problem (2.1). Intuitively, the solution of the regular perturbation problem should not differ significantly from that of the unperturbed problem. For example, for $n=1, m=0$, the solution of Equation (2.2) is of the form

$$\mathbf{x}_1(t,\varepsilon) = \mathbf{x}_{1,0}(t) + \varepsilon \mathbf{x}_{1,1}(t) + \cdots \tag{2.3}$$

The solution (2.3) is known as a regular perturbation expansion. $\mathbf{x}_{1,0}(t)$ is the solution of the original problem (2.1), and the higher-order terms $\mathbf{x}_{1,1}(t), \ldots$ are determined successively by substituting the regular expansion (2.3) into the original differential equation (2.1) (Haberman 1998).

Example 2.1. A storage tank of cross-sectional area $A_t = 1\ \text{m}^2$ (Figure 2.1) is fed at the top at a flow rate $F_0 = 0.221\,47\ \text{m}^3/\text{s}$ with a liquid of density $\rho = 1000\ \text{kg/m}^3$. The liquid drains under gravity via a pipe of cross-section $A_1 = 0.05\ \text{m}^2$ located at the bottom of the tank. We will compute the evolution of the tank level h, starting from an empty tank ($h = 0$ m) under the above conditions, comparing the results with a case in which the bottom of the tank leaks via a small fracture of cross-sectional area $A_2 = 0.0005\ \text{m}^2$.

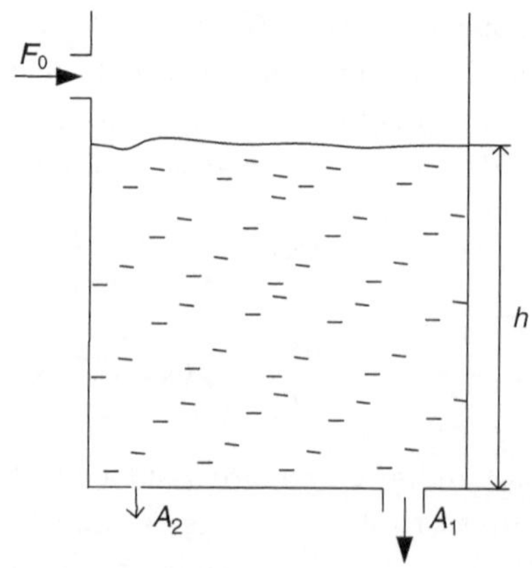

Figure 2.1 A gravity-drained tank with a leak.

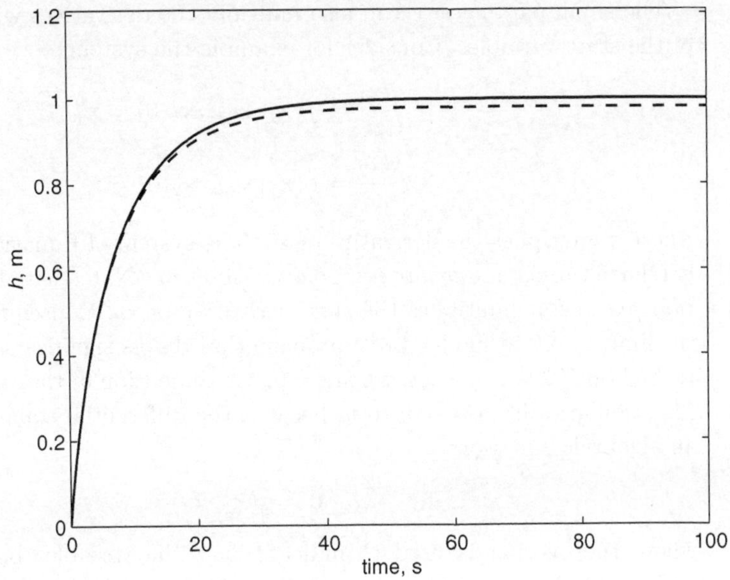

Figure 2.2 Time evolution of the tank level with (dashed) and without (solid) a leak at the bottom of the tank.

Assuming that the fracture is at the same height as the outlet pipe, an equation for the time evolution of the tank level h can easily be written as

$$\frac{dh}{dt} = \frac{1}{A_t}\left[F_0 - (A_1 + A_2)\sqrt{2gh}\right] \tag{2.4}$$

with $g = 9.81\ \mathrm{m/s^2}$ being the gravitational constant. Notice that, from the data above, the cross-sectional area of the fracture is much smaller than the cross-section of the pipe; we can thus define

$$\varepsilon = \frac{A_2}{A_1} = 0.01 \ll 1 \tag{2.5}$$

and rewrite (2.4) as

$$\frac{dh}{dt} = \frac{1}{A_t}\left(F_0 - A_1\sqrt{2gh}\right) - \varepsilon\frac{1}{A_t}A_1\sqrt{2gh} \tag{2.6}$$

The above equation is in the form of Equation (2.2), with the presence of the leak constituting a regular perturbation to the system dynamics. It is therefore to be expected that the solutions in the two cases differ by a small, $\mathcal{O}(\varepsilon)$ quantity.

Figure 2.2 shows the results of integrating Equation (2.4) numerically from $h(t=0)=0$ for 100 s. These results confirm the theoretical analysis presented above, insofar as the time evolution of the tank level in the two cases is very similar; in effect, the two trajectories differ by only 0.02 m $(=\mathcal{O}(\varepsilon))$ at steady state.

The small parameter ε can also multiply the derivatives with respect to time of the state variables. Consider for example the system

$$\frac{d\mathbf{x}_1}{dt} = f(\mathbf{x}_1, \mathbf{x}_2), \ \mathbf{x}_1(0) = \mathbf{x}_1^0 \tag{2.7}$$

$$\varepsilon\frac{d\mathbf{x}_2}{dt} = \mathbf{g}(\mathbf{x}_1, \mathbf{x}_2), \ \mathbf{x}_2(0) = \mathbf{x}_2^0 \tag{2.8}$$

where ε multiplies the derivative of $\mathbf{x}_2$. The system of Equations (2.7) and (2.8) is referred to as a *singular perturbation problem*. Note that the small perturbation parameter multiplies the time derivative of $\mathbf{x}_2$. Consequently, in this case, the limit $\varepsilon \to 0$ would lead to a problem that differs significantly from the unperturbed one (2.1). In effect, when $\varepsilon = 0$, the dimension of the state space of (2.7)–(2.8) collapses from $n + m$ to n, because the differential equation (2.8) becomes an algebraic equation:

$$\mathbf{0} = \mathbf{g}(\bar{\mathbf{x}}_1, \bar{\mathbf{x}}_2) \tag{2.9}$$

where the overbar is used to indicate that the variables belong to a system with $\varepsilon = 0$. The original system (2.7)–(2.8) collapses to a system of differential algebraic equations (DAEs):[1]

$$\frac{d\bar{\mathbf{x}}_1}{dt} = \mathbf{f}(\bar{\mathbf{x}}_1, \bar{\mathbf{x}}_2) \tag{2.10}$$

$$\mathbf{0} = \mathbf{g}(\bar{\mathbf{x}}_1, \bar{\mathbf{x}}_2) \tag{2.11}$$

Definition 2.1. *The model of Equations (2.7) and (2.8) is said to be in a* standard *singularly perturbed form if, in a domain of interest, Equation (2.9) has $k \geq 1$ distinct (isolated) roots*

$$\bar{\mathbf{x}}_2 = \bar{\mathbf{\Phi}}_i(\bar{\mathbf{x}}_1), \ i = 1, \ldots, k \tag{2.12}$$

The condition stated in Definition 2.1 assures that a well-defined n-dimensional reduced model will correspond to each solution (2.12); whenever this condition is violated, the system in Equations (2.7) and (2.8) is said to be in a *nonstandard singularly perturbed form.*

To obtain the (ith) reduced model, we substitute Equation (2.12) into Equation (2.10):

$$\frac{d\bar{\mathbf{x}}_1}{dt} = \mathbf{f}(\bar{\mathbf{x}}_1, \bar{\mathbf{\Phi}}_i(\bar{\mathbf{x}}_1)); \qquad \bar{\mathbf{x}}_1(t = 0) = \mathbf{x}_1^0 \tag{2.13}$$

which can be written in the more compact form

$$\frac{d\bar{\mathbf{x}}_1}{dt} = \bar{\mathbf{f}}(\bar{\mathbf{x}}_1); \qquad \bar{\mathbf{x}}_1(t = 0) = \mathbf{x}_1^0 \tag{2.14}$$

[1] A definition of differential algebraic equations and some related notions are presented in Appendix A.

Equation (2.14) is referred to as a quasi-steady-state model, because $\mathbf{x}_2$, whose rate of change $d\mathbf{x}_2/dt = (1/\varepsilon)\mathbf{g}$ can be large when ε is small, may rapidly converge to a solution (2.12), which is the quasi-steady-state form of Equation (2.8).

Remark 2.1. *For a standard singularly perturbed model, the DAE system (2.10) has an index $\nu = 1$, i.e., the variables $\bar{\mathbf{x}}_2$ can be solved for directly from the algebraic equations (2.9) and the reduced-order (equivalent ODE) representation (2.13) is obtained directly. For systems that are in the nonstandard singularly perturbed form, the DAE system (2.10) obtained in the limit as $\varepsilon \to 0$ has an index $\nu > 1$ and an equivalent ODE representation for the slow dynamics is not always readily available.*

The presence of a singular perturbation induces multiple-time-scale behavior in dynamical systems, which is characterized by the presence of both fast and slow transients in their time response. The slow response is approximated by the reduced model (2.14), while the difference between the response of the reduced model (2.14) and that of the full model (2.7)–(2.8) is the fast transient.

In the slow model, the variables $\mathbf{x}_2$ have been substituted by the "quasi-steady-state" $\bar{\mathbf{x}}_2$. In contrast to the original variable $\mathbf{x}_2$, which starts from the initial condition $\mathbf{x}_2^0$ at $t = 0$, the initial value of $\bar{\mathbf{x}}_2$ is $\bar{\mathbf{x}}_2(t = 0) = \bar{\mathbf{\Phi}}(\bar{\mathbf{x}}_1(t = 0), 0)$, and the discrepancy between $\mathbf{x}_2^0$ and $\bar{\mathbf{x}}_2(t = 0)$ may be large. Therefore, $\bar{\mathbf{x}}_2$ is not a uniform approximation of $\mathbf{x}_2$ on the entire time interval from $t \geq 0$, but it will be within $\mathcal{O}(\varepsilon)$ of $\mathbf{x}_2$ on a finite time interval $t \in [t_1, T]$ $t_1 > 0$, i.e.,

$$\mathbf{x}_2 = \bar{\mathbf{x}}_2(t) + \mathcal{O}(\varepsilon) \tag{2.15}$$

On the other hand, the quasi-steady-state $\bar{\mathbf{x}}_1$ can be constrained to start from the initial condition $\mathbf{x}_1^0$, and it is therefore possible that the approximation of $\mathbf{x}_1$ by $\bar{\mathbf{x}}_1$ can be uniform. Provided that $\bar{\mathbf{x}}_1$ exists at every time $t \in [0, T]$, we can write

$$\mathbf{x}_1 = \bar{\mathbf{x}}_1(t) + \mathcal{O}(\varepsilon), \ \forall t \in [0, T] \tag{2.16}$$

Equation (2.15) states that, during an initial time interval $[0, t_1]$ (frequently referred to as the "boundary layer"), the original variable $\mathbf{x}_2$ approaches $\bar{\mathbf{x}}_2$, and remains close to $\bar{\mathbf{x}}_2$ during $[t_1, T]$ in the interval $[t_1, T]$.

The rate at which $\mathbf{x}_2$ approaches $\bar{\mathbf{x}}_2$ can be very large, since $d\mathbf{x}_2/dt = (1/\varepsilon)\mathbf{g}$, and $\varepsilon \to 0$. Singular perturbation theory relies on defining a "stretched" time variable $\tau = t/\varepsilon$, with $\tau = 0$ at $t = 0$, to analyze such fast transient phenomena. The term "stretched" refers to the behavior of the new time variable τ, which tends to ∞ even for t only slightly larger than 0. Note that, while $\mathbf{x}_2$ and τ vary very rapidly, $\mathbf{x}_1$ stays near its initial value $\mathbf{x}_1^0$.

The behavior of $\mathbf{x}_2$ as a function of τ (i.e., its behavior in the boundary layer) is described using a *boundary-layer correction* $\hat{\mathbf{x}}_2 = \mathbf{x}_2 - \bar{\mathbf{x}}_2$ that satisfies

$$\frac{d\hat{\mathbf{x}}_2}{d\tau} = \mathbf{g}\left(\mathbf{x}_1^0, \hat{\mathbf{x}}_2(\tau) + \bar{\mathbf{x}}_2(t=0)\right), \text{ with } \hat{\mathbf{x}}_2(t=0) = \mathbf{x}_2^0 - \bar{\mathbf{x}}_2(t=0) \tag{2.17}$$

The solution $\hat{\mathbf{x}}_2$ of (2.17) is used as a correction of the expression in Equation (2.15), giving another approximate expression for $\mathbf{x}_2$ that is possibly uniform:

$$\mathbf{x}_2 = \bar{\mathbf{x}}_2(t) + \hat{\mathbf{x}}_2(\tau) + \mathcal{O}(\varepsilon) \tag{2.18}$$

Here, $\bar{\mathbf{x}}_2$ is the slow transient of $\mathbf{x}_2$ and $\hat{\mathbf{x}}_2$ is the fast transient. Note that, for the corrected approximation in Equation (2.18) to converge rapidly to the slow approximative solution (2.15), the term $\hat{\mathbf{x}}_2$ must decay as $t \to \infty$ to an $\mathcal{O}(\varepsilon)$ quantity. In the slow time scale t, this decay is fast, since

$$\frac{d\hat{\mathbf{x}}_2(\tau)}{dt} = \frac{1}{\varepsilon}\frac{d\hat{\mathbf{x}}_2(\tau)}{d\tau}$$

A very important result regarding the validity of the approximate solutions (2.18) and (2.16), and implicitly of the decomposition of the singularly perturbed system (2.7)–(2.8) into a slow model (2.14) and a fast model (2.17), is a theorem due to A. N. Tikhonov (Tikhonov 1948).

Theorem 2.1. *If*

(i) the equilibrium $\hat{\mathbf{x}}_2(\tau) = \mathbf{0}$ of (2.17) is asymptotically stable uniformly in $\mathbf{x}_1^0$ and $t=0$, and $\mathbf{x}_2^0 - \bar{\mathbf{x}}_2(t=0)$ belongs to its domain of attraction, so $\hat{\mathbf{x}}_2(\tau)$ exists for $\tau \geq 0$, and

(ii) the eigenvalues of $\partial\mathbf{g}/\partial\mathbf{x}_2$, evaluated, for $\varepsilon = 0$, along $\bar{\mathbf{x}}_1(t), \bar{\mathbf{x}}_2(t)$, have real parts smaller than a given negative number c,

then the approximations (2.18) and (2.16) are valid for any $t \in [0, T]$ and there exists a $t_1 \geq 0$ such that (2.15) is valid for any $t \in [t_1, T]$.

Remark 2.2. *Condition (i) in Theorem 2.1 states that*

$$\lim_{\tau \to \infty} \hat{\mathbf{x}}_2(\tau) = \mathbf{0}$$

uniformly in $\mathbf{x}_1^0$, $t=0$; that is, $\mathbf{x}_2$ will come close to the quasi-steady-state value $\bar{\mathbf{x}}_2$ at a time $t_1 > 0$, while condition (ii) assumes that, for any $t \in [t_1, T]$, $\mathbf{x}_2$ will stay close (within $\mathcal{O}(\varepsilon)$) to $\bar{\mathbf{x}}_2$.

Remark 2.3. *Conditions (i) and (ii) describe a strong stability property of the boundary layer (fast) system (2.17). Note that, if $\mathbf{x}_2^0$ is sufficiently close to $\bar{\mathbf{x}}_2(t=0)$, then condition (ii) encompasses condition (i). Also, the condition*

$$\det \frac{\partial \mathbf{g}}{\partial \mathbf{x}_2} \neq 0$$

implies that the roots $\bar{\mathbf{x}}_2(t)$ are distinct, as required in Definition 2.1.

Remark 2.4. *In a general nonlinear system, there may be several distinct solutions* $\bar{\mathbf{x}}_{2,i} = \bar{\mathbf{\Phi}}_i$, $i \in \{1, \ldots, k\}$. *In such a case, one focuses on a particular solution and the corresponding representation for the slow subsystem (2.13) in an appropriate neighborhood. The choice of a particular quasi-steady-state solution depends on the initial condition* $\mathbf{x}_1^0, \mathbf{x}_2^0$. *The solution* $\hat{\mathbf{x}}_2(\tau)$ *of the fast system will stabilize at the corresponding steady state* $\bar{\mathbf{x}}_{2,i} = \bar{\mathbf{\Phi}}_i(\mathbf{x}_1^0)$.

Remark 2.5. *Tikhonov's theorem holds only for bounded time intervals. Under the additional assumption that the slow system (2.14) is also locally exponentially stable, a similar result exists for infinite time intervals (Khalil 2002).*

It was previously shown that, in the limit as $\varepsilon \to 0$, the dimension of the state space of (2.7)–(2.8) collapses from $(n+m)$ to n. This allows a geometric interpretation: in the $(n+m)$-dimensional state space of $\mathbf{x}_1$ and $\mathbf{x}_2$, an n-dimensional subspace or *manifold* M_ε can be defined as

$$M_\varepsilon : \left\{ \mathbf{x}_2 = \mathbf{\Phi}(\mathbf{x}_1), \text{ with } \mathbf{x}_1 \in \mathbb{R}^n \text{ and } \mathbf{x}_2 \in \mathbb{R}^m \right\} \tag{2.19}$$

where $\mathbf{\Phi}(\mathbf{x}_1)$ is a sufficiently smooth function. The decrease in dimension of the state space of (2.7)–(2.8) is then due to the constraint that states $\mathbf{x}_2$ remain in M_ε. For instance, in $\mathbb{R}^2$, if $n=1$ and $m=1$, the manifold is a one-dimensional (1D) line; in $\mathbb{R}^3$, if $n=1$ and $m=2$, M_ε will be a curve. If (2.19) holds at time t^* and for any $t > t^*$, then the manifold M_ε is said to be *invariant*.

The discussion above was based on the limiting case $\varepsilon \to 0$. The manifold M_ε will, in this limiting case, be

$$M_0 : \left\{ \bar{\mathbf{x}}_2 = \bar{\mathbf{\Phi}}(\bar{\mathbf{x}}_1),\ 0 = \bar{\mathbf{g}}(\bar{\mathbf{x}}_1, \bar{\mathbf{\Phi}}(\bar{\mathbf{x}}_1)) \right\} \tag{2.20}$$

where the convention of distinguishing the variables in the case $\varepsilon = 0$ by an overbar has been maintained.

Example 2.2. A reaction system consists of a reactant R_1, which is transformed into product P_1 and intermediate R_2, the latter of which is subsequently transformed into product P_2:

$$R_1 \to P_1 + R_2 \tag{2.21}$$
$$R_2 \to P_2 \tag{2.22}$$

with rate constants $k_1 = 0.10\ \mathrm{s}^{-1}$ and $k_2 = 10\ \mathrm{s}^{-1}$, and initial conditions $x_1(t=0) = x_{1,0}$ and $x_2(t=0) = x_{2,0}$, respectively. The evolution of the molar concentrations x_1 and x_2 of R_1 and R_2 is described by

$$\frac{dx_1}{dt} = -k_1 x_1 \tag{2.23}$$
$$\frac{dx_2}{dt} = k_1 x_1 - k_2 x_2 \tag{2.24}$$

In this case, given the large difference in the reaction rate constants, we can define

$$\varepsilon = \frac{k_1}{k_2} = 0.01 \ll 1 \tag{2.25}$$

and rewrite (2.23) and (2.24) as

$$\frac{dx_1}{dt} = -k_1 x_1 \tag{2.26}$$

$$\varepsilon \frac{dx_2}{dt} = \varepsilon k_1 x_1 - k_1 x_2 \tag{2.27}$$

which is in the form of Equations (2.7) and (2.8).

Following the above, we consider the limit $\varepsilon \to 0$, obtaining the DAE system

$$\frac{dx_1}{dt} = -k_1 x_1 \tag{2.28}$$

$$0 = -k_1 x_2 \tag{2.29}$$

Equation (2.29) has one distinct root, $\bar{x}_2 = 0$, and hence the above singularly perturbed ODE system is in standard form. Proceeding with our analysis, we obtain a reduced-order, uniform approximation of the slow component of the dynamics as

$$\frac{dx_1}{dt} = -k_1 x_1 \tag{2.30}$$

$$0 = \bar{x}_2 \tag{2.31}$$

or, in integrated form,

$$x_1(t) = e^{-k_1 t} + x_{1,0} - 1 \tag{2.32}$$

Moving now to the fast dynamics of this reaction system, we define the fast time scale $\tau = t/\varepsilon$, obtaining

$$\frac{d\hat{x}_2}{d\tau} = -k_1 \hat{x}_2 \tag{2.33}$$

which can easily be solved for $\hat{x}_2$:

$$\hat{x}_2(\tau) = e^{-k_1 \tau} + \hat{x}_{2,0} - 1 \tag{2.34}$$

Finally, by combining the above results, we can derive a uniform approximation for x_2, as

$$x_2(t) = \bar{x}_2 + \hat{x}_2 = 0 + e^{-k_1 \tau} + \hat{x}_{2,0} - 1 \tag{2.35}$$

In computing the above solution, it is important to notice that, while $\bar{x}_2(t) = 0$ and $\bar{x}_2(t=0) = 0$, the initial condition for $\hat{x}_2$ is $\hat{x}_2 = x_{2,0}$.

Using the derivations above, we can infer the following.

- The reaction system (2.23)–(2.23) will have a dynamic behavior with two time scales, and the concentration x_2 of R_2 will evolve much faster than the concentration of R_1.

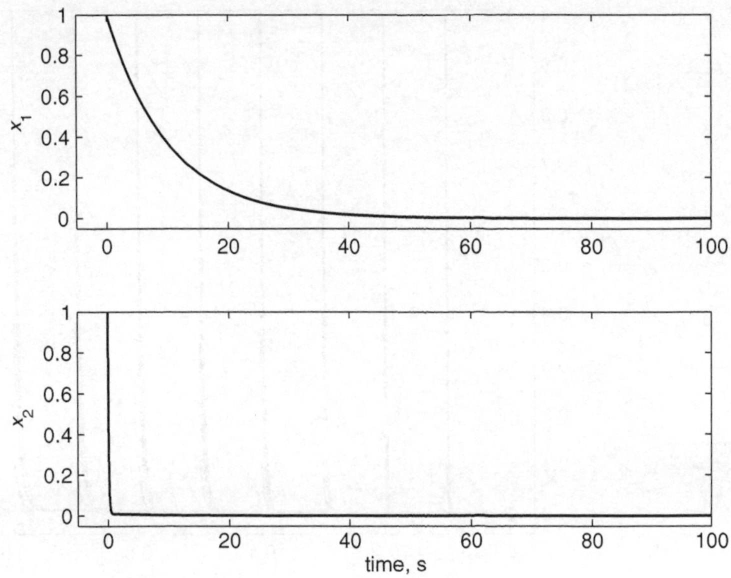

Figure 2.3 Evolution of the reactant concentrations from initial conditions $x_1(t=0)=x_2(t=0)=1\,\text{mol/l}$: numerical (solid) and approximate analytical (dashed) solutions.

- x_2 will quickly reach its quasi-steady-state value $\bar{x}_2=0$, and $x_2=0$ constitutes an equilibrium manifold of this system.
- Within the equilibrium manifold, x_1 will slowly evolve towards the equilibrium point $(0,0)$ of the entire system.

The above results are confirmed by numerical simulations: Figure 2.3 shows the evolution of the reactant concentrations from the initial values $x_1(t=0)=x_2(t=0)=1$ mol/l. As expected, x_2 quickly converges to its quasi-steady-state value, while the transient behavior of x_1 is much slower.

Figure 2.4 presents the trajectories of the two concentrations in the phase plane, revealing the presence of the equilibrium manifold: the phase trajectories starting from any initial condition $(x_{1,0}, x_{2,0}) \in [0,1] \times [0,1]$ approach the horizontal line $x_2=0$, followed by convergence towards the equilibrium point $(0,0)$.

In contrast, let us analyze the same reaction system in a second case, namely considering that the reaction rate constants have the same value, $k_1=k_2=1\ \text{s}^{-1}$. A comparison of the evolution of the compositions in the two cases is presented in Figure 2.5, and a phase plane of the second case is shown in Figure 2.6.

Clearly, in the second case, the rate of change of concentration of the two reactants is identical and the equilibrium manifold is not present in the phase portrait, confirming the absence of a two-time-scale behavior.

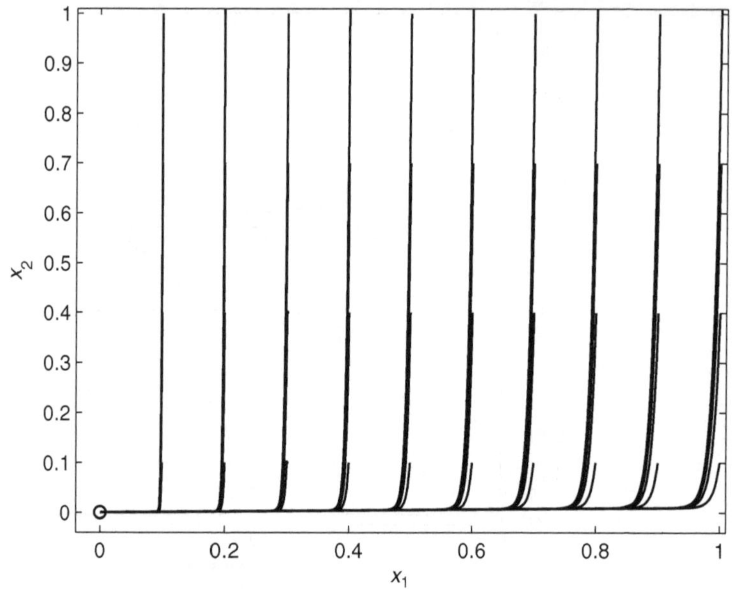

Figure 2.4 Phase trajectories of the reacting system starting from different initial conditions $(x_{1,0}, x_{2,0}) \in [0,1] \times [0,1]$; $(0,0)$ is a stable steady state.

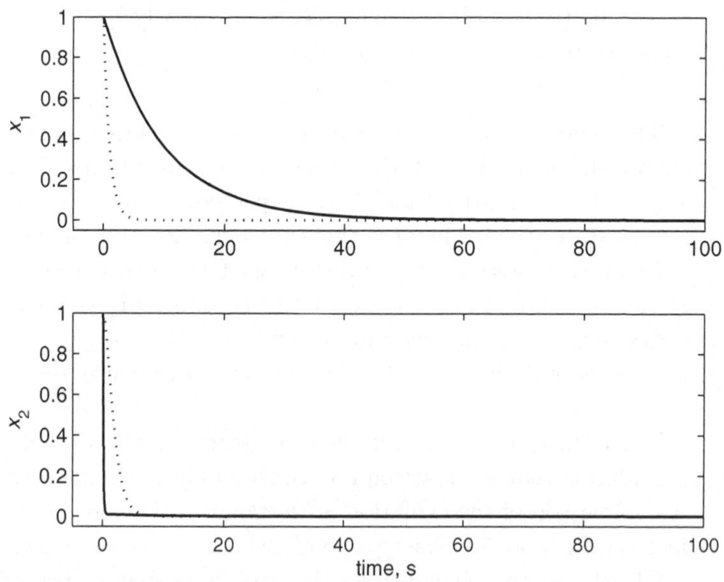

Figure 2.5 Evolution of the reactant concentrations from initial conditions $x_1(t=0) = x_2(t=0) = 1$ mol/l, with $k_1 = 0.1\ \mathrm{s}^{-1}$ and $k_2 = 10\ \mathrm{s}^{-1}$ (solid) and $k_1 = k_2 = 1\ \mathrm{s}^{-1}$ (dashed).

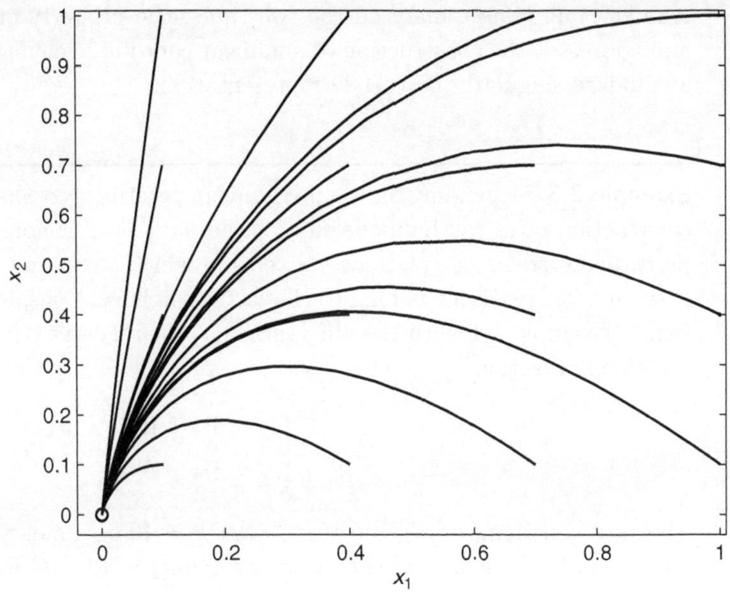

Figure 2.6 Phase trajectories of the reacting system with $k_1 = k_2 = 1\ \mathrm{s}^{-1}$ from different initial conditions $(x_{1,0}, x_{2,0}) \in [0, 1] \times [0, 1]$.

2.3 Nonstandard singularly perturbed systems with two time scales

In what follows, we will focus on a class of systems arising from the detailed dynamic models of chemical processes, which are characterized by the presence of large parameters in the explicit rate equations.[2] On defining the small parameter ε as the reciprocal of such a representative large parameter, it can be shown (Kumar and Daoutidis 1999a) that the dynamic models of these fast-rate processes are given by a system with the following general description:

$$\dot{\mathbf{x}} = \mathbf{f}(\mathbf{x}) + \mathbf{G}(\mathbf{x})\mathbf{u} + \frac{1}{\varepsilon}\mathbf{B}(\mathbf{x})\mathbf{k}(\mathbf{x}) \tag{2.36}$$

where $\mathbf{x} \subset \mathcal{X} \in \mathbb{R}^n$ is the vector of state variables, $\mathbf{f}(\mathbf{x})$ and $\mathbf{k}(\mathbf{x})$ are smooth vector fields of dimension n and p, $p < n$, and $\mathbf{G}(x)$ and $\mathbf{B}(x)$ are matrices of dimensions $n \times m$ and $n \times p$, respectively. In the rate-based models of Equation (2.36), the term $(1/\varepsilon)\mathbf{B}(\mathbf{x})\mathbf{k}(\mathbf{x})$ corresponds to the fast phenomena for which the rate expressions involve large parameters; the matrix $\mathbf{B}(\mathbf{x})$ and the Jacobian $\partial\mathbf{k}(\mathbf{x})/\partial\mathbf{x}$ are assumed to have full column and row rank p, respectively.

Clearly, the system in Equation (2.36) is not in a *standard* singularly perturbed form and therefore the results derived so far are not directly applicable.

[2] As will be seen throughout the rest of the text, typical examples include fast reactions and large heat and mass transfer rates.

Kumar *et al.* (1998) analyzed the two-time-scale property of the system (2.36) and addressed the construction of nonlinear coordinate changes that would yield a standard singularly perturbed representation.

Example 2.3. Depending on the mechanism, reacting systems with vastly different reaction rates can be modeled by either standard or nonstandard singularly perturbed systems of equations. Systems in which a reactant is involved in both slow and fast reactions belong to the latter category. Consider the reaction system in Example 2.2, with the difference that the reactant R_1 also participates in the second reaction:

$$R_1 \rightarrow P_1 + R_2 \tag{2.37}$$

$$R_1 + R_2 \rightarrow P_2 \tag{2.38}$$

The rate constants are $k_1 = 0.10\ \mathrm{s}^{-1}$ and $k_2 = 10\ \mathrm{s}^{-1}\,\mathrm{l\,mol}^{-1}$. In this case, the evolution of the molar concentrations x_1 and x_2 of R_1 and R_2 is described by

$$\frac{dx_1}{dt} = -k_1 x_1 - k_2 x_1 x_2 \tag{2.39}$$

$$\frac{dx_2}{dt} = k_1 x_1 - k_2 x_1 x_2 \tag{2.40}$$

Owing to the fact that R_1 is the sole feedstock, we can reasonably assume that $x_{1,0}$ is not insignificant. For simplicity, let $x_{1,0} = 1$ mol/l. We can thus define

$$\varepsilon = \frac{k_1}{k_2 x_{1,0}} = 0.01 \ll 1 \tag{2.41}$$

and rewrite (2.39)–(2.40) as

$$\frac{dx_1}{dt} = -k_1 x_1 - \frac{1}{\varepsilon}\frac{k_1}{x_{1,0}} x_1 x_2 \tag{2.42}$$

$$\frac{dx_2}{dt} = k_1 x_1 - \frac{1}{\varepsilon}\frac{k_1}{x_{1,0}} x_1 x_2 \tag{2.43}$$

which is in the form of Equation (2.36) (i.e., a *nonstandard* singularly perturbed ODE), with $\mathbf{x} = [x_1\ x_2]^{\mathrm{T}}$, $\mathbf{f}(\mathbf{x}) = k_1 x_1 [-1\ 1]^{\mathrm{T}}$, $\mathbf{G}(\mathbf{x}) = \mathbf{0}$, $\mathbf{B}(x) = -(k_1/x_{1,0})[1\ 1]^{\mathrm{T}}$, and $\mathbf{k}(\mathbf{x}) = x_1 x_2$.

Within the framework proposed in Kumar *et al.* (1998), a time-scale decomposition is initially used to derive separate representations of the slow and fast dynamics of (2.36) in the appropriate time scales and to provide some insight into the variables that should be used as part of the desired coordinate change. Specifically, by multiplying Equation (2.36) by ε and considering the limit $\varepsilon \rightarrow 0$, we obtain the following set of (linearly independent) constraints that must be satisfied in the slow time scale t:

$$k_i(\mathbf{x}) = 0,\ i = 1, \ldots, p \tag{2.44}$$

where $k_i(\mathbf{x})$ denotes the ith component of $\mathbf{k}(\mathbf{x})$.

For the system (2.36), in the limit $\varepsilon \to 0$, the term $(1/\varepsilon)\mathbf{k}(\mathbf{x})$ becomes indeterminate. For rate-based chemical and physical process models, this allows a physical interpretation: in the limit when the large parameters in the rate expressions approach infinity, the fast heat and mass transfer, reactions, etc., approach the quasi-steady-state conditions of phase and/or reaction equilibrium (specified by $\mathbf{k}(\mathbf{x}) = \mathbf{0}$). In this case, the rates of the fast phenomena, as given by the explicit rate expressions, become indeterminate (but, generally, remain different from zero; i.e., the fast reactions and heat and mass transfer *do still occur*).

Thus, defining

$$z_i = \lim_{\varepsilon \to 0} \frac{k_i(\mathbf{x})}{\varepsilon}$$

as the finite but unknown rates of the fast phenomena, the system (2.36) takes the following form:

$$\begin{aligned} \dot{\mathbf{x}} &= \mathbf{f}(\mathbf{x}) + \mathbf{G}(\mathbf{x})\mathbf{u} + \mathbf{B}(\mathbf{x})\mathbf{z} \\ \mathbf{0} &= \mathbf{k}(\mathbf{x}) \end{aligned} \tag{2.45}$$

which describes the slow dynamics of Equation (2.36). In the above DAE system, $\mathbf{x}$ is the vector of differential variables and $\mathbf{z} \in \mathbb{R}^p$ is a vector of algebraic variables.

Note that the system (2.45) is a DAE system of nontrivial index, since $\mathbf{z}$ cannot be evaluated directly from the algebraic equations. A solution for the variables $\mathbf{z}$ must be obtained by differentiating the constraints $\mathbf{k}(\mathbf{x}) = \mathbf{0}$. For most chemical processes, such as reaction networks (Gerdtzen *et al.* 2004), reactive distillation processes (Vora 2000), and complex chemical plants (Kumar and Daoutidis 1999a), the $\mathbf{z}$ variables can be obtained after just one differentiation of the algebraic constraints:

$$\mathbf{z} = -\left(L_{\mathbf{B}}\mathbf{k}(\mathbf{x})\right)^{-1} \left\{L_{\mathbf{f}}\mathbf{k}(\mathbf{x}) + L_{\mathbf{G}}\mathbf{k}(\mathbf{x})\mathbf{u}\right\} \tag{2.46}$$

since in such cases the $(p \times p)$ matrix $L_{\mathbf{B}}\mathbf{k}(\mathbf{x})$, denoting the Lie derivative[3] of function $\mathbf{k}$ along $\mathbf{B}$, is nonsingular. In the interest of preserving the simplicity of the discussion, this observation is generalized as follows.

Assumption 2.1. *For the systems under consideration, the matrix $L_{\mathbf{B}}\mathbf{k}(\mathbf{x})$ is nonsingular.*

[3] Please see Appendix A for a definition of the Lie derivative.

Assumption 2.1 fixes the index of the DAE system (2.45) to two, and the numbers of slow and fast variables as $(n-p)$ and p, respectively.

A state-space realization (ODE representation) of the DAE system (2.45) can readily be obtained as

$$\dot{\mathbf{x}} = \mathbf{f}(\mathbf{x}) + \mathbf{G}(\mathbf{x})\mathbf{u} - \mathbf{B}(\mathbf{x})(L_{\mathbf{B}}\mathbf{k}(\mathbf{x}))^{-1}\{L_{\mathbf{f}}\mathbf{k}(\mathbf{x}) + L_{\mathbf{G}}\mathbf{k}(\mathbf{x})\mathbf{u}\} \tag{2.47}$$

$$\mathbf{0} = \mathbf{k}(\mathbf{x}) \tag{2.48}$$

Similarly, a representation of the fast dynamics in the limit $\varepsilon \to 0$ is obtained in the "stretched" fast time scale $\tau = t/\varepsilon$ as

$$\frac{d\mathbf{x}}{d\tau} = \mathbf{B}(\mathbf{x})\mathbf{k}(\mathbf{x}) \tag{2.49}$$

Note that, though Equations (2.49) and (2.45) represent the fast and slow *dynamics*, the fast and slow *variables* are still not explicitly separated.

As mentioned before, obtaining an explicit variable separation for the system in Equation (2.36) requires a nonlinear coordinate transformation. The fact that $\mathbf{k}(\mathbf{x}) = \mathbf{0}$ in the slow time scale t and $\mathbf{k}(\mathbf{x}) \neq \mathbf{0}$ in the fast time scale τ indicates that the functions $k_i(\mathbf{x})$ should be used in such a coordinate transformation as fast variables. Then, it can be shown (see, e.g., Kumar and Daoutidis 1999a) that a coordinate change of the form

$$\begin{bmatrix} \boldsymbol{\zeta} \\ \boldsymbol{\eta} \end{bmatrix} = \mathbf{T}(\mathbf{x}) = \begin{bmatrix} \boldsymbol{\phi}(\mathbf{x}) \\ \mathbf{k}(\mathbf{x}) \end{bmatrix} \tag{2.50}$$

that results in a standard singularly perturbed representation of the system (2.36), where $\boldsymbol{\zeta} \in \mathbb{R}^{n-p}$ are the slow variables and $\boldsymbol{\eta} \in \mathbb{R}^{p}$ are the constraints associated with the quasi-steady state of the fast component of the dynamics, exists if and only if the matrix $L_{\mathbf{B}}\mathbf{k}(\mathbf{x})$ is nonsingular, and the p-dimensional distribution spanned by the columns of the matrix $\mathbf{B}(\mathbf{x})$ is involutive.[4]

Assuming that the above conditions are satisfied, under the coordinate change of Equation (2.50), we obtain the following standard singularly perturbed form for (2.36):

$$\dot{\boldsymbol{\zeta}} = \tilde{\mathbf{f}}(\boldsymbol{\zeta}, \boldsymbol{\eta}) + \tilde{\mathbf{G}}(\boldsymbol{\zeta}, \boldsymbol{\eta})\mathbf{u} \tag{2.51}$$

$$\varepsilon\dot{\boldsymbol{\eta}} = \varepsilon\bar{\mathbf{f}}(\boldsymbol{\zeta}, \boldsymbol{\eta}) + \varepsilon\bar{\mathbf{G}}(\boldsymbol{\zeta}, \boldsymbol{\eta})\mathbf{u} + \mathbf{Q}(\boldsymbol{\zeta}, \boldsymbol{\eta})\eta \tag{2.52}$$

where $\tilde{\mathbf{f}} = L_{\mathbf{f}}\boldsymbol{\phi}(\mathbf{x})$, $\bar{\mathbf{f}} = L_{\mathbf{f}}\mathbf{k}(\mathbf{x})$, $\tilde{\mathbf{G}} = \mathbf{L}_{\mathbf{G}}\boldsymbol{\phi}(\mathbf{x})$, $\bar{\mathbf{G}} = \mathbf{L}_{\mathbf{G}}\mathbf{k}(x)$, $\mathbf{Q} = \mathbf{L}_{\mathbf{B}}\mathbf{k}(\mathbf{x})$, evaluated at $\mathbf{x} = \mathbf{T}^{-1}(\boldsymbol{\zeta}, \boldsymbol{\eta})$, and $\mathbf{Q}(\boldsymbol{\zeta}, \boldsymbol{\eta})$ is nonsingular uniformly in $\boldsymbol{\zeta}, \boldsymbol{\eta}$, and the $(n-p)$-dimensional vector field $\boldsymbol{\phi}(\mathbf{x})$ is such that $L_{\mathbf{B}}\boldsymbol{\phi}(\mathbf{x}) \equiv \mathbf{0}$.

Considering now (2.51)–(2.52) in the limit $\varepsilon \to 0$, we obtain $\boldsymbol{\eta} = \mathbf{0}$ as the quasi-steady-state solution of the fast variables, and the following model of dimension $(n-p)$ is obtained for the slow dynamics:

$$\dot{\boldsymbol{\zeta}} = \tilde{\mathbf{f}}(\boldsymbol{\zeta}, \mathbf{0}) + \tilde{\mathbf{G}}(\boldsymbol{\zeta}, \mathbf{0})\mathbf{u} \tag{2.53}$$

[4] The notion of involutivity is defined in Appendix A.

On introducing the "stretched" fast time scale $\tau = t/\varepsilon$, and considering (2.51)–(2.52) in the limit $\varepsilon \to 0$, we also obtain the following description of the fast dynamics:

$$\begin{aligned} \frac{d\zeta}{d\tau} &= 0 \\ \frac{d\eta}{d\tau} &= \mathbf{Q}(\zeta, \eta)\eta \end{aligned} \tag{2.54}$$

Note that in this case, the variables $\zeta = \phi(\mathbf{x})$ and $\eta = \mathbf{k}(\mathbf{x})$ represent the "true" slow and fast variables, respectively, since the fast transients are observed only in the η variables.

Example 2.4. Two metal objects B_1 and B_2 (with constant masses m_1 and m_2 and constant heat capacities C_{p1} and C_{p2}, respectively), initially at different temperatures ($T_{1,0}$ and $T_{2,0}$), are brought into contact. Heat transfer occurs over a contact area A, with a heat-transfer coefficient U. The objects are assumed to be isolated from the environment; however, the insulation on B_2 is not perfect and heat is lost to the environment over a similar area A; the heat transfer coefficient U_e between B_2 and the environment is, however, much lower than U. The environment is assumed to act as a heat sink at a constant temperature T_e.

The energy balance for this system is described by

$$\frac{d(m_1 C_{p1} T_1)}{dt} = UA(T_2 - T_1) \tag{2.55}$$

$$\frac{d(m_2 C_{p2} T_2)}{dt} = -UA(T_2 - T_1) - U_e A(T_2 - T_e) \tag{2.56}$$

Since $U_e \ll U$, we can define

$$\varepsilon = \frac{U_e}{U} \ll 1 \tag{2.57}$$

and rewrite (2.55)–(2.56) as

$$\frac{dT_1}{dt} = \frac{1}{\varepsilon} \frac{U_e A}{m_1 C_{p1}} (T_2 - T_1) \tag{2.58}$$

$$\frac{dT_2}{dt} = -\frac{1}{\varepsilon} \frac{U_e A}{m_2 C_{p2}} (T_2 - T_1) - \frac{U_e A}{m_2 C_{p2}} (T_2 - T_e) \tag{2.59}$$

which is in the form of Equation (2.36), with

$$\mathbf{x} = \begin{bmatrix} T_1 \\ T_2 \end{bmatrix} \tag{2.60}$$

$$\mathbf{f}(\mathbf{x}) = \begin{bmatrix} 0 \\ -\dfrac{U_e A}{m_2 C_{p2}} (T_2 - T_e) \end{bmatrix} \tag{2.61}$$

$$\mathbf{B}(\mathbf{x}) = \begin{bmatrix} \dfrac{U_e A}{m_1 C_{p1}} \\ \\ -\dfrac{U_e A}{m_2 C_{p2}} \end{bmatrix} \tag{2.62}$$

$$\mathbf{k}(\mathbf{x}) = T_2 - T_1 \tag{2.63}$$

The nonstandard singularly perturbed form of the model of this system potentially indicates a dynamic behavior with two time scales. This is, in effect, quite intuitive, in view of the presence of different rates of heat transfer induced by the different heat-transfer coefficients U and U_e.

In order to capture the fast component of the dynamics, we define the stretched time scale $\tau = t/\varepsilon$ and consider the limit $\varepsilon \to 0$ (i.e., an infinitely high heat-transfer coefficient between B_1 and B_2). We thus obtain a description of the fast dynamics as

$$\frac{dT_1}{d\tau} = \frac{U_e A}{m_1 C_{p1}}(T_2 - T_1) \tag{2.64}$$

$$\frac{dT_2}{d\tau} = -\frac{U_e A}{m_2 C_{p2}}(T_2 - T_1) \tag{2.65}$$

with the corresponding quasi-steady-state constraint

$$0 = T_2 - T_1 \tag{2.66}$$

This result also lends itself to an intuitive interpretation: temperature equilibration is a fast phenomenon and $T_1 = T_2$ – a line in the (T_1, T_2) coordinate system – is the equilibrium manifold of the fast dynamics.

Turning now to the slow dynamics, we consider the same limit $\varepsilon \to 0$ in the original time scale t. On defining

$$z = \lim_{\varepsilon \to 0} \frac{T_2 - T_1}{\varepsilon} \tag{2.67}$$

we obtain

$$\frac{dT_1}{dt} = \frac{U_e A}{m_1 C_{p1}} z \tag{2.68}$$

$$\frac{dT_2}{dt} = -\frac{U_e A}{m_2 C_{p2}} z - \frac{U_e A}{m_2 C_{p2}}(T_2 - T_e) \tag{2.69}$$

$$0 = T_2 - T_1 \tag{2.70}$$

The algebraic variable z can be computed after differentiating the algebraic constraints (2.69):

$$z = -\left(\frac{U_e A}{m_1 C_{p1}} + \frac{U_e A}{m_2 C_{p2}}\right)^{-1} \frac{U_e A}{m_2 C_{p2}}(T_2 - T_e) \tag{2.71}$$

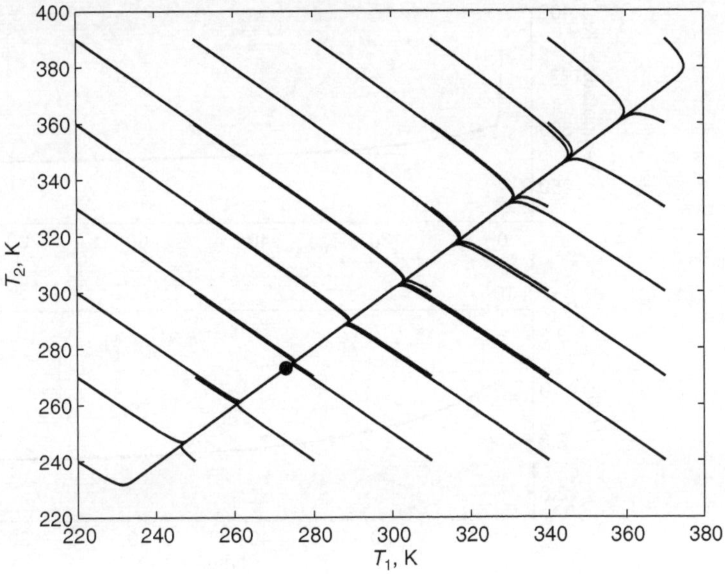

Figure 2.7 Phase portrait of the system (2.55)–(2.56).

Finally, the coordinate change

$$\zeta = m_1 C_{p1} T_1 + m_2 C_{p2} T_2 \tag{2.72}$$

$$\eta = T_2 - T_1 \equiv 0 \tag{2.73}$$

yields the 1D state-space representation of the slow dynamics:

$$\frac{d\zeta}{dt} = -U_{\mathrm{e}} A \left(\frac{\zeta}{m_1 C_{p1} + m_2 C_{p2}} - T_{\mathrm{e}} \right) \tag{2.74}$$

The representation (2.74) of the slow dynamics provides yet another valuable insight: while the temperatures of B_1 and B_2 exhibit a two-time-scale behavior, the *total enthalpy* of the system (captured by the variable ζ) is a true slow variable, evolving only in the slow time scale.

Figures 2.7 and 2.8 present a set of numerical simulations carried out on the system (2.55)–(2.56) using the following parameters: $A = 0.1\ \mathrm{m}^2$, $m_1 = m_2 = 0.1$ kg, $U = 1000\ \mathrm{W\,m^{-2}\,K^{-1}}$, $C_{p1} = C_{p2} = 1000\ \mathrm{J\,kg^{-1}\,K^{-1}}$ and $T_{\mathrm{e}} = 273$ K.

Figure 2.7 reveals the presence of the equilibrium manifold: phase trajectories approach the $T_1 = T_2$ line and converge along this line to the equilibrium point $T_1 = T_2 = T_{\mathrm{e}} = 273$ K.

Figure 2.8 presents the evolution of the temperatures starting from the initial state ($T_1 = 370$ K, $T_2 = 220$ K): thermal equilibrium between B_1 and B_2 is reached very quickly; subsequently, due to heat losses, the temperatures of the two objects slowly reach the temperature of the environment. Conversely, the total enthalpy of the system evolves slowly towards its value at $T_{\mathrm{e}} = 273$ K.

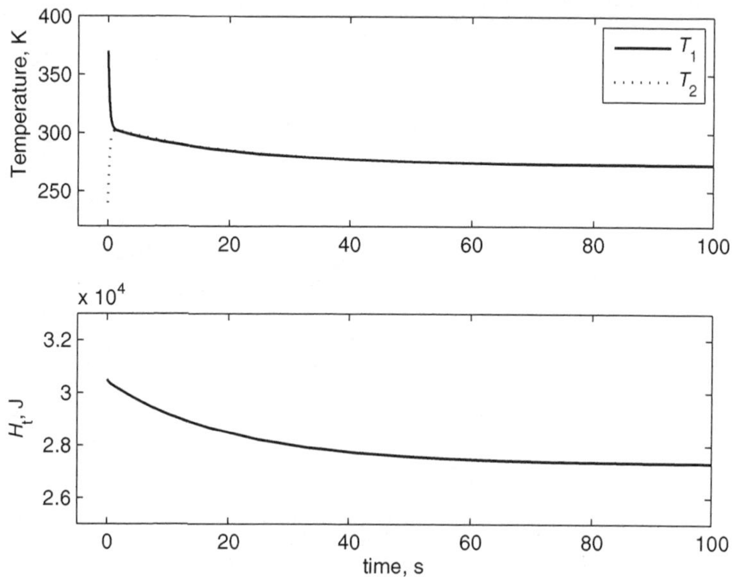

Figure 2.8 The system temperatures exhibit both fast and slow transients (top), while the total enthalpy evolves only in the slow time scale (bottom).

Remark 2.6. *Note that the above results require the involutivity of the distribution spanned by the columns of* $\mathbf{B}(\mathbf{x})$. *This condition, however, is quite restrictive, and is, in general, violated for nonlinear systems with* $p > 1$. *In such cases, it is possible to construct an* ε*-dependent coordinate transformation that is singular at* $\varepsilon = 0$ *to derive a standard singularly perturbed form. Specifically, under a coordinate transformation of the form*

$$\begin{bmatrix} \boldsymbol{\zeta} \\ \boldsymbol{\eta} \end{bmatrix} = \mathbf{T}(\mathbf{x}) = \begin{bmatrix} \boldsymbol{\phi}(\mathbf{x}) \\ \dfrac{\mathbf{k}(\mathbf{x})}{\varepsilon} \end{bmatrix} \tag{2.75}$$

where $\boldsymbol{\zeta} \in \mathbb{R}^{n-p}$ *and* $\boldsymbol{\eta} \in \mathbb{R}^{p}$, *the system of Equation (2.36) takes the following standard singularly perturbed form:*

$$\dot{\boldsymbol{\zeta}} = \tilde{\mathbf{f}}(\boldsymbol{\zeta}, \varepsilon\boldsymbol{\eta}) + \tilde{\mathbf{G}}(\boldsymbol{\zeta}, \varepsilon\boldsymbol{\eta})\,\mathbf{u} + \tilde{\mathbf{B}}(\boldsymbol{\zeta}, \varepsilon\boldsymbol{\eta})\boldsymbol{\eta} \tag{2.76}$$

$$\varepsilon\dot{\boldsymbol{\eta}} = \bar{\mathbf{f}}(\boldsymbol{\zeta}, \varepsilon\boldsymbol{\eta}) + \bar{\mathbf{G}}(\boldsymbol{\zeta}, \varepsilon\boldsymbol{\eta})\,\mathbf{u} + \mathbf{Q}(\boldsymbol{\zeta}, \varepsilon\boldsymbol{\eta})\boldsymbol{\eta} \tag{2.77}$$

where $\tilde{\mathbf{f}} = L_{\mathbf{f}}\boldsymbol{\phi}(\mathbf{x})$, $\tilde{\mathbf{G}} = L_{\mathbf{G}}\boldsymbol{\phi}(\mathbf{x})$, $\tilde{\mathbf{B}} = L_{\mathbf{B}}\boldsymbol{\phi}(\mathbf{x})$, $\bar{\mathbf{f}} = L_{\mathbf{f}}\mathbf{k}(\mathbf{x})$, $\bar{\mathbf{G}} = L_{\mathbf{G}}\mathbf{k}(\mathbf{x})$, *and* $\mathbf{Q} = L_{\mathbf{B}}\mathbf{k}(\mathbf{x})$ *are evaluated at* $\mathbf{x} = \mathbf{T}^{-1}(\boldsymbol{\zeta}, \varepsilon\boldsymbol{\eta})$, *and* $\mathbf{Q}(\boldsymbol{\zeta}, \mathbf{0})$ *is nonsingular uniformly in* $\boldsymbol{\zeta}$. *In the limit* $\varepsilon \to 0$, *a reduced-order model of dimension* $(n-p)$ *is obtained for the slow dynamics:*

$$\dot{\boldsymbol{\zeta}} = \tilde{\mathbf{f}}(\boldsymbol{\zeta}, \mathbf{0}) + \tilde{\mathbf{G}}(\boldsymbol{\zeta}, \mathbf{0})\mathbf{u} - \tilde{\mathbf{B}}(\boldsymbol{\zeta}, \mathbf{0})\mathbf{Q}(\boldsymbol{\zeta}, \mathbf{0})^{-1}\left[\bar{\mathbf{f}}(\boldsymbol{\zeta}, 0) + \bar{\mathbf{G}}(\boldsymbol{\zeta}, \mathbf{0})\mathbf{u}\right] \tag{2.78}$$

$$\boldsymbol{\eta} = -\mathbf{Q}(\boldsymbol{\zeta}, \mathbf{0})^{-1}\left[\bar{\mathbf{f}}(\boldsymbol{\zeta}, \mathbf{0}) + \bar{\mathbf{G}}(\boldsymbol{\zeta}, \mathbf{0})\mathbf{u}\right] \tag{2.79}$$

On introducing the "stretched" fast time scale $\tau = t/\varepsilon$ and considering the limit $\varepsilon \to 0$ in Equation (2.77), we also obtain the following description of the fast dynamics:

$$\begin{aligned}\frac{d\boldsymbol{\zeta}}{d\tau} &= \tilde{\mathbf{B}}(\boldsymbol{\zeta}, \varepsilon\boldsymbol{\eta})\hat{\mathbf{k}}(\boldsymbol{\zeta}, \boldsymbol{\eta}) \\ \frac{d\boldsymbol{\eta}}{d\tau} &= \mathbf{Q}(\boldsymbol{\zeta}, \boldsymbol{\eta})\hat{\mathbf{k}}(\boldsymbol{\zeta}, \boldsymbol{\eta})\end{aligned} \tag{2.80}$$

where $\hat{\mathbf{k}}(\boldsymbol{\zeta}, \boldsymbol{\eta}) = \mathbf{k}(\mathbf{x})_{\mathbf{x}=\mathbf{T}^{-1}(\boldsymbol{\zeta},\boldsymbol{\eta})}$. Note that, because $\hat{\mathbf{k}}(\boldsymbol{\zeta}, \boldsymbol{\eta})$ is nonzero in the fast time scale, the variables $\boldsymbol{\zeta}$ exhibit both fast and slow transients and hence, strictly speaking, are not "true" slow variables. Therefore, in this case, the system (2.51)–(2.52) exhibits a two-time-scale behavior only in a restricted subspace (Kumar and Daoutidis 1999a) where $\mathbf{k}(\mathbf{x})$ is $\mathcal{O}(\varepsilon)$, i.e.,

$$\bar{\mathcal{M}}(\varepsilon) = \{\mathbf{x} \in \mathcal{X} : \mathbf{k}(\mathbf{x}) = \mathcal{O}(\varepsilon)\} \tag{2.81}$$

2.4 Singularly perturbed systems with three or more time scales

The developments presented above have been limited to the case of a single small, singular perturbation parameter being present in the system description. However, in practical applications, e.g., the analysis of complex reaction networks (Vora 2000, Gerdtzen *et al.* 2004) or of processes with physical and chemical phenomena occurring at different rates (Vora and Daoutidis 2001), it is possible that several such parameters ε_i, $i = 1, \ldots, k$, are present. Typically, the values of these parameters are themselves of very different magnitudes, with

$$\frac{\varepsilon_{j+1}}{\varepsilon_j} \to 0 \text{ as } \varepsilon_j \to 0 \tag{2.82}$$

In such cases, the system is said to be in a *multiply singularly perturbed form*, and has the potential to exhibit a dynamic behavior featuring more than two time scales.

As in the case of two-time-scale systems, research work on systems exhibiting more than two time scales has mostly focused on the standard singularly perturbed form (Hoppensteadt 1971). Comparatively, however, such systems (and, in particular, nonstandard multiply singularly perturbed ones) have received far less attention than their two-time-scale counterparts (Vora *et al.* 2006).

Time-scale decomposition and model reduction methods for multiply singularly perturbed systems typically involve the nested application of the procedures discussed so far. For the interested reader, an overview of existing research results concerning the dynamic behavior of multiply singularly perturbed systems is presented in Appendix B.

2.5 Control of singularly perturbed systems

Let us now consider an augmented representation of the standard and nonstandard singularly perturbed systems discussed above, namely

$$\frac{d\mathbf{x}_1}{dt} = \mathbf{f}_1(\mathbf{x}_1, \mathbf{x}_2) + \mathbf{G}_1(\mathbf{x}_1, \mathbf{x}_2)\mathbf{u} \tag{2.83}$$

$$\varepsilon\frac{d\mathbf{x}_2}{dt} = \mathbf{f}_2(\mathbf{x}_1, \mathbf{x}_2) + \mathbf{G}_2(\mathbf{x}_1, \mathbf{x}_2)\mathbf{u} \tag{2.84}$$

$$\mathbf{y} = \mathbf{h}(\mathbf{x}_1, \mathbf{x}_2) \tag{2.85}$$

and

$$\frac{d\mathbf{x}}{dt} = \mathbf{f}(\mathbf{x}) + \mathbf{G}(\mathbf{x})\mathbf{u} + \frac{1}{\varepsilon}\mathbf{B}(\mathbf{x})\mathbf{k}(\mathbf{x}) \tag{2.86}$$

$$\mathbf{y} = \mathbf{h}(\mathbf{x}) \tag{2.87}$$

respectively. Here $\mathbf{u} \in \mathbb{R}^m$ is a vector of manipulated inputs, or "handles" that can be used to change the behavior of the system, $\mathbf{y}$ is a vector of system outputs that singles out the states or combinations of states which can be measured (and need to be controlled), and $\mathbf{G}_1(\mathbf{x}), \mathbf{G}_2(\mathbf{x})$, and $\mathbf{G}(\mathbf{x})$ are matrix functions of appropriate dimensions.

In very generic terms, controller design seeks to use the (inverse of the) model (2.83)–(2.85) or (2.86)–(2.87) to compute the inputs $\mathbf{u}$ as a function of the outputs $\mathbf{y}$ (or the states $\mathbf{x}$), so as to minimize the difference between the latter and a given value, or setpoint.

The majority of the existing literature on the control of singularly perturbed systems considers the two-time-scale, standard form (see, e.g., Kokotović *et al.* 1986, Christofides and Daoutidis 1996a, 1996b). Nevertheless, the methods available for standard singularly perturbed systems can be extended to systems in nonstandard form, since these can be transformed into an equivalent standard form as mentioned above.

For two-time-scale systems, it is well established that inversion-based controllers designed without explicitly accounting for the time-scale multiplicity are ill-conditioned and can lead to closed-loop instability. In order to avoid such issues, controller design must be addressed on the basis of the reduced-order representations of the slow and fast dynamics, an approach referred to as *composite control* (see, e.g., Chow and Kokotović 1976, 1978, Saberi and Khalil 1985, Kokotović *et al.* 1986, Christofides and Daoutidis 1996a, 1996b).

Composite control relies on the use of a single controller consisting of a *fast component* and a *slow component*, which are designed *separately* on the basis of the reduced-order models for the dynamics in the respective time scales (Figure 2.9).

Whether considering linear or nonlinear systems, and regardless of the controller design method employed, the fast component of a composite control system is typically designed to stabilize the fast dynamics (if it is unstable), while the

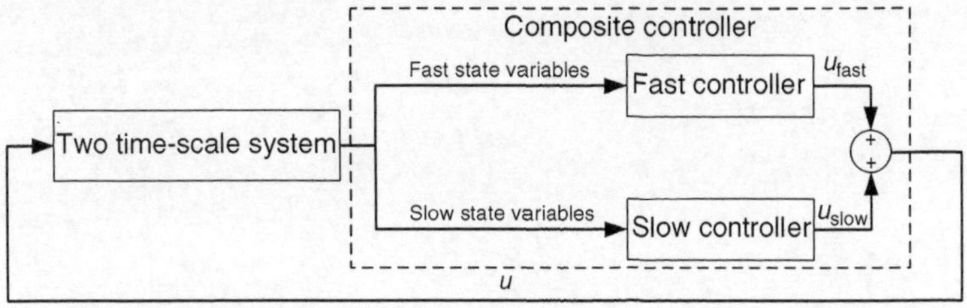

Figure 2.9 Composite control relies on separate, coordinated fast and slow controllers, designed on the basis of the respective reduced-order models, to compute a control action that is consistent with the dynamic behavior of two-time-scale systems.

slow component aims to achieve the desired closed-loop performance objectives on the basis of the slow subsystem, which practically governs the input–output behavior of the overall two-time-scale system.

2.6 Synopsis

This chapter has reviewed existing results in addressing the analysis and control of multiple-time-scale systems, modeled by singularly perturbed systems of ODEs. Several important concepts were introduced, amongst which the classification of perturbations to ODE systems into regular and singular, with the latter subdivided into standard and nonstandard forms. In each case, we discussed the derivation of reduced-order representations for the fast dynamics (in a newly defined stretched time scale, or boundary layer) and the corresponding equilibrium manifold, and for the slow dynamics. Illustrative examples were provided in each case.

We also introduced the idea of composite control, which is based on the use of separate, coordinated controllers for the fast and slow components of the dynamics.

These concepts will be applied extensively in the remainder of the book as the basis for further theoretical developments concerning singularly perturbed systems, and as a support for analyzing the dynamic and control implications of process integration. Specifically, we will demonstrate that many salient design features of integrated processes translate into dynamic models that are in a singularly perturbed form. Subsequently, we will exploit these findings in well-conditioned, hierarchical controller designs that naturally account for the process dynamics and lead to excellent performance at the process level.

Part II

Process systems with material integration

3 Process systems with significant material recycling

3.1 Introduction

Modern chemical plant designs favor "lean" configurations featuring material recycling, fewer units, and diminished material inventory. Together with the elimination of provision for intermediary storage (buffer tanks), these traits intuitively result in significant dynamic interactions between the process units, leading to an intricate dynamic behavior at the process level. Consequently, the design and implementation of advanced, model-based control systems aimed at improving plant performance is a difficult matter, with the complexity, large dimension, and ill conditioning of the process models being major hindrances.

In this chapter, we begin to explore the dynamic impact of process integration. Following closely the arguments in (Kumar and Daoutidis 2002), we consider a specific category of process systems, namely processes with significant (compared with the feed flow rate) material recovery and recycling. We will establish a connection between this steady-state design feature and the transient behavior of the process, followed by postulating a hierarchical control framework that exploits our findings in the realm of process dynamics in the design of well-conditioned, low-dimensional controllers.

3.2 Modeling of process systems with large recycle streams

We consider a generic class of reaction–separation process systems, such as the one in Figure 3.1, consisting of N units (modeled as lumped parameter systems) in series, with one material recycle stream.

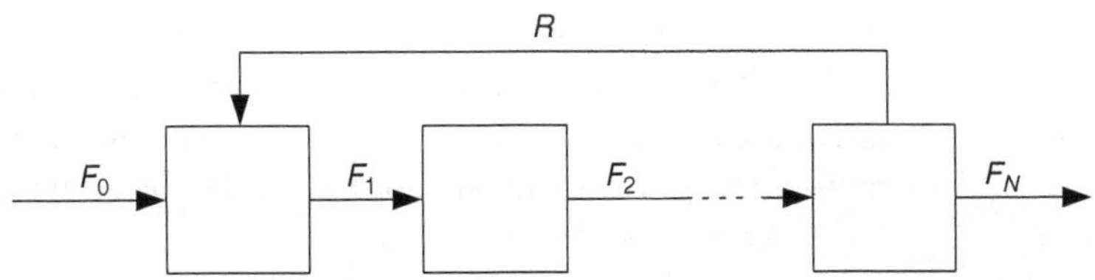

Figure 3.1 A generic integrated process system, featuring a material recycle loop.

In its general form, the mathematical model describing the overall and component material balances of this process can be written as

$$\dot{\mathbf{x}} = \mathbf{f}(\mathbf{x}) + \sum_{j=0,N} \mathbf{g}_j(\mathbf{x})u_j + \mathsf{Rc}\sum_{j=1}^{N-1} k_j\mathbf{g}_j(\mathbf{x})u_j + \mathsf{Rc}\mathbf{g}_{\mathrm{R}}(\mathbf{x})u_{\mathrm{r}} \tag{3.1}$$

where $\mathbf{x}$ is the vector of state variables (i.e., total and component inventories in each unit), the term $\mathbf{f}(\mathbf{x})$ captures the impact of chemical reactions on the material balance, $u_j = (F_j/F_{j,\mathrm{s}})$ represent (possibly manipulated) dimensionless variables that correspond to the flow rates of the process streams, $k_j = F_{j,\mathrm{s}}/F_{\mathrm{R,s}}, j = 1, \ldots, N$, and $\mathbf{g}_j(\mathbf{x})$ and $\mathbf{g}_{\mathrm{R}}(\mathbf{x})$ are vector functions of appropriate dimensions. The subscript s denotes steady-state values.

Note that the model explicitly identifies the terms that involve the feed and product flow rates ($j = 0, N$), the internal flow rates ($j = 1, \ldots, N-1$), and the recycle ($j = \mathrm{R}$).

In order to investigate the impact of the presence and magnitude of the recycle stream on the process dynamics, Equation (3.1) also makes use of the recycle number **Rc**, which we define below.

Definition 3.1. *The **recycle number** of a material recycle loop in an integrated process is a process-wide dimensionless number, expressed as the ratio of the (steady-state) flow rates of the recycle stream and the process throughput, as captured by the (total) flow rate of the process feed stream(s):*

$$\mathsf{Rc} = \frac{R_{\mathrm{s}}}{F_{0\mathrm{s}}} \tag{3.2}$$

Evidently, for a given production rate (set by F_0), the value of **Rc** varies with the flow rate of the recycle stream. Two limit cases are of interest.

Case I. $\mathsf{Rc} \ll 1$: in this case, the recycle flow rate is much smaller than the feed flow rate, the internal flow rates are comparable to the feed flow rate, and $k_j \gg 1$, $\mathsf{Rc}k_j = \tilde{k}_j = \mathcal{O}(1), j = 1, \ldots, N-1$. In the limit as the recycle flow rate tends to zero ($\mathsf{Rc} \to 0$), the last term in Equation (3.1) vanishes, and the model reduces to the model of a series system:

$$\dot{\mathbf{x}} = \mathbf{f}(\mathbf{x}) + \sum_{j=0,N} \mathbf{g}_j(\mathbf{x})u_j + \sum_{j=1}^{N-1} \tilde{k}_j\mathbf{g}_j(\mathbf{x})u_j \tag{3.3}$$

In this case, the model of the process with recycle is a regular perturbation of the nominal (no recycle) series system. In light of the concepts introduced in Chapter 2, we can expect that the presence of the (small) material recycle stream will not have a significant impact on the dynamics of the process.

Case II. $\mathsf{Rc} \gg 1$. in this case the recycle flow rate is much larger than the feed flow rate, the internal flow rates are equally large and $k_j = \mathcal{O}(1)$,

$j = 1, \ldots, N-1$. In this context, the model of Equation (3.1) can be rewritten as

$$\dot{\mathbf{x}} = \mathbf{f}(\mathbf{x}) + \sum_{j=0,N} \mathbf{g}_j(\mathbf{x})u_j + \frac{1}{\varepsilon}\left[\sum_{j=1}^{N-1} k_j \mathbf{g}_j(\mathbf{x})u_j + \mathbf{g}_{\mathrm{R}}(\mathbf{x})u_{\mathrm{r}}\right] \tag{3.4}$$

with

$$\varepsilon = \frac{1}{\mathsf{Rc}}, \quad 0 < \varepsilon \ll 1 \tag{3.5}$$

The model in (3.4) is a system of equations in singularly perturbed form. Considering again the theory presented in Chapter 2, we can expect the dynamic behavior of systems with large recycle to differ significantly from the dynamics of the nominal system with N units in series.

Example 3.1.

A process consisting of an isothermal continuously stirred tank reactor of volume V and an ideal separator (Figure 3.2) converts a feed stream of flow rate F_0, containing the reactant A (concentration C_{A0}) to product B in the first-order reaction

$$\mathrm{A} \rightarrow \mathrm{B} \tag{3.6}$$

with the reaction rate

$$r = kC_{\mathrm{A}} \tag{3.7}$$

Complete conversion in the outlet stream is desired (i.e., $C_{\mathrm{Aout}} = 0$) and the reactor effluent is separated, with the unreacted A being recycled to the reactor at a flow rate R. It is assumed for simplicity that $C_{\mathrm{AR}} = C_{\mathrm{A0}}$.

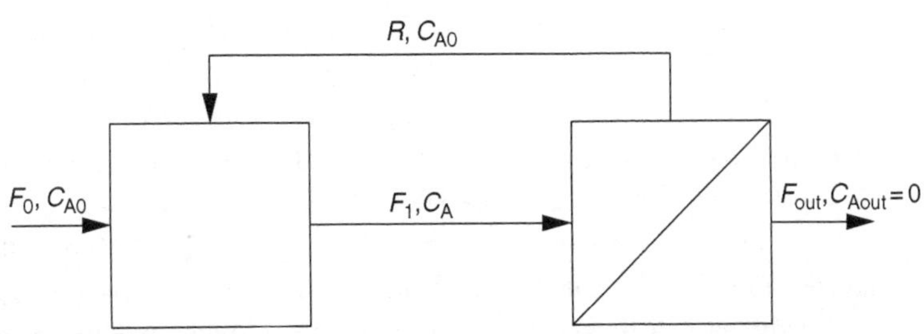

Figure 3.2 A process system consisting of a reactor and an ideal separator, with recycling of the unreacted feed A.

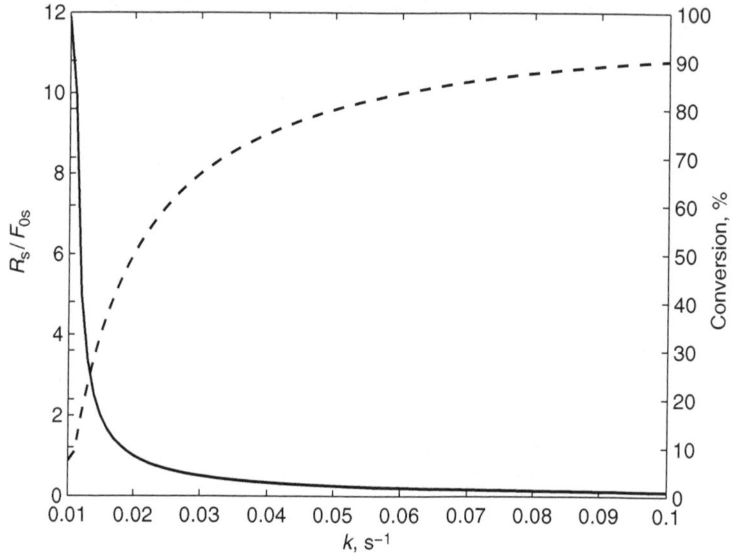

Figure 3.3 Recycle flow rates (presented in terms of the recycle number $\mathbf{Rc} = R_s/F_{0s}$) required to achieve complete conversion at the system outlet (solid line, left ordinate) and corresponding reactor conversion (dashed, right ordinate) as a function of the reaction rate constant.

The flow rate of the recycle stream, R, required to achieve complete conversion at the process outlet can be computed as a function of the reaction rate constant and the reactor volume from the steady-state material balance equations:

$$0 = \frac{V}{F_0 + R} - \frac{C_{A0} - C_A}{kC_A} \tag{3.8}$$

$$0 = F_1 C_A - RC_{A0} \tag{3.9}$$

On solving the above equations for $C_{A0} = 1000\ \text{mol/m}^3$, $V = 1\ \text{m}^3$, and $F_0 = 0.01\ \text{m}^3/\text{s}$, for different values of k, we obtain the dependences represented in Figure 3.3.

The results in Figure 3.3 are quite intuitive: for the selected reactor size, the conversion in the reactor is low at low values of the reaction rate constant k. The reactor effluent will therefore contain a significant amount of reactant A, which must be separated and recycled, and in this case the flow rate of the material recycle stream is very high compared with the flow rate of the feed stream (Case II). Conversely, at high single-pass conversions, the required recycle flow rate drops sharply (Case I).

Evidently, changes in the reactor size impact on the above findings: allowing an increase in the reactor holdup leads to an increase in the single-pass conversion and reduces the flow rate of the material recycle stream. While plant configurations with low reactor capacity are preferred in processes featuring multiple reactions with valuable intermediate products (Luyben 1993b), the optimal sizes of the reactor and separation equipment can be determined only by considering the tradeoff between capital and operating costs[1] and the profit generated from the products.

Example 3.1 highlights the fact that the presence of (and need for) material recycle streams with significant flow rates is entirely a **steady-state design** feature of a process. In what follows, we will focus on investigating the profound impact of this feature on the **dynamics and control** of the processes under consideration.

To this end, let us rewrite the model in Equation (3.4) in a more general form:

$$\dot{\mathbf{x}} = \mathbf{f}(\mathbf{x}) + \mathbf{G}^{\mathrm{s}}(\mathbf{x})\mathbf{u}^{\mathrm{s}} + \frac{1}{\varepsilon}\mathbf{G}^{\mathrm{l}}(\mathbf{x})\mathbf{u}^{\mathrm{l}} \tag{3.10}$$

where, as above, $\mathbf{x}$ is the vector of state variables, $\mathbf{u}^{\mathrm{s}} \in \mathbb{R}^{m^{\mathrm{s}}}$ is the vector of scaled input variables that correspond to the small flow rates, $\mathbf{u}^{\mathrm{l}} \in \mathbb{R}^{m^{\mathrm{l}}}$ is the vector of scaled input variables that correspond to the large flow rates of the internal streams, and $\mathbf{G}^{\mathrm{s}}(\mathbf{x})$ and $\mathbf{G}^{\mathrm{l}}(\mathbf{x})$ are matrices of appropriate dimensions.

According to the developments in Section 2.3, the model of Equation (3.10) is in a *nonstandard* singularly perturbed form. We thus expect its dynamics (and, consequently, the dynamics of integrated process systems with large material recycle) to feature two distinct time scales. However, the analysis of the system dynamics is complicated by the presence of the term $\mathbf{u}^{\mathrm{l}}$, which, as we will see below, precludes the direct application of the methods presented in Chapter 2 for deriving representations of the slow and fast components of the system dynamics.

3.3 Model reduction

3.3.1 Fast dynamics

We define the fast, "stretched" time scale $\tau = t/\varepsilon$. On rewriting Equation (3.10) in this time scale and considering the limit case $\varepsilon \to 0$ (which physically corresponds to an infinitely large recycle number or, equivalently, an infinitely high recycle flow rate), we obtain a description of the fast dynamics of the process:

$$\frac{d\mathbf{x}}{d\tau} = \mathbf{G}^{\mathrm{l}}(\mathbf{x})\mathbf{u}^{\mathrm{l}} \tag{3.11}$$

Note that the above model of the fast dynamics involves only the large recycle and internal flow rates $\mathbf{u}^{\mathrm{l}}$, and does not involve the small feed/product flow rates $\mathbf{u}^{\mathrm{s}}$. Examining the material balance equations in (3.1), it is intuitive that the flow rates of the internal streams do not affect the *total holdup* of any component in the process, and that total holdups are affected only by the flow rates $\mathbf{u}^{\mathrm{s}}$ of the

[1] Associated, respectively, with increasing the reactor and separation equipment size, and with the amount of energy required by daily operations.

feed/product streams. In other words, Equation (3.11) effectively describes the dynamics of the individual units in the recycle loop – note that, by definition, τ is of the order of magnitude of the residence time of an individual process unit – and does not capture the overall (process-level) changes in the material balance of the process. We can use this observation to further infer the following.

- The differential equations (3.11) are not linearly independent. Consequently, as indicated in Section 2.3, the steady-state condition

$$\mathbf{0} = \mathbf{G}^{\mathrm{l}}(\mathbf{x})\mathbf{u}^{\mathrm{l}} \tag{3.12}$$

  for the fast dynamics (3.11) does not specify a set of isolated equilibrium points, but rather a low-dimensional equilibrium subspace (manifold), in which a slow component of the system dynamics evolves.
- The slow component of the process dynamics is associated with the evolution of the total material holdup of the process and with the total holdups of the chemical components present in the process.
- From physical considerations, at most $\mathcal{C}$ equations (with $\mathcal{C}$ being the number of chemical components) are required in order to completely capture the above overall, process-level material balance. Thus, we can expect the dimension of the system of equations describing the slow dynamics of the process to be at most $\mathcal{C}$, and the equilibrium manifold (3.12) of the fast dynamics to be at most $\mathcal{C}$-dimensional.

Remark 3.1. *In contrast to the theory presented thus far (Section 2.3), the algebraic constraints of (3.12) incorporate a set of (unknown) manipulated inputs,* $\mathbf{u}^{\mathrm{l}}$*. The equilibrium manifold described by (3.12) is thus referred to as* ***control-dependent****.*

3.3.2 Slow dynamics

In order to obtain the description of the slow dynamics, and using the developments in Section 2.3, we make the following, typically true (see, e.g., Kumar and Daoutidis 2002, Contou-Carrère *et al.* 2004, Baldea *et al.* 2006, Baldea and Daoutidis 2007), assumption.

Assumption 3.1. *The matrix* $\mathbf{G}^{\mathrm{l}}(\mathbf{x})$ *can be decomposed as* $\mathbf{G}^{\mathrm{l}}(\mathbf{x}) = \mathbf{B}(\mathbf{x})\tilde{\mathbf{G}}^{\mathrm{l}}(\mathbf{x})$*, with* $\mathbf{B}(\mathbf{x}) \in \mathbb{R}^{n\times(n-\mathcal{C})}$ *being a full column rank matrix and the matrix* $\tilde{\mathbf{G}}^{\mathrm{l}}(\mathbf{x}) \in \mathbb{R}^{(n-\mathcal{C})\times m^{\mathrm{l}}}$ *having linearly independent rows.*

We now multiply Equation (3.10) by ε and consider the limit of an infinitely high recycle flow rate (i.e., $\varepsilon \to 0$) in the original time scale t. In this limit, we obtain the constraints in Equation (3.12), or equivalently, the linearly independent constraints

$$\tilde{\mathbf{G}}^{\mathrm{l}}(\mathbf{x})\mathbf{u}^{\mathrm{l}} = \mathbf{0} \tag{3.13}$$

which must be satisfied in the slow time scale.

Also in the limit $\varepsilon \to 0$, the terms

$$\frac{\bar{\mathbf{G}}^{\mathrm{l}}(\mathbf{x})\mathbf{u}^{\mathrm{l}}}{\varepsilon} \tag{3.14}$$

which correspond to the *differences* of large flow rates present in the material balance equations of every process unit become indeterminate. On defining

$$\mathbf{z} = \lim_{\varepsilon \to 0} \frac{\bar{\mathbf{G}}^{\mathrm{l}}(\mathbf{x})\mathbf{u}^{\mathrm{l}}}{\varepsilon} \tag{3.15}$$

as the vector of these finite, but unknown terms, the system in Equation (3.10) becomes

$$\begin{aligned} \dot{\mathbf{x}} &= \mathbf{f}(\mathbf{x}) + \mathbf{G}^{\mathrm{s}}(\mathbf{x})\mathbf{u}^{\mathrm{s}} + \bar{\mathbf{B}}(\mathbf{x})\mathbf{z} \\ \mathbf{0} &= \bar{\mathbf{G}}^{\mathrm{l}}(\mathbf{x})\mathbf{u}^{\mathrm{l}} \end{aligned} \tag{3.16}$$

which represents the model of the slow dynamics of the process induced by the large recycle flow rate.

As anticipated in Remark 3.1, the constraints in Equation (3.13) depend on $\mathbf{u}^{\mathrm{l}}$. In other words, the slow dynamics of the process cannot be completely characterized (in the sense of obtaining a reduced-order ODE representation of the type (2.48)) prior to defining $\mathbf{u}^{\mathrm{l}}$ as a function of the process state variables (or measured outputs) via an appropriate control law. These issues are addressed in the following section.

Remark 3.2. *We can regard the developments above from a converse perspective. Namely, if we consider the model of each individual unit (preserving the input and output flow structure of the process) in the fast time scale τ, we can write a simplified model for unit i in the form*

$$\frac{d\mathbf{x}}{d\tau} = \varepsilon \mathbf{f}_i(\mathbf{x}) + \varepsilon \mathbf{G}_i^{\mathrm{s}}(\mathbf{x})\mathbf{u}_i^{\mathrm{s}} + \mathbf{G}_i^{\mathrm{l}}(\mathbf{x})\mathbf{u}_i^{\mathrm{l}} \tag{3.17}$$

with $\mathbf{f}_i(\mathbf{x})$ being a vector field, and $\mathbf{G}_i^{\mathrm{s}}$ and $\mathbf{G}_i^{\mathrm{l}}$ matrices of appropriate dimensions. $\mathbf{u}_i^{\mathrm{s}}$ denote the inputs corresponding to any small mass flows and $\mathbf{u}_i^{\mathrm{l}}$ the inputs corresponding to any large mass flows into the process unit. The model of the process unit written in the form (3.17) is a regularly perturbed *model. In the limit case of an infinitely small recycle number, $\varepsilon \to 0$ (which, in this context, can intuitively be viewed as the limit in which the flow rates of the small streams $\mathbf{u}_i^{\mathrm{s}}$ are reduced to zero), the model reduces to*

$$\frac{d\mathbf{x}}{d\tau} = \mathbf{G}_i^{\mathrm{l}}(\mathbf{x})\mathbf{u}_i^{\mathrm{l}} \tag{3.18}$$

This result further confirms that the dynamics at the unit level are strongly influenced by the flow rates of the large internal material streams $\mathbf{u}_i^{\mathrm{l}}$, while the flow rates of the small process input and output streams, $\mathbf{u}_i^{\mathrm{s}}$, have little impact on the fast dynamics.

Moreover, the fact that, at the unit level, the presence of flow rates of vastly different magnitudes is modeled as a regular *perturbation, while at the process level it is modeled as a* singular *perturbation, establishes that it is the* use of recycle streams *of large flow rates that induces the time-scale multiplicity at the process level.*

3.4 Control of integrated processes with large recycle

3.4.1 Hierarchical controller design

The two-time-scale behavior of the material balance of integrated processes with large recycle suggests the use of a hierarchical control structure with two tiers of control action:

- distributed control, addressing control objectives for individual process units in the fast time scale
- supervisory control, addressing control objectives for the overall process in the slow time scale.

To this end, let us complete the description of Equation (3.10) with a vector of output variables $\mathbf{y}$:

$$\begin{aligned} \dot{\mathbf{x}} &= \mathbf{f}(\mathbf{x}) + \mathbf{G}^{\mathrm{s}}(\mathbf{x})\mathbf{u}^{\mathrm{s}} + \frac{1}{\varepsilon}\mathbf{G}^{\mathrm{l}}(\mathbf{x})\mathbf{u}^{\mathrm{l}} \\ \mathbf{y} &= \mathbf{h}(\mathbf{x}) \end{aligned} \tag{3.19}$$

and let us further define

$$\mathbf{y} = \begin{bmatrix} \mathbf{y}^{\mathrm{l}} \\ \mathbf{y}^{\mathrm{s}} \end{bmatrix} \tag{3.20}$$

where $\mathbf{y}^{\mathrm{l}}$ denotes the subset of the output variables which are associated with control objectives for the individual process units (typically involving the control of local inventories, i.e., the stabilization of liquid holdups or gas pressures) and $\mathbf{y}^{\mathrm{s}}$ those associated with control objectives for the overall process, e.g., production rate, total inventory, and product quality.

The above time-scale decomposition provides a transparent framework for the selection of manipulated inputs that can be used for control in the two time scales. Specifically, it establishes that output variables $\mathbf{y}^{\mathrm{l}}$ need to be controlled in the fast time scale, using the large flow rates $\mathbf{u}^{\mathrm{l}}$, while the control of the variables $\mathbf{y}^{\mathrm{s}}$ is to be considered in the slow time scale, using the variables $\mathbf{u}^{\mathrm{s}}$. Moreover, the reduced-order approximate models for the fast (Equation (3.11)) and slow (the state-space realization of Equation (3.16)) dynamics can serve as a basis for the synthesis of well-conditioned nonlinear controllers in each time scale.

3.4.2 Control of the fast dynamics

Owing to the dependence of the state space where the slow dynamics evolves on the large flow rates (Remark 3.1), the fast-controller design must precede the slow-controller design.

The design of the fast distributed controllers for the individual units can, in general, be addressed as a collection of individual control problems, where the strictness of the operational requirements for each unit dictates the complexity of the corresponding controller; typical applications rely on the use of simple linear controllers, e.g., proportional (P), proportional–integral (PI) or proportional–integral–derivative (PID).

Tikhonov's theorem (Theorem 2.1) indicates a further requirement that must be fulfilled by the controllers in the fast time scale: in order for the time-scale decomposition developed above to remain valid, these controllers must ensure the exponential stability of the fast dynamics. From a practical point of view, this is an intuitive requirement: one cannot expect stability and control performance at the process level if the operation of the process units is not stable.

3.4.3 Control in the slow time scale

The design and implementation of the fast controllers allows the derivation of a minimal-order realization of the DAE model of the slow dynamics in Equation (3.16). For illustration purposes, let us assume that the fast controllers are defined by the static state-feedback control law

$$\mathbf{u}^{\mathrm{l}} = \mathbf{K}\left(\mathbf{y}^{\mathrm{l}} - \mathbf{y}^{\mathrm{l}}_{\mathrm{sp}}\right) \tag{3.21}$$

with $\mathbf{y}^{\mathrm{l}}_{\mathrm{sp}}$ denoting the setpoint for the outputs $\mathbf{y}^{\mathrm{l}}$, which stabilizes the fast dynamics and induces the desired closed-loop response in the fast time scale. Under the control law (3.21), the slow model in Equation (3.16) becomes

$$\begin{aligned} \dot{\mathbf{x}} &= \mathbf{f}(\mathbf{x}) + \mathbf{G}^{\mathrm{s}}(\mathbf{x})\mathbf{u}^{\mathrm{s}} + \bar{\mathbf{B}}(\mathbf{x})\mathbf{z} \\ \mathbf{0} &= \bar{\mathbf{G}}^{\mathrm{l}}(\mathbf{x})\mathbf{K}\left(\mathbf{y}^{\mathrm{l}} - \mathbf{y}^{\mathrm{l}}_{\mathrm{sp}}\right) \end{aligned} \tag{3.22}$$

Using the methods presented in Chapter 2, the above formulation can be used to derive a state-space realization of the slow dynamics of the type in Equation (2.48). The resulting low-dimensional model should subsequently form the basis for formulating and solving the control problems associated with the slow time scale, i.e., stabilization, output tracking, and disturbance rejection at the process level.

From a practical perspective, this is the model that should be used to design a (multivariable) controller that manipulates the inputs $\mathbf{u}^{\mathrm{s}}$ to fulfill the control objectives $\mathbf{y}^{\mathrm{s}}$. It is important to note that the availability of a low-order ODE model of the process-level dynamics affords significant flexibility in designing the supervisory control system, since any of the available inversion- or optimization-based (e.g., Kravaris and Kantor 1990, Mayne *et al.* 2000, Zavala and Biegler 2009) control methods for nonlinear ODE systems can be applied. Using the reduced-order model for controller synthesis is also beneficial from an implementation point of view: the reduced dimensions and improved conditioning (reduced stiffness) compared with those of the original model (3.10) will result in shorter online calculation times and less sensitivity to noise and disturbances.

3.4.4 Cascaded control configurations

Practical considerations in implementing the hierarchical control framework developed above concern the availability of manipulated inputs to address the control objectives in the slow time scale (it is possible that $\dim(\mathbf{u}^{\mathrm{s}}) < \dim(\mathbf{y}^{\mathrm{s}})$), as well as achieving a tighter coordination between the distributed and supervisory control layers. Both issues are effectively addressed by using a cascaded control configuration, which extends the choice of controlled variables in the slow time scale to include the *setpoints* $\mathbf{y}^{\mathrm{l}}_{\mathrm{sp}}$ of the distributed controllers.

In this case, however, the algebraic constraints in the DAE system describing the slow dynamics (3.22) will explicitly involve manipulated input variables (i.e., $\mathbf{y}^{\mathrm{l}}_{\mathrm{sp}}$), and the direct application of the methods in Section 2.3 for the derivation of a state-space realization of the slow dynamics is not possible.[2]

It can be shown (Kumar and Daoutidis 1996, Contou-Carrère *et al.* 2004) that, under some mild assumptions (including Assumption 3.1), the models of the process systems under consideration can be transformed into regular DAE systems by introducing an additional set of appropriately defined differential variables, i.e., by constructing a dynamic extension of the process model. Within this framework, considering that a subset $\bar{\mathbf{y}}^{\mathrm{l}}_{\mathrm{sp}} \subseteq \mathbf{y}^{\mathrm{l}}_{\mathrm{sp}}$ of the setpoints of the fast controllers are used as manipulated inputs in the slow time scale, the dynamic extension

$$\bar{\mathbf{y}}^{\mathrm{l}}_{\mathrm{sp}} = \mathbf{w} \tag{3.23}$$

$$\dot{\mathbf{w}} = \mathbf{v}_1 \tag{3.24}$$

$$\mathbf{u}^{\mathrm{s}} = \mathbf{v}_2 \tag{3.25}$$

transforms the (nonregular) DAE model of Equation (3.22) into an index-2 DAE system whose state space is independent of the new vector of manipulated inputs $\mathbf{v} = \left[\mathbf{v}_1^{\mathrm{T}}\ \mathbf{v}_2^{\mathrm{T}}\right]^{\mathrm{T}}$, which can subsequently be used for obtaining an ODE representation of the slow dynamics as described above.

Remark 3.3. *The hierarchical control structure proposed in this chapter (illustrated in Figure 3.4) is dissimilar to the composite control approach reviewed in*

[2] Differential algebraic equation systems whose algebraic equations explicitly involve manipulated input variables are referred to as *nonregular* (Kumar and Daoutidis 1999a). See also Definition A.6.

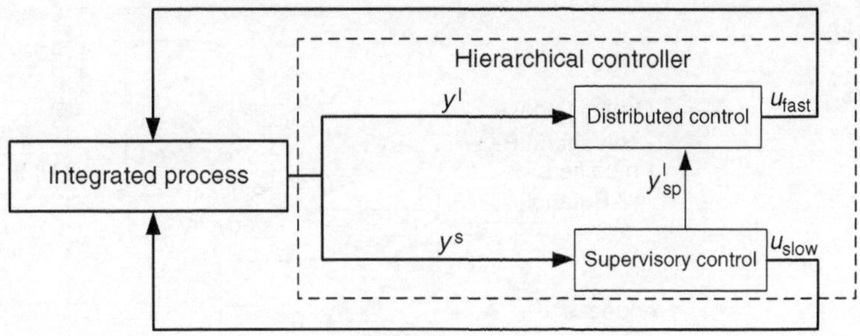

Figure 3.4 Hierarchical control relies on separate, but coordinated, fast and slow controllers, designed on the basis of the respective reduced-order models, to compute the values of the separate inputs that influence the fast and slow dynamics of the process. Tighter coordination between the distributed and supervisory control layers is achieved by using a cascaded configuration.

Section 2.2, in that two layers *of control action involving* separate controllers *are proposed, whereas composite control relies on a* single *(possibly multivariable)* controller *with* two components, *a fast one and a slow one. Thus, the hierarchical control structure accounts for the separation of the flow rates of the process streams into* two groups of inputs *that act upon the dynamics in the different time scales. On the other hand, composite controller design (Figure 2.9) presupposes that the available manipulated inputs impact both the fast and the slow dynamics and relies on* one set of inputs *to regulate both components of the system dynamics.*

Remark 3.4. *In the context of the present chapter (and of the remainder of the book), the term* hierarchical *control structure reflects the use of two (or multiple) coordinated tiers of control action, and should not be confused with "hierarchical plant-wide controller design strategies" (see, e.g. Ponton and Laing 1993, Luyben* et al. *1997, Zheng* et al. *1999, Antelo* et al. *2007, Scattolini 2009, and references therein), which use the term "hierarchy" to denote a set of guidelines, to be followed in sequence, for designing the control system for a chemical plant.*

Example 3.2. Non-aqueous solvents are used as a reaction medium when an aqueous synthesis route is not possible or economically feasible. Organic solvents (alkanes, ethers, etc.) can facilitate product and enzyme recovery in enzyme-catalyzed processes (Zaks and Klibanov 1985, Illanes 2008, Kobayashi 2009, Adams *et al.* 2009). Ionic-liquids-based processes have garnered attention in recent years owing to improved environmental safety and enhanced solvent recovery (Wheeler *et al.* 2001, van Rantwijk *et al.* 2003, Yang and Pan 2005, Sureshkumar and Lee 2009). All such processes require an effective solvent-recovery strategy; incorporating solvent recycling into the process design reduces the need for expensive make-up solvents and helps meet environmental regulations.

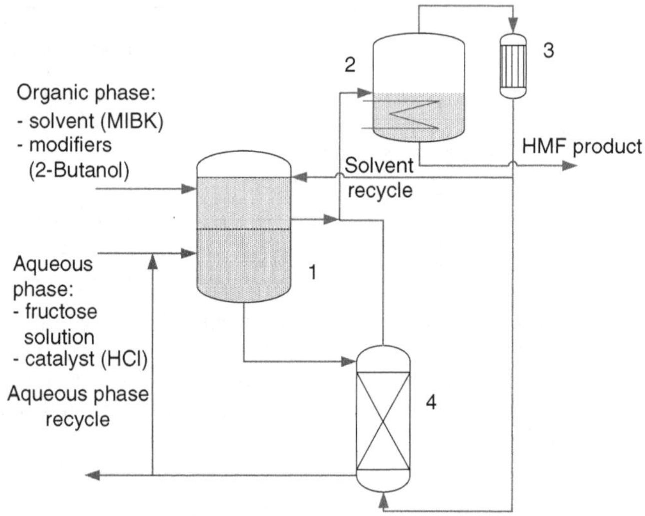

Figure 3.5 HMF production using a solvent-based process (1: two-phase reactor, 2: evaporator, 3: condenser, 4: extractor).

5-hydroxymethylfurfural (HMF), a furanic compound, is currently regarded as a key intermediate in the production of biomass-derived fuels and chemicals. In particular, HMF can be used to synthesize building blocks for the production of polymer analogs that are at present produced from petroleum (e.g., polyethylene terephthalate, polybutylene terephthalate).

HMF can be produced through the acid-catalyzed dehydration of fructose in aqueous media. This synthesis path is highly non-selective, leading to several soluble and insoluble products besides HMF (Kuster and Temmink 1977). As an additional complication, HMF reacts with water yielding levulinic and formic acids, and fructose–fructose and fructose–HMF oligomers. A recently developed (Roman-Leshkov *et al.* 2006, Torres *et al.* 2010) process for the production of HMF relies on solvent extraction to continuously remove HMF from the reaction system. The process (Figure 3.5) consists of a biphasic reactor coupled with an extractor and an evaporator. A fructose solution and a soluble acid catalyst are fed into a two-phase continuously stirred tank reactor. Fructose dehydration takes place in the aqueous phase, while the organic phase, composed of 7:3 methyl iso-butylketone (MIBK):2-butanol, selectively extracts the HMF produced, thus minimizing its decomposition. The aqueous stream from the reactor enters a liquid–liquid extractor, where any residual aqueous HMF is recovered. The aqueous stream is subsequently recycled to the reactor. A purge stream is used to prevent the accumulation of byproducts in the aqueous phase. The streams exiting the organic phase of the reactor and the extractor are sent to the evaporator, where purified HMF is obtained as a liquid product. The evaporated solvent is condensed and cycled back to the extractor and the reactor. A small make-up of fresh solvent compensates for any solvent losses.

An optimally designed process (Torres *et al.* 2010) is capable of producing HMF of 95% purity at the evaporator liquid exit. Solvent losses are therefore very small (about 5% of the product flow rate) and the required solvent make-up flow is of comparable magnitude. Process performance is enhanced by operating all the units (except the evaporator) at high dilution, which minimizes the extent of secondary reactions. This, in turn, requires that a significant amount of solvent be used (and recycled) in the system: the solvent recycle flow rate is two orders of magnitude (about 600 times) larger than the fresh-solvent feed flow, and the recycled solvent stream is largely devoid of HMF.

Following the path of the MIBK–butanol mixture from the feed to the HMF product stream, we notice that the process in Figure 3.5 is in effect a process with significant material recycling. Specifically, it features a large *solvent* recycle stream.

This observation can be used to make some important inferences concerning the dynamics and control of solvent-based HMF production. In light of the arguments developed in this chapter, we can expect that the dynamics of the process in Figure 3.5 exhibit a fast component, related to the dynamics of the individual process units, and a slow component, associated with the total material holdup of the process and the total holdup of solvent in the solvent recycle loop. More specifically, we should expect the solvent content (and, evidently, the actual product content) of the product stream to change slowly in response to changes in the manipulated inputs. Controlling the inventory at the process level and product purity control should therefore be addressed over a longer time horizon, using a separate controller designed on the basis of the model of the slow dynamics (Jogwar *et al.* 2011).

3.5 Case study: control of a reactor–distillation–recycle process

3.5.1 Process description

We consider the process of Figure 3.6, consisting of a continuously stirred tank reactor and a distillation column. A series of irreversible first-order reactions transforms the reactant A, fed to the CSTR at a flow rate F_0, into product P_1 and undesired byproduct P_2:

$$\mathrm{A} \xrightarrow{k_1} \mathrm{P}_1 \xrightarrow{k_2} \mathrm{P}_2, \quad k_1 = k_2 \tag{3.26}$$

The reactor effluent, consisting of a mixture of unreacted A with the products P_1 and P_2, is separated in the distillation column. The reactor effluent is fed to the column on the feed stage N_f (stages are numbered $1, \ldots, N$ from top to bottom) at a flow rate F. The reactant A is the lightest component of the reactor-effluent mixture and separates at the top of the column, being subsequently completely

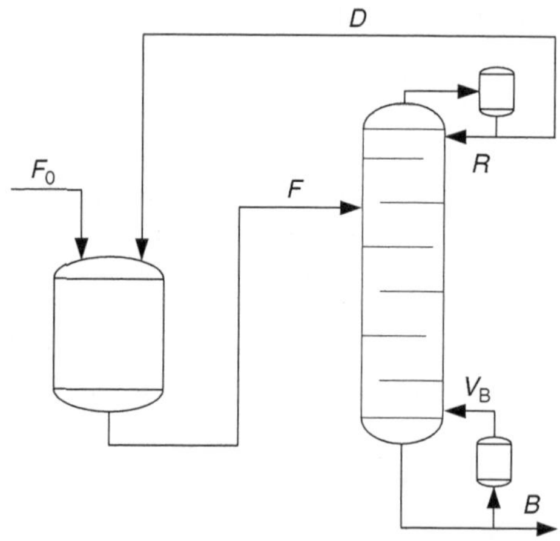

Figure 3.6 A reactor-separator process, consisting of a CSTR and a distillation column. The unreacted feed material is recycled to the reactor.

recycled to the reactor, at a flow rate D. The bottom stream of the column (of flow rate B) consists of the heavier products P_1 and P_2, which are separated in a downstream column.

The operating targets for this process call for a high conversion of A, as well as high selectivity in the desired product P_1. From steady-state considerations, both objectives can be achieved by maintaining a low single-pass conversion in the reactor and using a recycle flow rate D much larger than the fresh-reactant feed flow rate F_0. A large recycle flow rate also implies that F, the flow rate of the reactor-effluent stream, and V, the column vapor flow rate, are large, while R and B are comparable to F_0. The nominal values of the process variables are presented in Table 3.1. For this process, we assume that

- the thermal effects of the reactions are negligible
- the latent heats of vaporization of A, P_1, and P_2 are comparable
- the relative volatilities α_1 of A, α_2 of P_1, and α_3 of P_2 are such that $\alpha_1 > \alpha_2 > \alpha_3 = 1$
- a trayed column is used, with each tray being equal to a theoretical stage
- the molar overflow from the column trays is constant
- the flow rates of the material streams F, V, B, R, and D are available as manipulated inputs.

The control objectives of the process are to

- stabilize the liquid levels/holdups in the reactor, condenser, and column reboiler, and
- control the concentration of P_1, $x_{2,B}$, in the column bottom (product) stream.

Table 3.1. Nominal values of the process variables (adapted from (Luyben 1993b))

F_0	100.00 mol/h	F	1880.00 mol/h
D	1780.00 mol/h	R	290.00 mol/h
V_{B}	2070.00 mol/h	B	100.00 mol/h
k_1	1.10 h^{-1}	k_2	1.10 h^{-1}
α_1	4.50	α_2	2.10
M_{R}	101.53 mol	M_{D}	173.00 mol
M_i	175.00 mol	M_{B}	181.00 mol
$x_{1,\mathrm{R}}$	0.8934	$x_{2,\mathrm{R}}$	0.0998
$x_{1,\mathrm{D}}$	0.9434	$x_{2,\mathrm{D}}$	0.0556
$x_{1,\mathrm{B}}$	0.00221	$x_{2,\mathrm{B}}$	0.8863
N	15 trays	N_{f}	4

Under the above modeling assumptions, the dynamic model of the reactor–column–recycle system consists of the material balance for the total molar holdup of the reactor, condenser, and reboiler, and component-wise balances for the reactant A and product P_1 in the reactor, condenser, reboiler, and column trays, having a total of $2N+9$ differential equations. Specifically,

$$\left.\begin{aligned} \dot{M}_{\mathrm{R}} &= F_0 - F + D \\ \dot{x}_{1,\mathrm{R}} &= \frac{1}{M_{\mathrm{R}}}\left[F_0(1-x_{1,\mathrm{R}}) + D(x_{1,\mathrm{D}} - x_{1,\mathrm{R}})\right] - k_1 x_{1,\mathrm{R}} \\ \dot{x}_{2,\mathrm{R}} &= \frac{1}{M_{\mathrm{R}}}\left[-F_0 x_{2,\mathrm{R}} + D(x_{2,\mathrm{D}} - x_{2,\mathrm{R}})\right] + k_1 x_{1,\mathrm{R}} - k_2 x_{2,\mathrm{R}} \end{aligned}\right\} \text{reactor}$$

$$\left.\begin{aligned} \dot{M}_{\mathrm{C}} &= V_{\mathrm{B}} - R - D \\ \dot{x}_{1,\mathrm{D}} &= \frac{V_{\mathrm{B}}}{M_{\mathrm{C}}}(y_{1,1} - x_{1,\mathrm{D}}) \\ \dot{x}_{2,\mathrm{D}} &= \frac{V_{\mathrm{B}}}{M_{\mathrm{C}}}(y_{2,1} - x_{2,\mathrm{D}}) \end{aligned}\right\} \text{condenser}$$

$$\left.\begin{aligned} \dot{x}_{1,i} &= \frac{1}{M_i}\left[V_{\mathrm{B}}(y_{1,i+1} - y_{1,i}) + R(x_{1,i-1} - x_{1,i})\right] \\ \dot{x}_{2,i} &= \frac{1}{M_i}\left[V_{\mathrm{B}}(y_{2,i+1} - y_{2,i}) + R(x_{2,i-1} - x_{2,i})\right] \end{aligned}\right\} \text{tray } i < N_{\mathrm{f}}$$

$$\left.\begin{aligned}\dot{x}_{1,i} &= \frac{1}{M_i}[V_B(y_{1,i+1} - y_{1,i}) + R(x_{1,i-1} - x_{1,i})\\ &\quad + F(x_{1,f} - x_{1,i})]\\ \dot{x}_{2,i} &= \frac{1}{M_i}[V_B(y_{2,i+1} - y_{2,i}) + R(x_{2,i-1} - x_{2,i})\\ &\quad + F(x_{2,f} - x_{2,i})]\end{aligned}\right\} \text{feed tray } i = N_f \tag{3.27}$$

$$\left.\begin{aligned}\dot{x}_{1,i} &= \frac{1}{M_i}[V_B(y_{1,i+1} - y_{1,i}) + R(x_{1,i-1} - x_{1,i})\\ &\quad + F(x_{1,i-1} - x_{1,i})]\\ \dot{x}_{2,i} &= \frac{1}{M_i}[V_B(y_{2,i+1} - y_{2,i}) + R(x_{2,i-1} - x_{2,i})\\ &\quad + F(x_{2,i-1} - x_{2,i})]\end{aligned}\right\} \text{tray } i > N_f$$

$$\left.\begin{aligned}\dot{M}_B &= R - V_B + F - B\\ \dot{x}_{1,B} &= \frac{1}{M_B}[R(x_{1,N} - x_{1,B}) - V_B(y_{1,B} - x_{1,B}) + F(x_{1,N} - x_{1,B})]\\ \dot{x}_{2,B} &= \frac{1}{M_B}[R(x_{2,N} - x_{2,B}) - V_B(y_{2,B} - x_{2,B}) + F(x_{2,N} - x_{2,B})]\end{aligned}\right\} \text{reboiler}$$

where M_R, M_C, M_B, and M_i denote, respectively, the liquid molar holdup in the reactor, condenser, reboiler, and trays i; and $x_{1,i}$, $x_{2,i}$ and $y_{1,i}$, $y_{2,i}$ are the corresponding liquid-phase and vapor-phase mole fractions for components A and P_1. The vapor–liquid equilibrium relating $x_{j,i}$ and $y_{j,i}$ is described by

$$y_{1,i} = \frac{\alpha_1 x_{1,i}}{1 + (\alpha_1 - 1)x_{1,i} + (\alpha_2 - 1)x_{2,i}} \tag{3.28}$$

$$y_{2,i} = \frac{\alpha_2 x_{2,i}}{1 + (\alpha_1 - 1)x_{1,i} + (\alpha_2 - 1)x_{2,i}} \tag{3.29}$$

Note that the model in Equation (3.27) does not include the secondary column required to separate P_1 and P_2. This unit is not part of the material recycle loop and has no dynamic interaction with the reactor or the first column. Consequently, the control problem for this column can be formulated and addressed independently and will not be considered in the remainder of the present study.

3.5.2 Model reduction and hierarchical controller design

Let us employ the ratio of the nominal, steady-state values (denoted by the subscript s) of the feed-stream and recycle-stream flow rates to define the recycle number

$$\mathsf{Rc} = \frac{D_{\mathrm{s}}}{F_{0\mathrm{s}}} \tag{3.30}$$

We also define $\varepsilon = 1/\mathsf{Rc}$, the $\mathcal{O}(1)$ ratios $\kappa_1 = F_{\mathrm{s}}/D_{\mathrm{s}}$ and $\kappa_2 = V_{\mathrm{Bs}}/D_{\mathrm{s}}$, and the scaled ($\mathcal{O}(1)$) input functions $u_1 = F/F_{\mathrm{s}}$, $u_2 = V_{\mathrm{B}}/V_{\mathrm{Bs}}$, $u_3 = B/B_{\mathrm{s}}$, $u_4 = R/R_{\mathrm{s}}$, and $u_{\mathrm{R}} = D/D_{\mathrm{s}}$. Using the above definitions, the dynamic model in Equation (3.27) becomes

$$\begin{aligned}
\dot{M}_{\mathrm{R}} &= F_0 + \frac{1}{\varepsilon} F_{0,\mathrm{s}}(u_{\mathrm{R}} - \kappa_1 u_1) \\
\dot{x}_{1,\mathrm{R}} &= \frac{1}{M_{\mathrm{R}}}\left[F_0(1 - x_{1,\mathrm{R}}) + \frac{1}{\varepsilon} F_{0,\mathrm{s}}(x_{1,\mathrm{D}} - x_{1,\mathrm{R}})u_{\mathrm{R}}\right] - k_1 x_{1,\mathrm{R}} \\
\dot{x}_{2,\mathrm{R}} &= \frac{1}{M_{\mathrm{R}}}\left[-F_0 x_{2,\mathrm{R}} + \frac{1}{\varepsilon} F_{0,\mathrm{s}}(x_{2,\mathrm{D}} - x_{2,\mathrm{R}})u_{\mathrm{R}}\right] + k_1 x_{1,\mathrm{R}} - k_2 x_{2,\mathrm{R}} \\
\dot{M}_{\mathrm{C}} &= R_{\mathrm{s}} u_4 + \frac{1}{\varepsilon} F_{0,\mathrm{s}}(\kappa_2 u_2 - u_{\mathrm{R}}) \\
\dot{x}_{1,\mathrm{D}} &= \frac{1}{\varepsilon}\frac{F_{0,\mathrm{s}}}{M_{\mathrm{C}}}\kappa_2 u_2(y_{1,1} - x_{1,\mathrm{D}}) \\
\dot{x}_{2,\mathrm{D}} &= \frac{1}{\varepsilon}\frac{F_{0,\mathrm{s}}}{M_{\mathrm{C}}}\kappa_2 u_2(y_{2,1} - x_{2,\mathrm{D}}) \\
&\ \ \vdots \\
\dot{x}_{1,i} &= \frac{1}{M_i}\left[R_{\mathrm{s}} u_4(x_{1,i-1} - x_{1,i}) + \frac{1}{\varepsilon} F_{0,\mathrm{s}}\kappa_2 u_2(y_{1,i+1} - y_{1,i})\right] \\
& \qquad\qquad\qquad\qquad\qquad\qquad\qquad\qquad\qquad\qquad i < N_{\mathrm{f}} \\
\dot{x}_{2,i} &= \frac{1}{M_i}\left[R_{\mathrm{s}} u_4(x_{2,i-1} - x_{2,i}) + \frac{1}{\varepsilon} F_{0,\mathrm{s}}\kappa_2 u_2(y_{2,i+1} - y_{2,i})\right] \\
&\ \ \vdots
\end{aligned}$$

$$
\left.
\begin{aligned}
\dot{x}_{1,i} &= \frac{1}{M_i}\Bigg[R_s u_4(x_{1,i-1} - x_{1,i}) \\
&\qquad + \frac{1}{\varepsilon}F_{0,\mathrm{s}}\left[\kappa_2 u_2(y_{1,i+1} - y_{1,i}) + \kappa_1 u_1(x_{1,\mathrm{f}} - x_{1,i})\right]\Bigg] \\
\dot{x}_{2,i} &= \frac{1}{M_i}\Bigg[R_s u_4(x_{2,i-1} - x_{2,i}) \\
&\qquad + \frac{1}{\varepsilon}F_{0,\mathrm{s}}\left[\kappa_2 u_2(y_{2,i+1} - y_{2,i}) + \kappa_1 u_1(x_{2,\mathrm{f}} - x_{2,i})\right]\Bigg]
\end{aligned}
\right\} \quad i = N_{\mathrm{f}}
$$

$$
\vdots \tag{3.31}
$$

$$
\left.
\begin{aligned}
\dot{x}_{1,i} &= \frac{1}{M_i}\Bigg[R_{\mathrm{s}} u_4(x_{1,i-1} - x_{1,i}) \\
&\qquad + \frac{1}{\varepsilon}F_{0,\mathrm{s}}\left[\kappa_2 u_2(y_{1,i+1} - y_{1,i}) + \kappa_1 u_1(x_{1,i-1} - x_{1,i})\right]\Bigg] \\
\dot{x}_{2,i} &= \frac{1}{M_i}\Bigg[R_{\mathrm{s}} u_4(x_{2,i-1} - x_{2,i}) \\
&\qquad + \frac{1}{\varepsilon}F_{0,\mathrm{s}}\left[\kappa_2 u_2(y_{2,i+1} - y_{2,i}) + \kappa_1 u_1(x_{2,i-1} - x_{2,i})\right]\Bigg]
\end{aligned}
\right\} \quad \imath > N_{\mathrm{f}}
$$

$$
\vdots
$$

$$
\begin{aligned}
\dot{M}_{\mathrm{B}} &= R_{\mathrm{s}} u_4 - B_{\mathrm{s}} u_3 + \frac{1}{\varepsilon}F_{0,\mathrm{s}}(\kappa_1 u_1 - \kappa_2 u_2) \\
\dot{x}_{1,\mathrm{B}} &= \frac{1}{M_{\mathrm{B}}}\Bigg[R_{\mathrm{s}} u_4(x_{1,N} - x_{1,\mathrm{B}}) \\
&\qquad + \frac{1}{\varepsilon}F_{0,\mathrm{s}}\left[\kappa_1 u_1(x_{1,N} - x_{1,\mathrm{B}}) - \kappa_2 u_2(y_{1,\mathrm{B}} - x_{1,\mathrm{B}})\right]\Bigg] \\
\dot{x}_{2,\mathrm{B}} &= \frac{1}{M_{\mathrm{B}}}\Bigg[R_{\mathrm{s}} u_4(x_{2,N} - x_{2,\mathrm{B}}) \\
&\qquad + \frac{1}{\varepsilon}F_{0,\mathrm{s}}\left[\kappa_1 u_1(x_{2,N} - x_{2,\mathrm{B}}) - \kappa_2 u_2(y_{2,\mathrm{B}} - x_{2,\mathrm{B}})\right]\Bigg]
\end{aligned}
$$

The model in Equation (3.31) conforms to the general form in Equation (3.10), with $\mathbf{u}^{\mathrm{s}} = [u_3\ u_4]^{\mathrm{T}}$, $\mathbf{u}^{\mathrm{l}} = [u_1\ u_2\ u_{\mathrm{R}}]^{\mathrm{T}}$, and

$$\mathbf{f} = \begin{bmatrix} F_0 \\ (F_0/M_{\mathrm{R}})(1 - x_{1,\mathrm{R}}) - k_1 x_{1,\mathrm{R}} \\ -(F_0/M_{\mathrm{R}})x_{2,\mathrm{R}} + k_1 x_{1,\mathrm{R}} - k_2 x_{2,\mathrm{R}} \\ 0 \\ 0 \\ 0 \\ \vdots \\ 0 \\ 0 \\ 0 \end{bmatrix} \tag{3.32}$$

$$\mathbf{G}^{\mathrm{s}} = \begin{bmatrix} 0 & 0 \\ 0 & 0 \\ 0 & 0 \\ 0 & R_{\mathrm{s}} \\ 0 & 0 \\ 0 & 0 \\ \vdots & \\ 0 & \frac{1}{M_i}\{R_{\mathrm{s}}(x_{1,\mathrm{D}} - x_{1,1})\} \\ 0 & \frac{1}{M_i}\{R_{\mathrm{s}}(x_{2,\mathrm{D}} - x_{2,1})\} \\ \vdots & \\ 0 & \frac{1}{M_i}\{R_{\mathrm{s}}(x_{1,i-1} - x_{1,i})\} \\ 0 & \frac{1}{M_i}\{R_{\mathrm{s}}(x_{2,i-1} - x_{2,i})\} \\ \vdots & \\ 0 & \frac{1}{M_i}\{R_{\mathrm{s}}(x_{1,N-1} - x_{N,i})\} \\ 0 & \frac{1}{M_i}\{R_{\mathrm{s}}(x_{2,N-1} - x_{N,i})\} \\ -B_{\mathrm{s}} & R_{\mathrm{s}} \\ 0 & \frac{1}{M_{\mathrm{B}}}\{R_{\mathrm{s}}(x_{1,N} - x_{1,\mathrm{B}})\} \\ 0 & \frac{1}{M_{\mathrm{B}}}\{R_{\mathrm{s}}(x_{2,N} - x_{2,\mathrm{B}})\} \end{bmatrix} \tag{3.33}$$

$$\mathbf{G}^{\mathrm{l}} = F_{0,\mathrm{s}} \begin{bmatrix} -\kappa_1 & 0 & 1 \\ 0 & 0 & (1/M_{\mathrm{R}})(x_{1,\mathrm{D}} - x_{1,\mathrm{R}}) \\ 0 & 0 & (1/M_{\mathrm{R}})(x_{2,\mathrm{D}} - x_{2,\mathrm{R}}) \\ 0 & 1 & -1 \\ 0 & (\kappa_2/M_{\mathrm{C}})(y_{1,1} - x_{1,\mathrm{D}}) & 0 \\ 0 & (\kappa_2/M_{\mathrm{C}})(y_{2,1} - x_{2,\mathrm{D}}) & 0 \\ \vdots & & \\ 0 & (\kappa_2/M_i)(y_{1,i+1} - y_{1,i}) & 0 \\ 0 & (\kappa_2/M_i)(y_{2,i+1} - y_{2,i}) & 0 \\ \vdots & & \\ (\kappa_1/M_{\mathrm{f}})(x_{1,\mathrm{f}} - x_{1,i}) & (\kappa_2/M_{\mathrm{f}})(y_{1,i+1} - y_{1,i}) & 0 \\ (\kappa_1/M_{\mathrm{f}})(x_{2,\mathrm{f}} - x_{2,i}) & (\kappa_2/M_{\mathrm{f}})(y_{2,i+1} - y_{2,i}) & 0 \\ \vdots & & \\ (\kappa_1/M_i)(x_{1,i-1} - x_{1,i}) & (\kappa_2/M_i)(y_{1,i+1} - y_{1,i}) & 0 \\ (\kappa_1/M_i)(x_{2,i-1} - x_{2,i}) & (\kappa_2/M_i)(y_{2,i+1} - y_{2,i}) & 0 \\ \vdots & & \\ \kappa_1 & -\kappa_2 & 0 \\ (\kappa_1/M_{\mathrm{B}})(x_{1,N} - x_{1,\mathrm{B}}) & -(\kappa_2/M_{\mathrm{B}})(y_{1,\mathrm{B}} - x_{1,\mathrm{B}}) & 0 \\ (\kappa_1/M_{\mathrm{B}})(x_{2,N} - x_{2,\mathrm{B}}) & -(\kappa_2/M_{\mathrm{B}})(y_{2,\mathrm{B}} - x_{2,\mathrm{B}}) & 0 \end{bmatrix} \tag{3.34}$$

Ignoring the two-time-scale behavior, a single $j \times 4$ $(1 \le j \le 5)$ multivariable optimization- or inversion-based controller could be designed for the overall system, using any combination of the flow rates of the five available material streams (F, V, B, R, and D) to address the four control objectives stated above (M_{R}, M_{C}, M_{B}, and $x_{2,\mathrm{B}}$).

However, the stiffness/ill conditioning of the model (3.31) will strongly impact on the implementation of optimization controllers (e.g., a model predictive controller) (Baldea *et al.* 2010). On the other hand, for any choice of four flow rates as manipulated inputs (keeping the remaining one constant at its nominal value), the system is non-minimum phase (Kumar and Daoutidis 2002) and thus potentially closed-loop unstable with an inversion-based controller.[3] As discussed in the previous section, a more systematic controller-design approach would

[3] A *linear* system is referred to as minimum phase if all the zeros of its transfer function lie in the left-hand plane; else, the system is non-minimum phase. Naturally, the inverse of the transfer function of a non-minimum phase system will be unstable. In the case of *nonlinear* systems, the concept of transfer function zeros is replaced by the *zero dynamics* (Isidori 1995); a nonlinear system is minimum phase if its zero dynamics are stable.

therefore entail a time-scale decomposition of the system dynamics and the design of separate distributed and supervisory controllers. This is addressed below.

Distributed control in the fast time scale

The distributed control objectives for this process involve the stabilization of the individual unit holdups (M_{R}, M_{C}, and M_{B}), which, according to our prior analysis, should be addressed in the fast time scale. The design of the distributed controllers for the stabilization of the three holdups can easily be achieved, using the large flow rates F, D, and V as manipulated inputs and employing simple proportional controllers – note that only these three flow rates (i.e., the components of $\mathbf{u}^{\mathrm{l}}$) affect the fast dynamics. More specifically, the proportional control laws

$$u_1 = 1 - K_1(M_{\mathrm{Rsp}} - M_{\mathrm{R}}) \tag{3.35}$$

$$u_{\mathrm{R}} = 1 - K_2(M_{\mathrm{Dsp}} - M_{\mathrm{D}}) \tag{3.36}$$

$$u_2 = 1 - K_3(M_{\mathrm{Bsp}} - M_{\mathrm{B}}) \tag{3.37}$$

can be used to stabilize the holdups M_{R}, M_{C}, and M_{B} at their setpoints M_{Rsp}, M_{Dsp}, and M_{Bsp}, respectively, with gains $K_1 = 0.0266\ \mathrm{mol}^{-1}$, $K_2 = 0.0281\ \mathrm{mol}^{-1}$, and $K_3 = 0.0242\ \mathrm{mol}^{-1}$.

The implementation of the above distributed controllers stabilizes the fast dynamics and affords us the opportunity to carry out a numerical experiment: Figure 3.7 shows the evolution of the product purity $x_{2,\mathrm{B}}$ for the original system with the above proportional stabilizing controllers and starting from an initial condition slightly perturbed from the nominal steady state. Clearly, $x_{2,\mathrm{B}}$ exhibits an initial fast transient ("boundary layer") followed by a slow dynamics, highlighting the two-time-scale behavior of this process.

The remaining state variables in Equation (3.27) display a similar behavior. The fast and slow dynamics are thus not associated with any distinct subsets of the state variables, which is consistent with the statement that the model of the process under consideration is a *nonstandard* singularly perturbed system of equations.

Supervisory control in the slow time scale

The supervisory control objective to be addressed in the slow time scale is the regulation of the product purity. Additionally, we must consider the control of the total material holdup $M_{\mathrm{t}} = M_{\mathrm{R}} + M_{\mathrm{C}} + M_{\mathrm{B}}$, which is not affected by changes in the flow rates of the large internal material streams F, D, and V, as can easily be verified from the corresponding mass-balance equations.

In order to address these objectives, we follow the procedure outlined in Section 3.4.3 to obtain a reduced-order model of the dynamics in the slow time scale. Specifically, we consider the limit of an infinitely high recycle flow rate

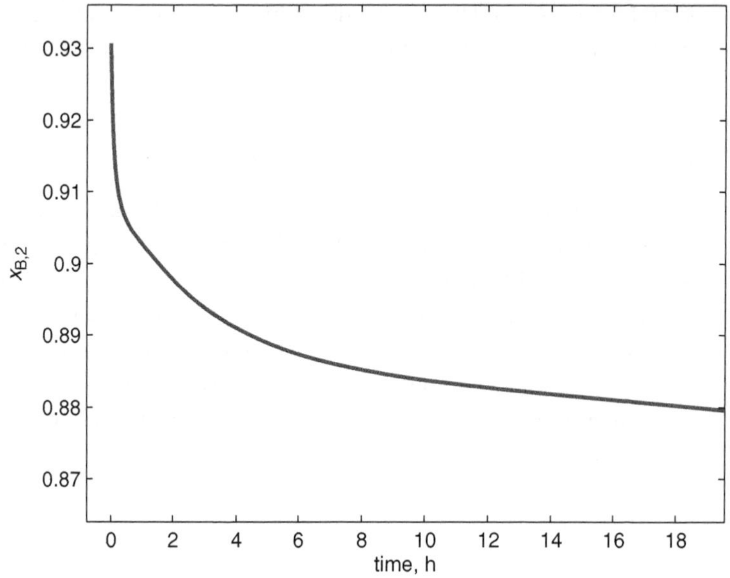

Figure 3.7 Two-time-scale behavior of the reactor–distillation-column–recycle process.

(i.e., $\varepsilon \to 0$), in which case the system in Equation (3.31) yields the following set of quasi-steady-state algebraic constraints:

$$
\begin{aligned}
0 &= u_{\mathrm{R}} - \kappa_1 u_1 \\
0 &= x_{1,\mathrm{D}} - x_{1,\mathrm{R}} \\
0 &= x_{2,\mathrm{D}} - x_{2,\mathrm{R}} \\
0 &= \kappa_2 u_2 - u_{\mathrm{R}} \\
0 &= y_{1,1} - x_{1,\mathrm{D}} \\
0 &= y_{2,1} - x_{2,\mathrm{D}} \\
&\vdots \\
0 &= y_{1,i+1} - y_{1,i} \qquad i < N_{\mathrm{f}} \\
0 &= y_{2,i+1} - y_{2,i} \\
&\vdots \\
0 &= \kappa_2 u_2 (y_{1,i+1} - y_{1,i}) + \kappa_1 u_1 (x_{1f} - x_{1,i}) \qquad i = N_{\mathrm{f}} \\
0 &= \kappa_2 u_2 (y_{2,i+1} - y_{2,i}) + \kappa_1 u_1 (x_{2f} - x_{2,i}) \\
&\vdots \\
0 &= \kappa_2 u_2 (y_{1,i+1} - y_{1,i}) + \kappa_1 u_1 (x_{1,i-1} - x_{1,i}) \qquad i > N_{\mathrm{f}} \\
0 &= \kappa_2 u_2 (y_{2,i+1} - y_{2,i}) + \kappa_1 u_1 (x_{2,i-1} - x_{2,i}) \\
&\vdots \\
0 &= \kappa_1 u_1 - \kappa_2 u_2 \\
0 &= \kappa_1 u_1 (x_{1,N} - x_{1,\mathrm{B}}) - \kappa_2 u_2 (y_{1,\mathrm{B}} - x_{1,\mathrm{B}}) \\
0 &= \kappa_1 u_1 (x_{2,N} - x_{2,\mathrm{B}}) - \kappa_2 u_2 (y_{2,\mathrm{B}} - x_{2,\mathrm{B}})
\end{aligned}
\tag{3.38}
$$

where u_1, u_2, and u_R are given by the proportional controllers in Equation (3.35). The constraints in Equation (3.38) are not linearly independent; it is easy to verify that the last three constraints can be expressed as linear combinations of the other $2N+6$ algebraic equations. For example, we can write

$$\kappa_1 u_1 - \kappa_2 u_2 = -[u_\mathrm{R} - \kappa_1 u_1] - [\kappa_2 u_2 - u_\mathrm{R}] \tag{3.39}$$

Consequently, the slow dynamics of the process are described by a DAE system with $2N+6$ algebraic equations (specifically, the first $2N+6$ equations in (3.38)) and $2N+6$ algebraic variables $\mathbf{z}$, defined as the ratios of the right-hand sides of the above $2N+6$ equations and ε, in the limit as $\varepsilon \to 0$. The index of this DAE system is exactly two, i.e., a single differentiation of the algebraic constraints yields a solution for the algebraic variables, and an ODE representation of the slow dynamics can be obtained using the methods described above. In this case, the ODE system that describes the evolution of the process in the slow time scale is of dimension three (the number of components present in the system is $\mathcal{C}=3$). Furthermore, it is important to note that the dimension of the reduced-order model of the slow dynamics is independent of the number of stages in the distillation column.

A potential choice of manipulated inputs to address the control objectives in the slow time scale is $[u_3 \;\; M_\mathrm{Rsp}]^\mathrm{T}$, i.e., the product flow rate from the column reboiler, and the *setpoint* for the reactor holdup used in the proportional feedback controller of Equation (3.35). This cascade control configuration is physically meaningful as well: intuitively, the regulation of the product purity $x_{2,\mathrm{B}}$ is associated with the conversion and selectivity achieved by the reactor, which in turn are affected by the reactor residence time.

Figure 3.8 shows the evolution of the product purity $x_{2,\mathrm{B}}$ (the main performance indicator of the process) for different step changes in M_Rsp. The operation of the process is stabilized by the three proportional controllers in Equation (3.35). Clearly, these responses indicate that the slow dynamics of the process is nonlinear; the implementation of a nonlinear supervisory controller for this process is thus highly desirable.

Note, however, that due to this choice of manipulated inputs, the DAE system describing the slow dynamics is nonregular, i.e., the underlying algebraic constraints in the DAE system explicitly involve the input M_Rsp. A dynamic extension (Section 3.4.4) was therefore used to obtain a state-space realization (ODE description of the slow dynamics). The state-space realization was subsequently used in the synthesis of a nonlinear input–output linearizing controller (Daoutidis and Kravaris 1992) for the slow system. The controller was designed to enforce a first-order decoupled response in the nominal slow system:

$$x_{2,\mathrm{B}} + \gamma_1 \dot{x}_{2,\mathrm{B}} = y_{1\mathrm{sp}} \tag{3.40}$$

$$M_\mathrm{B} + \gamma_2 \dot{M}_\mathrm{B} = y_{2\mathrm{sp}} \tag{3.41}$$

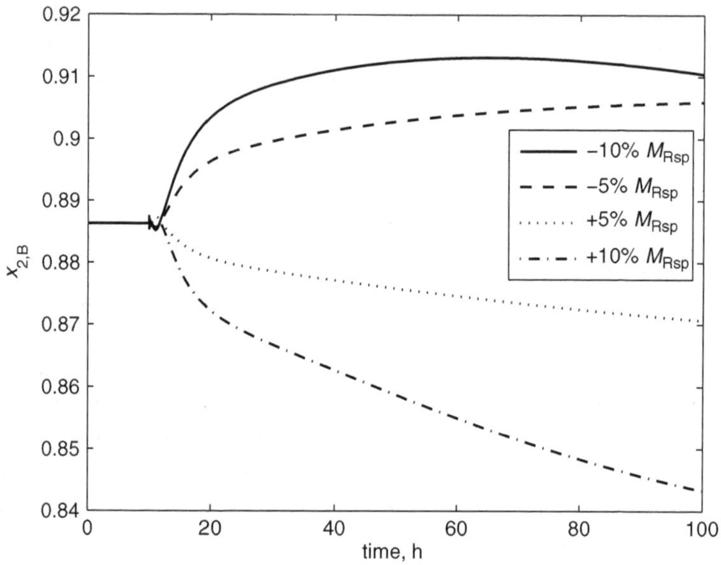

Figure 3.8 Nonlinear behavior of the slow dynamics of the integrated process.

where y_{1sp} and y_{2sp} denote the setpoints for the two outputs and $\gamma_1 = 5h$ and $\gamma_2 = 6h$ are the time constants.

Note that, owing to the underlying algebraic constraints in the DAE system that describes the slow dynamics, the holdups M_B, M_C, and M_R are not independent (there are only two linearly independent constraints among the three holdups, i.e., $0 = u_R - \kappa_1 u_1$ and $0 = \kappa_2 u_2 - u_R$, where u_1, u_2, and u_R are determined by the proportional control laws in Equation (3.35)). Thus, controlling one of the holdups (e.g., M_B) amounts to regulating the total material holdup in the process.

3.5.3 Simulation results and discussion

The hierarchical, cascade control structure described above was tested in several simulation scenarios. Figures 3.9 and 3.10 show the closed-loop profiles for a 15% increase in process throughput (via a direct increase in the feed flow rate F_0), imposed at $t = 15$ h, and a 1.5% decrease in the setpoint for $x_{2,B}$. The proposed hierarchical control framework clearly yields the desired closed-loop performance. Figures 3.11 and 3.12 show the corresponding profiles for the same setpoint change, in the presence of a plant–model mismatch in the model parameters, namely a +10% error in α_A and a −10% error in α_B, demonstrating that the controller performance is very robust with respect to these modeling errors.

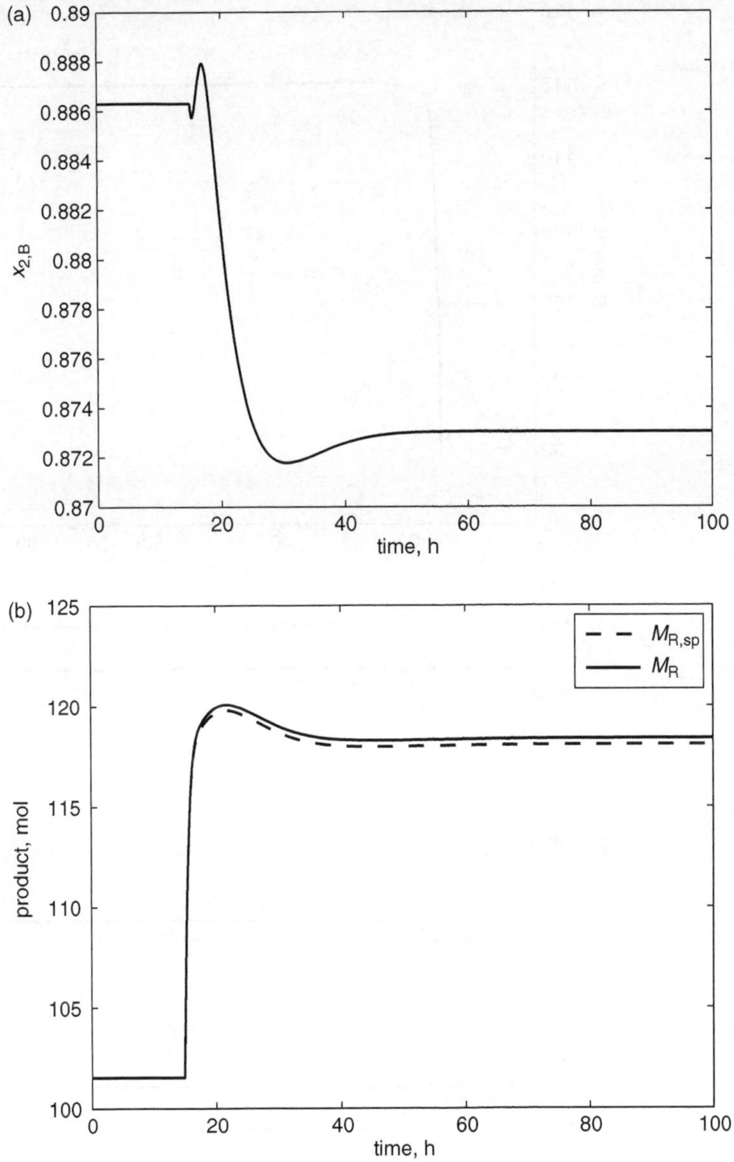

Figure 3.9 Closed-loop response for a 15% increase in production rate and 1.5% decrease of the product-purity setpoint. (a) Product composition and (b) reactor holdup and setpoint in the nominal case with no modeling errors.

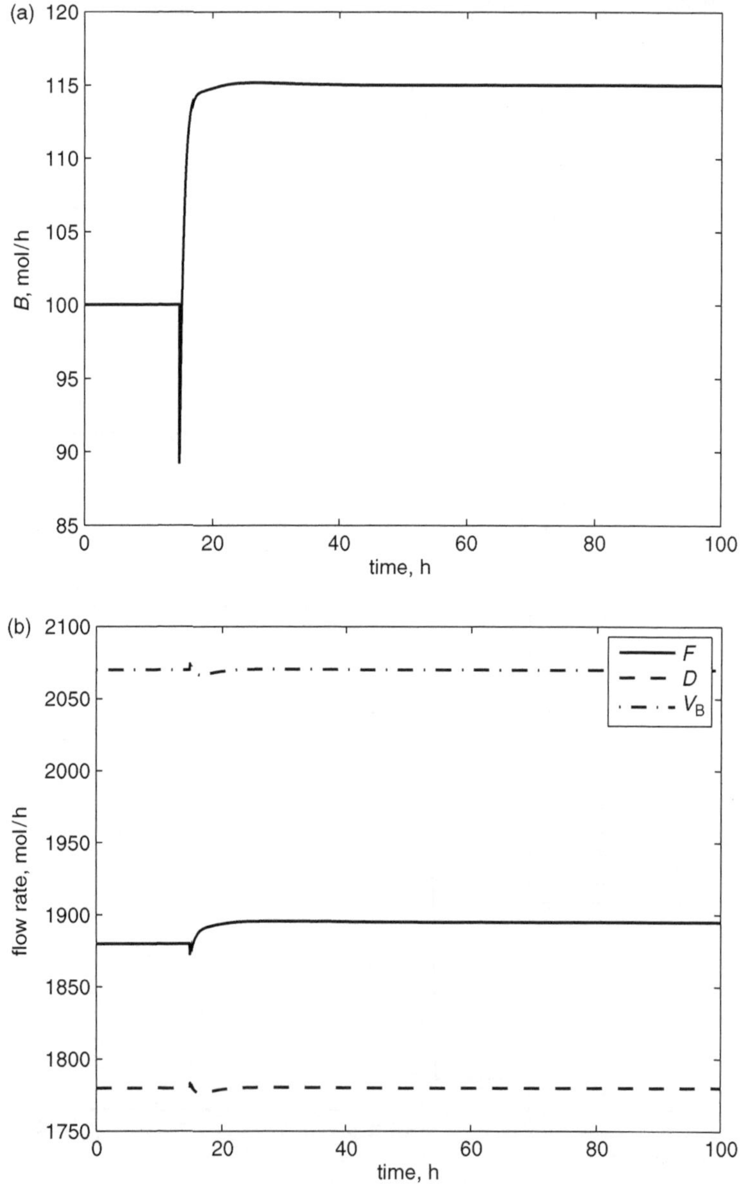

Figure 3.10 Closed-loop response for a 15% increase in production rate and 1.5% decrease of the product-purity setpoint. Process flow rates in the nominal case with no modeling errors: (a) product flow rate and (b) reactor effluent, recycle, and column boilup flow rates.

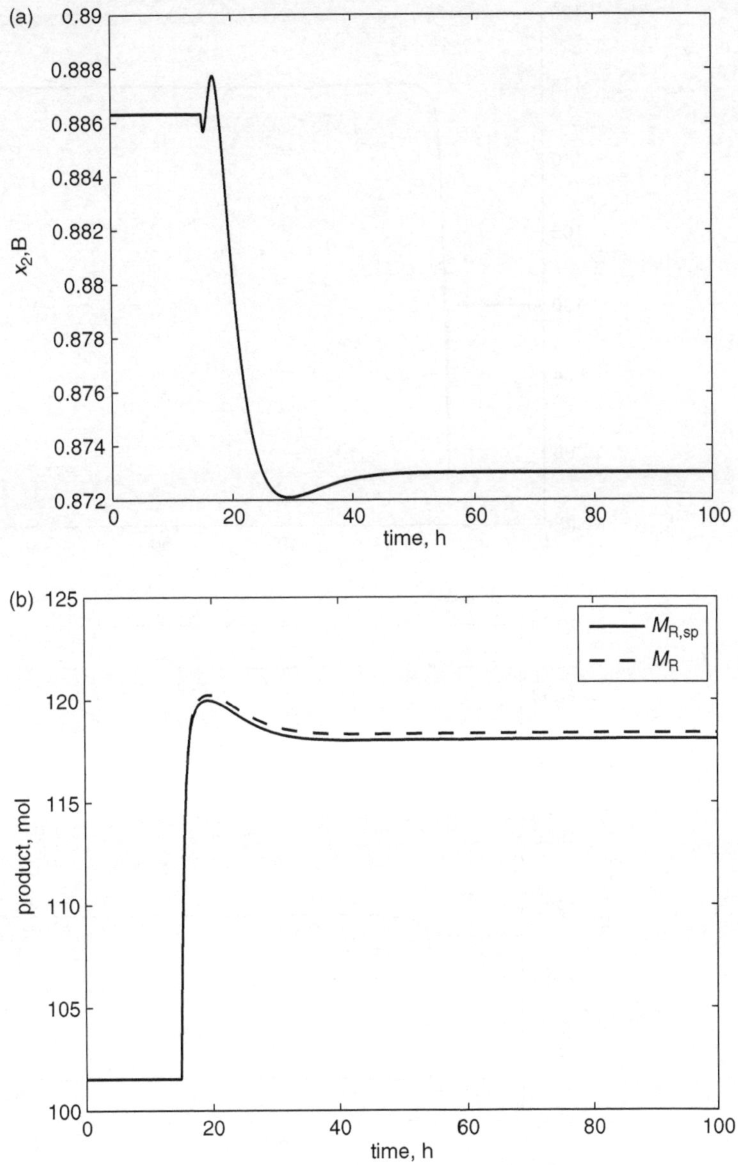

Figure 3.11 Closed-loop response of the product composition and reactor holdup for a 15% increase in production rate and 1.5% decrease of the product-purity setpoint, in the presence of plant–model mismatch. (a) Product stream composition and (b) reactor holdup and setpoint.

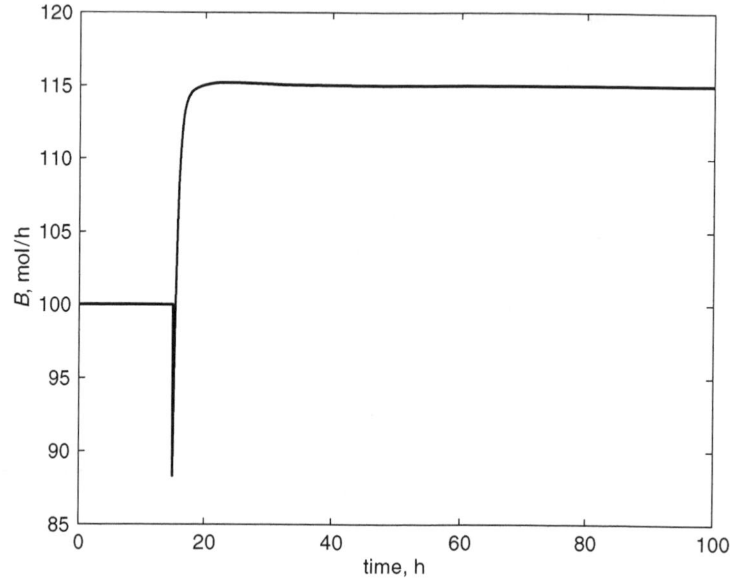

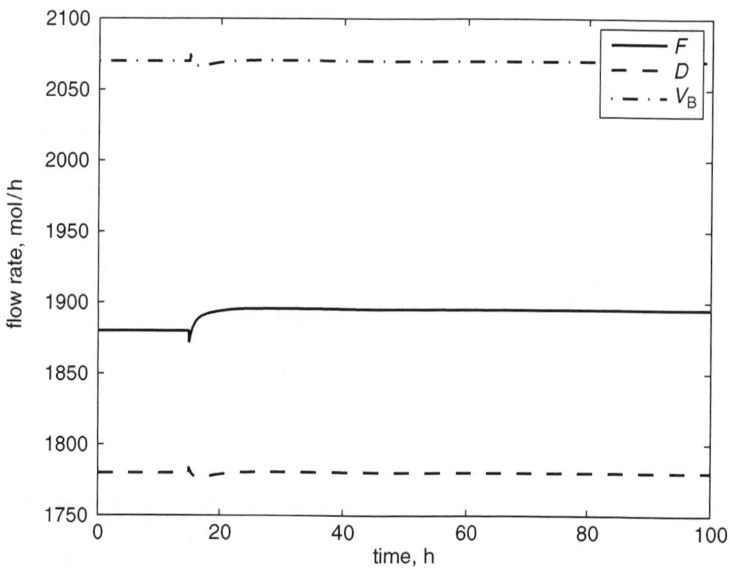

Figure 3.12 Closed-loop response of the process flow rates for a 15% increase in production rate and 1.5% decrease of the product-purity setpoint, in the presence of plant–model mismatch. (a) Product flow rate and (b) reactor effluent, recycle, and column boilup flow rates.

3.6 Synopsis

This chapter addressed the dynamics and control of process systems with material recycling. We established that whenever the flow rate of the recycle stream is significantly larger than the flow rates of the feed/product streams, the overall process exhibits a time-scale separation in its dynamics.

Specifically, the individual units of the process exhibit a fast dynamics in the fast time scale; the response times in the fast time scale are typically of the order of magnitude of the time constants of the individual units. In the fast time scale, the dynamic coupling between the units (induced by the recycle stream) is weak and can be ignored.

The interactions between units do, however, become significant over long periods of time: processes with recycle exhibit a slow, core dynamic component that must be addressed in any effective process-wide control strategy. This chapter presented an approach for systematically exploiting this two-time-scale behavior in a well-coordinated hierarchical controller design. The proposed framework relies on the use of simple distributed controllers to address unit-level control objectives in the fast time scale and a multivariable supervisory controller to accomplish process-wide control objectives over an extended time horizon.

4 Process systems with purge streams

4.1 Introduction

The present chapter focuses on the dynamics and control of integrated process systems in which impurities are present in small quantities, e.g., introduced as trace components in feed streams or generated as reaction byproducts. When such impurities are not readily removed by the product streams (e.g., if they are inert or noncondensable), they will accumulate in the process due to material recycling. The accumulation of impurities is detrimental to process operation (causing, for example, catalyst poisoning in the reactor) and process economics (owing to an increase in compression and recirculation costs) (Belanger and Luyben 1998, Luyben 2000, Dimian *et al.* 2001). Understanding the impact of the presence of impurities on the process dynamics is therefore critical and controlling the level of such components in the recycle structure can be a key operational objective.

In almost all such processes, the flow rate of the purge stream(s) is kept significantly smaller than the process throughput, with the evident goal of minimizing raw-material and product losses and the impact of releasing potentially hazardous chemicals into the environment; the difference between the purge flow rate and the flow rates of other process streams can span a few orders of magnitude. This discrepancy suggests the possibility of a "core" dynamics of the impurity levels in the process evolving in a much slower time scale than the dynamics of the individual process units and possibly the overall process. Developing an explicit nonlinear model of this slow dynamics can be beneficial both for analysis and evaluation purposes, and for model-based control.

In what follows, we begin by introducing two examples of process systems with recycle and purge. First, we analyze the case of a reactor with gas effluent connected via a gas recycle stream to a condenser, and a purge stream used to remove the light impurity present in the feed. In the second case, the products of a liquid-phase reactor are separated by a distillation column. The bottoms of the column are recycled to the reactor, and the trace heavy impurity present in the feed stream is removed via a liquid purge stream. We show that, in both cases, the dynamics of the system is modeled by a system of stiff ODEs that can, potentially, exhibit a two-time-scale behavior.

Using the concepts introduced in (Baldea *et al.* 2006), we will subsequently define a generic process structure consisting of a reaction–separation sequence in which impurities are present and a purge stream is used for their removal. We will show that, in the general case, the dynamics of such processes is described by a system of ODEs in a nonstandard singularly perturbed form. We investigate the dynamic behavior of the class of process systems considered within the framework of singular perturbations developed in Chapter 2, demonstrating that they do indeed exhibit a two-time-scale dynamics and deriving explicit reduced-order, non-stiff models for the dynamics in each time scale. The key result of this chapter is establishing that the slow dynamics of processes with purge streams is *one-dimensional* and is associated with the total impurity holdup in the recycle loop, which represents a true slow variable of the process. We will also highlight the control implications of this finding.

4.2 Motivating examples

4.2.1 Processes with light impurities

We consider the system of a gas-phase reactor and a condenser shown in Figure 4.1. The reactant A is fed at a molar flow rate F_0 to the reactor, where a first-order irreversible reaction A $\to$ B takes place with a reaction rate constant k_1. The reactor outlet stream is fed to a partial condenser that separates the light unconverted reactant A from the heavy product B. The gas phase, rich in A, is recycled to the reactor. A volatile inert impurity I is present in the feed stream in small quantities and a (small) purge stream P is used to prevent its accumulation in the recycle loop.

The interphase mass-transfer rates for the components A, B, and I in the condenser are governed by rate expressions of the form:

$$N_j = \mathcal{K}_j \mathcal{A} \left(y_j - \frac{\mathcal{P}_j^{\mathrm{S}}}{\mathcal{P}} x_j \right) \frac{M_{\mathrm{L}}}{\rho_{\mathrm{L}}} \tag{4.1}$$

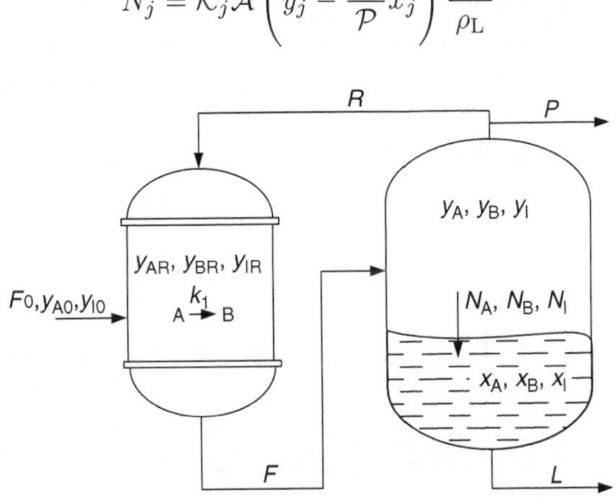

Figure 4.1 Process system with recycle and purge.

where $\mathcal{K}_j\mathcal{A}$ denotes a mass-transfer coefficient, y_j the mole fraction in the gas phase, x_j the mole fraction in the liquid phase, $\mathcal{P}_j^{\mathrm{S}}$ the saturation vapor pressure of the component j, $\mathcal{P}$ the total pressure in the condenser, M_{L} the total molar liquid holdup in the condenser, and ρ_{L} the molar density of the liquid. Assuming that the temperature effects in the reactor are negligible and that the operation of the condenser is isothermal, the dynamic model of the system has the form

$$\begin{aligned}
\dot{M}_{\mathrm{R}} &= F_0 + R - F \\
\dot{y}_{\mathrm{A,R}} &= \frac{1}{M_{\mathrm{R}}}[F_0(y_{\mathrm{A},0} - y_{\mathrm{A,R}}) + R(y_{\mathrm{A}} - y_{\mathrm{A,R}})] - r_{\mathrm{A}} \\
\dot{y}_{\mathrm{I,R}} &= \frac{1}{M_{\mathrm{R}}}[F_0(y_{\mathrm{I},0} - y_{\mathrm{I,R}}) + R(y_{\mathrm{I}} - y_{\mathrm{I,R}})] \\
\dot{M}_{\mathrm{V}} &= F - R - N - P \\
\dot{y}_{\mathrm{A}} &= \frac{1}{M_{\mathrm{V}}}[F(y_{\mathrm{A,R}} - y_{\mathrm{A}}) - N_{\mathrm{A}} + y_{\mathrm{A}}N] \\
\dot{y}_{\mathrm{I}} &= \frac{1}{M_{\mathrm{V}}}[F(y_{\mathrm{I,R}} - y_{\mathrm{I}}) - N_{\mathrm{I}} + y_{\mathrm{I}}N] \\
\dot{M}_{\mathrm{L}} &= N - L \\
\dot{x}_{\mathrm{A}} &= \frac{1}{M_{\mathrm{L}}}[N_{\mathrm{A}} - x_{\mathrm{A}}N] \\
\dot{x}_{\mathrm{I}} &= \frac{1}{M_{\mathrm{L}}}[N_{\mathrm{I}} - x_{\mathrm{I}}N]
\end{aligned} \tag{4.2}$$

where $N = N_{\mathrm{A}} + N_{\mathrm{B}} + N_{\mathrm{I}}$, M_{R}, M_{V}, and M_{L} denote the molar holdups in the reactor, vapor phase in the condenser, and liquid phase in the condenser, respectively, and $r_{\mathrm{A}} = k_1 y_{\mathrm{A,R}}$ represents the reaction rate.

In practical applications, for economic and operational reasons, the flow rate of the purge stream is very small compared with the throughput of the process. Hence, we can assume that the ratio of the purge flow rate to the feed flow rate under steady-state conditions is very small, i.e., $P_s/F_{0,s} = \epsilon \ll 1$. We will also consider that the mole fraction of the impurity in the feed (and, consequently, the rate at which the impurity enters the system) is very small, or $y_{\mathrm{I}0} = \beta_1\epsilon$, where β_1 is an $\mathcal{O}(1)$ quantity.

Owing to its high volatility, the impurity does not separate readily in the separation unit. Equivalently, the mass transfer rate for component I is very small:

$$\mathcal{K}_{\mathrm{I}}\mathcal{A} = \beta_2\epsilon^2 \tag{4.3}$$

where β_2 is $\mathcal{O}(1)$, and the vapor pressure of component I is high, i.e.,

$$\frac{\mathcal{P}_{\mathrm{I}}^{\mathrm{S}}}{\mathcal{P}} = \beta_3\frac{1}{\epsilon} \tag{4.4}$$

where β_3 is $\mathcal{O}(1)$.

Note that, from steady-state considerations, in order to remove an appreciable amount of impurity from the recycle loop via the purge stream (whose flow rate is small), the *mole fraction* of the impurity in the vapor phase in the condenser, y_I, has to be $\mathcal{O}(1)$. This implies that $O(\epsilon)$ moles of impurity enter and leave the system through the feed and purge streams. Note also that our assumption concerning the mass-transfer properties of the component I implies that a negligible amount of impurity leaves the recycle loop through condensation, exiting the process with the liquid stream from the bottom of the condenser.

Using the assumptions above, the dynamic model of the system takes the form

$$
\begin{aligned}
\dot{M}_\mathrm{R} &= F_0 + R - F \\
\dot{y}_\mathrm{AR} &= \frac{1}{M_\mathrm{R}}[F_0(1 - \beta_1\epsilon - y_\mathrm{A,R}) + R(y_\mathrm{A} - y_\mathrm{A,R})] - r_\mathrm{A} \\
\dot{y}_\mathrm{IR} &= \frac{1}{M_\mathrm{R}}[F_0(\beta_1\epsilon - y_\mathrm{I,R}) + R(y_\mathrm{I} - y_\mathrm{I,R})] \\
\dot{M}_\mathrm{V} &= F - R - (N_\mathrm{A} + N_\mathrm{B}) - \beta_2\epsilon^2 y_\mathrm{I} + \beta_2\beta_3\epsilon x_\mathrm{I} - \epsilon F_{0,\mathrm{s}}\frac{P}{P_\mathrm{s}} \\
\dot{y}_\mathrm{A} &= \frac{1}{M_\mathrm{V}}[F(y_\mathrm{A,R} - y_\mathrm{A}) - N_\mathrm{A} \\
&\qquad + y_\mathrm{A}(N_\mathrm{A} + N_\mathrm{B}) + y_\mathrm{A}(\beta_2\epsilon^2 y_\mathrm{I} - \beta_2\beta_3\epsilon x_\mathrm{I})] \\
\dot{y}_\mathrm{I} &= \frac{1}{M_\mathrm{V}}[F(y_\mathrm{I,R} - y_\mathrm{I}) - (\beta_2\epsilon^2 y_\mathrm{I} - \beta_2\beta_3\epsilon x_\mathrm{I}) \\
&\qquad + y_\mathrm{I}(N_\mathrm{A} + N_\mathrm{B}) + y_\mathrm{I}(\beta_2\epsilon^2 y_\mathrm{I} - \beta_2\beta_3\epsilon x_\mathrm{I})] \\
\dot{M}_\mathrm{L} &= (N_\mathrm{A} + N_\mathrm{B}) + \beta_2\epsilon^2 y_\mathrm{I} - \beta_2\beta_3\epsilon x_\mathrm{I} - L \\
\dot{x}_\mathrm{A} &= \frac{1}{M_\mathrm{L}}[N_\mathrm{A} - x_\mathrm{A}(N_\mathrm{A} + N_\mathrm{B}) - x_\mathrm{A}(\beta_2\epsilon^2 y_\mathrm{I} - \beta_2\beta_3\epsilon x_\mathrm{I})] \\
\dot{x}_\mathrm{I} &= \frac{1}{M_\mathrm{L}}[\beta_2\epsilon^2 y_\mathrm{I} - \beta_2\beta_3\epsilon x_\mathrm{I} - x_\mathrm{I}(N_\mathrm{A} + N_\mathrm{B}) - x_\mathrm{I}(\beta_2\epsilon^2 y_\mathrm{I} - \beta_2\beta_3\epsilon x_\mathrm{I})]
\end{aligned}
\tag{4.5}
$$

4.2.2 Processes with heavy impurities

Let us now consider a process consisting of a reactor and a distillation column with N stages, as in Figure 4.2. The first-order reaction

$$\mathrm{E} \rightarrow \mathrm{A} \tag{4.6}$$

takes place in the reactor. The feed F_0 contains the reactant E and a small quantity of a nonvolatile, heavy impurity I, and the effluent of the reactor is fed to a distillation column. The product A is light and is removed at the top of the column, while the heavy reactant E is removed as bottoms and recycled to the reactor. In order to prevent the accumulation of the heavy, inert impurity in the process, a small purge stream of flow rate P is used.

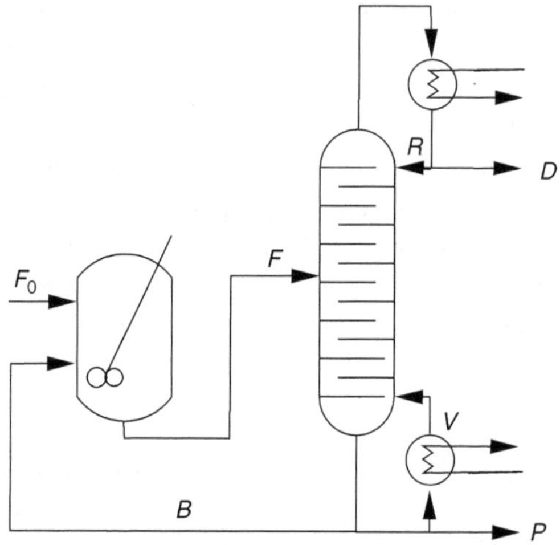

Figure 4.2 A reactor–distillation-column process system with recycle and purge.

Assuming that the relative volatilities of the components are constant and that the vapor and liquid phases on the column stages are at equilibrium, we can write

$$y_{\mathrm{A},i} = \frac{\alpha_{\mathrm{A}} x_{\mathrm{A},i}}{1 + (\alpha_{\mathrm{I}} - 1)x_{\mathrm{I},i} + (\alpha_{\mathrm{A}} - 1)x_{\mathrm{A},i}} \tag{4.7}$$

$$y_{\mathrm{I},i} = \frac{\alpha_I x_{I,i}}{1 + (\alpha_{\mathrm{I}} - 1)x_{\mathrm{I},i} + (\alpha_{\mathrm{A}} - 1)x_{\mathrm{A},i}} \tag{4.8}$$

where $x_{j,i}$ and $y_{j,i}$ represent, respectively, the liquid and vapor mole fractions of component j on tray i, with α_j being the relative volatility of component j, the $(2N + 9)$-dimensional model of the process is

$$\begin{aligned}
\dot{M}_{\mathrm{R}} &= F_0 + B - F \\
\dot{x}_{\mathrm{A,R}} &= (1/M_{\mathrm{R}})[F_0(x_{\mathrm{A},0} - x_{\mathrm{A,R}}) + B(x_{\mathrm{A,B}} - x_{\mathrm{A,R}}) + k(1 - x_{\mathrm{A,R}} - x_{\mathrm{I,R}})M_{\mathrm{R}}] \\
\dot{x}_{\mathrm{I,R}} &= (1/M_{\mathrm{R}})[F_0(x_{\mathrm{I},0} - x_{\mathrm{I,R}}) + B(x_{\mathrm{I,B}} - x_{\mathrm{I,R}})] \\
\dot{M}_{\mathrm{D}} &= V - R - D \\
\dot{x}_{\mathrm{A,D}} &= (V/M_{\mathrm{D}})(y_{\mathrm{A},1} - x_{\mathrm{A,D}}) \\
\dot{x}_{\mathrm{I,D}} &= (V/M_{\mathrm{D}})(y_{\mathrm{I},1} - x_{\mathrm{I,D}}) \\
&\vdots \\
\dot{x}_{\mathrm{A},i} &= (1/M_i)[V(y_{\mathrm{A},i+1} - y_{\mathrm{A},i}) + R(x_{\mathrm{A},i-1} - x_{\mathrm{A},i})] \\
\dot{x}_{\mathrm{I},i} &= (1/M_i)[V(y_{\mathrm{I},i+1} - y_{\mathrm{I},i}) + R(x_{\mathrm{I},i-1} - x_{\mathrm{I},i})] \\
&\vdots
\end{aligned} \tag{4.9}$$

$$
\begin{aligned}
\dot{x}_{A,f} &= (1/M_f)[V(y_{A,f+1} - y_{A,f}) + R(x_{A,f-1} - x_{A,f}) + F(x_{A,R} - x_{A,f})] \\
\dot{x}_{I,f} &= (1/M_f)[V(y_{I,f+1} - y_{I,f}) + R(x_{I,f-1} - x_{I,f}) + F(x_{I,R} - x_{I,f})] \\
&\vdots \\
\dot{x}_{A,i} &= (1/M_i)[V(y_{A,i+1} - y_{A,i}) + (R+F)(x_{A,i-1} - x_{A,i})] \\
\dot{x}_{I,i} &= (1/M_i)[V(y_{I,i+1} - y_{I,i}) + (R+F)(x_{I,i-1} - x_{I,i})] \\
&\vdots \\
\dot{M}_B &= R + F - B - P \\
\dot{x}_{A,B} &= (1/M_B)[(R+F)(x_{A,N} - x_{A,B}) + V(x_{A,B} - y_{A,B})] \\
\dot{x}_{I,B} &= (1/M_B)[(R+F)(x_{I,N} - x_{I,B}) + V(x_{I,B} - y_{I,B})]
\end{aligned}
$$

As in the previous case, we capture the fact that the flow rate of the purge stream is very small compared with the system throughput by defining

$$
\varepsilon = \frac{P_s}{F_{0,s}} \ll 1 \tag{4.10}
$$

and we assume that the mole fraction of the impurity in the feed stream is also very small, so that

$$
y_{I0} = \beta_1 \varepsilon \tag{4.11}
$$

with β_1 being $\mathcal{O}(1)$ and the subscript s denoting nominal values. Finally, it is reasonable to assume that the heavy impurity has a high boiling point, or, equivalently, a low relative volatility, which can be described as a function of the small parameter ε as

$$
\alpha_I = \beta_2 \varepsilon^2 \tag{4.12}
$$

with β_2 being a $\mathcal{O}(1)$ term.

Under the above considerations, the phase-equilibrium equations (4.7) become

$$
\begin{aligned}
y_{A,i} &= \frac{\alpha_A x_{A,i}}{\mathcal{N}} \\
y_{I,i} &= \frac{\varepsilon^2 \beta_2 x_{I,i}}{\mathcal{N}}
\end{aligned} \tag{4.13}
$$

with

$$
\mathcal{N} = 1 + (\varepsilon^2 \beta_2 - 1) x_{I,i} + (\alpha_A - 1) x_{A,i}
$$

and the process model takes the form

$$
\begin{aligned}
\dot{M}_{\mathrm{R}} &= F_0 + B - F \\
\dot{x}_{\mathrm{A,R}} &= (1/M_{\mathrm{R}})[F_0(1 - \beta_1\varepsilon - x_{\mathrm{A,R}}) + B(x_{\mathrm{A,B}} - x_{\mathrm{A,R}}) \\
&\qquad + k(1 - x_{\mathrm{A,R}} - x_{\mathrm{I,R}})M_{\mathrm{R}}] \\
\dot{x}_{\mathrm{I,R}} &= (1/M_{\mathrm{R}})[F_0(\beta_1\varepsilon - x_{\mathrm{I,R}}) + B(x_{\mathrm{I,B}} - x_{\mathrm{I,R}})] \\
\dot{M}_{\mathrm{D}} &= V - R - D \\
\dot{x}_{\mathrm{A,D}} &= (V/M_{\mathrm{D}})(y_{\mathrm{A},1} - x_{\mathrm{A,D}}) \\
\dot{x}_{\mathrm{I,D}} &= (V/M_{\mathrm{D}})(\varepsilon^2\beta_2 x_{\mathrm{I},1}/\mathcal{N} - x_{\mathrm{I,D}}) \\
&\vdots \\
\dot{x}_{\mathrm{A},i} &= (1/M_i)[V(y_{\mathrm{A},i+1} - y_{\mathrm{A},i}) + R(x_{\mathrm{A},i-1} - x_{\mathrm{A},i})] \\
\dot{x}_{\mathrm{I},i} &= (1/M_i)[V(\varepsilon^2\beta_2 x_{\mathrm{I},i+1}/\mathcal{N} - \varepsilon^2\beta_2 x_{\mathrm{I},i}/\mathcal{N}) + R(x_{\mathrm{I},i-1} - x_{\mathrm{I},i})] \\
&\vdots \\
\dot{x}_{\mathrm{A},f} &= (1/M_f)[V(y_{\mathrm{A},f+1} - y_{\mathrm{A},f}) \\
&\qquad + R(x_{\mathrm{A},f-1} - x_{\mathrm{A},f}) + F(x_{\mathrm{A,R}} - x_{\mathrm{A},f})] \\
\dot{x}_{\mathrm{I},f} &= (1/M_f)[V(\varepsilon^2\beta_2 x_{\mathrm{I},f+1}/\mathcal{N} - \varepsilon^2\beta_2 x_{\mathrm{I},f}/\mathcal{N}) \\
&\qquad + R(x_{\mathrm{I},f-1} - x_{\mathrm{I},f}) + F(x_{\mathrm{I,R}} - x_{\mathrm{I},f})] \\
&\vdots \\
\dot{x}_{\mathrm{A},i} &= (1/M_i)[V(y_{\mathrm{A},i+1} - y_{\mathrm{A},i}) + (R + F)(x_{\mathrm{A},i-1} - x_{\mathrm{A},i})] \\
\dot{x}_{\mathrm{I},i} &= (1/M_i)[V(\varepsilon^2\beta_2 x_{\mathrm{I},i+1}/\mathcal{N} - \varepsilon^2\beta_2 x_{\mathrm{I},i}/\mathcal{N}) + (R + F)(x_{\mathrm{I},i-1} - x_{\mathrm{I},i})] \\
&\vdots \\
\dot{M}_{\mathrm{B}} &= R + F - B - V - \varepsilon F_{0,\mathrm{s}}P/P_{\mathrm{s}} \\
\dot{x}_{\mathrm{A,B}} &= (1/M_{\mathrm{B}})[(R + F)(x_{\mathrm{A,N}} - x_{\mathrm{A,B}}) + V(x_{\mathrm{A,B}} - y_{\mathrm{A,B}})] \\
\dot{x}_{\mathrm{I,B}} &= (1/M_{\mathrm{B}})[(R + F)(x_{\mathrm{I,N}} - x_{\mathrm{I,B}}) + V(x_{\mathrm{I,B}} - \varepsilon^2\beta_2 x_{\mathrm{I,B}}/\mathcal{N})]
\end{aligned}
\tag{4.14}
$$

It is evident that the above models (Equations (4.5) and (4.14)) have terms of $\mathcal{O}(1)$ and $O(\epsilon)$ and are in a singularly perturbed form. This suggests, potentially, a two-time-scale behavior for the process systems with recycle and purge streams that they describe. In the next section, we will develop a generic modeling framework for such systems that captures this feature and allows a more general analysis of their dynamic behavior.

4.3 Modeling of process systems with recycle and purge

To generalize the findings in the examples presented above, let us consider the class of process systems presented in Figure 4.3. These systems consist of N

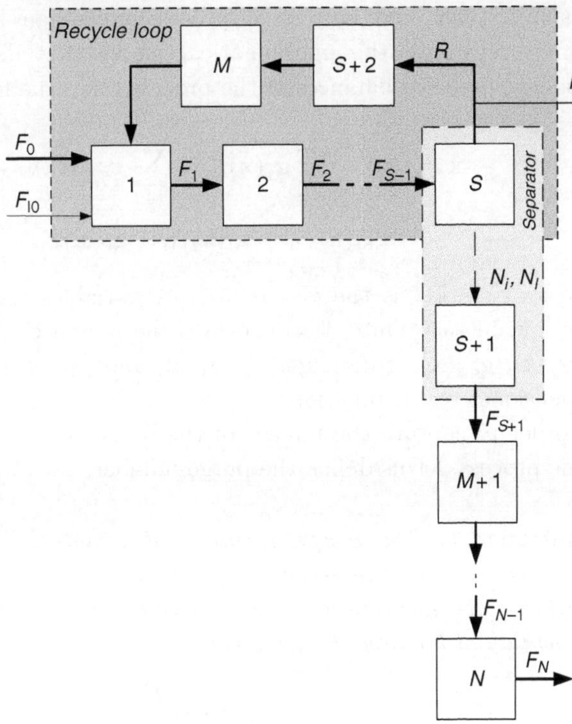

Figure 4.3 A generic reactor–separator process system with recycle and purge.

units with a total of C components and a single recycle loop. At least one (possibly multi-stage) separation unit is present and we denote by S the part of the separator that is included in the recycle loop, and by $S+1$ the part that is outside the loop. We consider that one of the output streams of the separator is at least partially recycled (possibly through units $S+2,\ldots,M$), while the other output stream leaves the system as a product stream, potentially after being processed in units $M+1,\ldots,N$. An impurity I is introduced into the process at a small rate, and we assume that it does not separate readily in the separator. A purge stream prevents the accumulation of the impurity in the recycle loop.

Let F_0 denote the feed flow rate to the first unit, $F_{\mathrm{I}0}$ the rate at which the impurity is input to the system,[1] F_j, $j=1,\ldots,N$, the outlet flow rate from the jth unit, F_{R} the recycle flow rate, and F_{P} the purge flow rate. Also, let N_i, $i=1,\ldots,C-1$ and N_{I} denote the net rates at which the ith component and, respectively, the impurity, are separated from the recycle loop.

[1] The impurity feed stream $F_{\mathrm{I}0}$ has no physical equivalent but is a convenient means to increase the generality of the model, since it can capture multiple practical scenarios concerning the origin of the impurities, e.g., as part of the feed stream F_0, leaks from the environment into the process or as the product of undesired secondary reactions.

Assuming that the individual process units can be modeled as lumped-parameter systems, the mathematical model that describes the overall and component material balances of the process takes the form

$$\dot{\mathbf{x}} = \mathbf{f}(\mathbf{x}) + \sum_{j=1}^{N} \mathbf{g}_j(\mathbf{x})F_j + \sum_{i=1}^{C-1} \mathbf{g}_{c,i}(\mathbf{x})N_i + \mathbf{g}_{\mathrm{R}}(\mathbf{x})F_{\mathrm{R}}$$
$$+ \mathbf{g}_{\mathrm{I0}}(\mathbf{x})F_{\mathrm{I0}} + \mathbf{g}_{\mathrm{I}}(\mathbf{x})N_{\mathrm{I}} + \mathbf{g}_{\mathrm{P}}(\mathbf{x})F_{\mathrm{p}} \tag{4.15}$$

where $\mathbf{x} \subset \mathcal{X} \in \mathbb{R}^n$ is the vector of state variables (i.e., total and component inventories in each unit), $\mathbf{f}(\mathbf{x})$ captures the presence of any chemical reactions, and $\mathbf{g}_j(\mathbf{x})$, $\mathbf{g}_{c,i}(\mathbf{x})$, $\mathbf{g}_{\mathrm{R}}(\mathbf{x})$, $\mathbf{g}_{\mathrm{I0}}(\mathbf{x})$, $\mathbf{g}_{\mathrm{I}}(x)$, and $\mathbf{g}_{\mathrm{P}}(x)$ are appropriately defined n-dimensional vector functions.

In order to capture the impact of the purge stream on the dynamics of this generic process, let us define the purge number.

Definition 4.1. *The* ***purge number*** *of a material recycle loop in an integrated process is a process-wide dimensionless number expressed as the ratio of the steady-state flow rate of the purge and the process throughput, as captured by the flow rate of the process feed stream:*

$$\mathsf{Pu} = \frac{F_{\mathrm{P,s}}}{F_{0,\mathrm{s}}} \tag{4.16}$$

Using the insights gained from the examples presented in Section 4.2, we also make the following assumptions.

Assumption 4.1. *The nominal flow rate of the purge stream is much smaller than that of the feed stream, i.e., the purge number of the process is very small:*

$$\mathsf{Pu} \ll 1$$

Assumption 4.2. *The rate at which the impurity is input to the process is very small and (naturally) comparable to the flow rate of the purge stream. Therefore, we have*

$$F_{\mathrm{I0,s}}/F_{0,\mathrm{s}} = \beta_1 \mathsf{Pu}$$

with β_1 being an $\mathcal{O}(1)$ quantity.

Assumption 4.3. *The net rate of impurity removal from the recycle loop by the product stream (along the path $N_{\mathrm{I}}, F_{S+1}, \ldots, F_N$) is much smaller than the rate at which the impurity is input to the system:*

$$N_{\mathrm{I,s}}/F_{0,\mathrm{s}} = \beta_2 \mathsf{Pu}^2$$

where β_2 is an $\mathcal{O}(1)$ quantity.

Note that Assumption 4.3 effectively captures the need to use a purge stream to prevent the buildup of the impurity in the units of the recycle loop.

Assumption 4.4. *The flow rates* $F_0, \ldots, F_N, F_{\mathrm{R}}$ *and* $N_1, \ldots, N_{C-1}$ *are of comparable magnitude, that is*

$$\frac{F_{j,\mathrm{s}}}{F_{0,\mathrm{s}}} = \mathcal{O}(1); \qquad \frac{N_{i,\mathrm{s}}}{F_{0,\mathrm{s}}} = \mathcal{O}(1)$$

We denote $u_{\mathrm{c},i} = N_i/N_{i,\mathrm{s}}$, $i = 1, \ldots, C-1, \mathrm{I}$. *Also, we define* $u_j = F_j/F_{j\mathrm{s}}$ *to be the scaled (possibly manipulated) inputs that correspond to the flow rates* F_j *of the material streams.*

Under Assumptions 4.1–4.4, the generic model in Equation (4.15) becomes

$$\dot{\mathbf{x}} = \mathbf{f}(\mathbf{x}) + \sum_{j=1}^{N} \mathbf{g}_j(\mathbf{x}) F_{j,\mathrm{s}} u_j + \sum_{i=1}^{C-1} \mathbf{g}_{\mathrm{c},i}(\mathbf{x}) N_{i,\mathrm{s}} u_{\mathrm{c},i} + \mathbf{g}_{\mathrm{R}}(\mathbf{x}) F_{\mathrm{R,s}} u_{\mathrm{R}}$$
$$+ \mathbf{Pu}\left[\mathbf{g}_{\mathrm{I0}}(\mathbf{x}) F_{0,\mathrm{s}} \beta_1 + \mathbf{Pu}\, \mathbf{g}_{\mathrm{I}}(\mathbf{x}) F_{0,\mathrm{s}} u_{\mathrm{c,I}} \beta_2 + \mathbf{g}_{\mathrm{P}}(\mathbf{x}) F_{0,\mathrm{s}} u_{\mathrm{p}}\right] \tag{4.17}$$

Equation (4.17) can be written in a more compact and mathematically convenient form as

$$\dot{\mathbf{x}} = \bar{\mathbf{f}}(\mathbf{x}, \mathbf{u}^{\mathrm{l}}) + \varepsilon[\mathbf{g}^{\mathrm{I0}}(\mathbf{x}) + \varepsilon \mathbf{g}^{\mathrm{I}}(\mathbf{x}) + \mathbf{g}^{\mathrm{P}}(\mathbf{x}) u_{\mathrm{p}}] \tag{4.18}$$

with

$$\varepsilon = \mathbf{Pu} \tag{4.19}$$

$\mathbf{u}^{\mathrm{l}}$ is the vector of scaled input variables corresponding to the "large" flow rates $F_0, \ldots, F_N, F_{\mathrm{R}}$, and $\bar{\mathbf{f}}(\mathbf{x}, \mathbf{u}^{\mathrm{l}})$, $\mathbf{g}^{\mathrm{I0}}(\mathbf{x})$, $\mathbf{g}^{\mathrm{I}}(\mathbf{x})$, and $\mathbf{g}^{\mathrm{P}}(\mathbf{x})$ are n-dimensional vector functions.

The mass-transfer rates represented in Equation (4.17) by $u_{\mathrm{c},i}$ are generally functions of the physical parameters of the system, i.e., $u_{\mathrm{c},i} = u_{\mathrm{c},i}(\mathbf{x})$, and cannot be manipulated independently. Hence, for notational convenience, the corresponding terms have been included in the expressions for $\bar{\mathbf{f}}(\mathbf{x}, \mathbf{u}^{\mathrm{l}})$ and $\mathbf{g}^{\mathrm{I}}(\mathbf{x})$.

Referring back to the theory introduced in Chapter 2, we can expect that the presence of terms of very different magnitudes (i.e., $\mathcal{O}(1)$ and $\mathcal{O}(\varepsilon)$) in the model (4.18) reflects a two-time-scale behavior in the dynamics of typical processes with recycle and purge. In what follows, we will show that this is indeed the case. Also, we will address the derivation of reduced-order models of the fast and slow dynamics, provide a physical interpretation of this dynamic behavior, and highlight its control implications.

4.4 Dynamic analysis and model reduction

Let us concentrate on the model in Equation (4.18) and consider the limit case of a zero purge number, i.e., $\varepsilon \to 0$. Physically, this limit corresponds to setting the flow rate of the purge stream, the rate at which impurity is input to the process, as well as the rate at which the impurity is separated from the recycle loop in the separator, to zero. In other words, this limit case assumes that there is no inflow or outflow of impurity to or from the process. Note, however, that this limit case does not necessarily assume that impurities are no longer present in the system (i.e., the impurity will be still present in the material being recycled, and, as mentioned in the motivating examples above, its concentration in the recycle loop will be $\mathcal{O}(1)$).

The effect of the impurity inlet and outlet thus eliminated, we obtain a description of the fast dynamics of the process:

$$\dot{\mathbf{x}} = \bar{\mathbf{f}}(\mathbf{x}, \mathbf{u}^{\mathrm{l}}) \tag{4.20}$$

Note that the expression above involves only $\mathbf{u}^{\mathrm{l}}$, which corresponds to the flow rates of all material streams *other than* the impurity input and output flows. It is easy to verify (and quite intuitive) that $\mathbf{u}^{\mathrm{l}}$ does *not* affect the *total* holdup of the impurity in the recycle loop.

It is also easy to verify that the total inventory of I *is* influenced by the inflow of impurity, its net removal rate in the separator, and the flow rate of the purge stream. In the limit of an infinitely low purge number, these are set to zero and, as can be seen from Equation (4.20), clearly have no influence on the dynamics in the fast time scale.

These observations imply that there exists a process variable – namely the total impurity holdup – whose dynamics are slow (in the sense defined in Chapter 2) and whose evolution is thus not captured by the fast model (4.20). As a consequence, one of the differential equations in (4.20) is redundant (i.e., these equations are not linearly independent). The steady-state conditions corresponding to the fast dynamics,

$$\mathbf{0} = \bar{\mathbf{f}}(\mathbf{x}, \mathbf{u}^{\mathrm{l}}) \tag{4.21}$$

will therefore specify a 1D equilibrium manifold.

As in the previous chapters, in order to proceed with our analysis, we will make the following assumption.

Assumption 4.5. *The vector* $\bar{\mathbf{f}}$ *can be decomposed as*

$$\bar{\mathbf{f}}(\mathbf{x}, \mathbf{u}^{\mathrm{l}}) = \mathbf{B}(\mathbf{x})\tilde{\mathbf{f}}(\mathbf{x}, \mathbf{u}^{\mathrm{l}}) \tag{4.22}$$

with $\mathbf{B}(\mathbf{x}) \in \mathbb{R}^{n\times(n-1)}$ *being a full column rank matrix and the vector* $\tilde{\mathbf{f}}(\mathbf{x}, \mathbf{u}^{\mathrm{l}}) \in \mathbb{R}^{n-1}$ *having linearly independent terms; that is, the linearly independent equations corresponding to (4.21) can be isolated.*

Next, in order to obtain a description of the slow dynamics, we define a slow, *compressed*, time scale $\tau = \varepsilon t$, in which the model of the process becomes

$$\varepsilon \frac{d\mathbf{x}}{d\tau} = \bar{\mathbf{f}}(\mathbf{x}, \mathbf{u}^{\mathrm{l}}) + \varepsilon[\mathbf{g}^{\mathrm{I0}}(\mathbf{x}) + \varepsilon \mathbf{g}^{\mathrm{I}}(\mathbf{x}) + \mathbf{g}^{\mathrm{P}}(\mathbf{x})u_{\mathrm{p}}] \tag{4.23}$$

Let us consider again the limit of an infinitely small purge number ($\varepsilon \to 0$), this time in the newly defined slow time scale. This yields the algebraic equations

$$0 = \bar{\mathbf{f}}(\mathbf{x}, \mathbf{u}^{\mathrm{l}}) \tag{4.24}$$

or, equivalently, the linearly independent equations

$$0 = \tilde{\mathbf{f}}(\mathbf{x}, \mathbf{u}^{\mathrm{l}}) \tag{4.25}$$

which are the quasi-steady-state conditions for the fast dynamics in Equation (4.21) and, respectively, (4.22). They describe the equilibrium manifold in which the slow dynamics of the process evolves and therefore constitute constraints that must be satisfied in the slow time scale.

Finally, on dividing Equation (4.23) by ε, i.e.,

$$\frac{d\mathbf{x}}{d\tau} = \frac{1}{\varepsilon}\tilde{\mathbf{f}}(\mathbf{x}, \mathbf{u}^{\mathrm{l}}) + \mathbf{g}^{\mathrm{IO}}(\mathbf{x}) + \varepsilon\mathbf{g}^{\mathrm{I}}(\mathbf{x}) + \mathbf{g}^{\mathrm{P}}(\mathbf{x})u_{\mathrm{P}} \tag{4.26}$$

and considering the limit $\varepsilon \to 0$ under the constraints above, we obtain the description of the slow dynamics of the system:

$$\begin{aligned} \frac{d\mathbf{x}}{d\tau} &= \mathbf{g}^{\mathrm{IO}}(\mathbf{x}) + \mathbf{g}^{\mathrm{P}}(\mathbf{x})u_{\mathrm{P}} + \mathbf{B}(\mathbf{x})\mathbf{z} \\ 0 &= \tilde{\mathbf{f}}(\mathbf{x}, \mathbf{u}^{\mathrm{l}}) \end{aligned} \tag{4.27}$$

Note that, in the limit $\varepsilon \to 0$, the term $\tilde{\mathbf{f}}(\mathbf{x}, \mathbf{u}^{\mathrm{l}})/\varepsilon$ in Equation (4.26) becomes indeterminate. Thus, in Equation (4.27), we defined

$$\mathbf{z} = \lim_{\varepsilon \to 0} \frac{\tilde{\mathbf{f}}(\mathbf{x}, \mathbf{u}^{\mathrm{l}})}{\varepsilon} \tag{4.28}$$

with $\mathbf{z} \in \mathrm{IR}^{n-1}$, to account for this finite but unknown term.

The model of the slow dynamics of the system consists therefore of a set of coupled DAEs of nontrivial index, since the variables $\mathbf{z}$ (that physically correspond to the net material flows of the system in the slow time scale) are implicitly fixed by the quasi-steady-state constraints, rather than explicitly specified in the dynamic model. Also, note that the DAE model (4.27) has a well-defined index only if the flow rates $\mathbf{u}^{\mathrm{l}}$ which appear in the algebraic constraints are specified as functions of the state variables $\mathbf{x}$. This is typically accomplished via a control law $\mathbf{u}^{\mathrm{l}}(\mathbf{x})$.

Once the flow rates $\mathbf{u}^{\mathrm{l}}$ have been specified, it is possible to differentiate the constraints in Equation (4.27) to obtain (after differentiating a sufficient number of times) a solution for the algebraic variables $\mathbf{z}$. One differentiation in time will yield

$$\mathbf{z} = -[L_{\mathbf{B}}\tilde{\mathbf{f}}(\mathbf{x})]^{-1}(L_{\mathbf{g}^{\mathrm{IO}}}\tilde{\mathbf{f}}(\mathbf{x}) + L_{\mathbf{g}^{\mathrm{P}}}\tilde{\mathbf{f}}(\mathbf{x})u_{\mathrm{P}}) \tag{4.29}$$

with

$$L_{\mathbf{B}}\tilde{\mathbf{f}}(\mathbf{x}) = \frac{\partial \tilde{\mathbf{f}}}{\partial \mathbf{x}}\mathbf{B}(\mathbf{x}) \tag{4.30}$$

$$L_{\mathbf{g}^{\mathrm{I0}}}\tilde{\mathbf{f}}(\mathbf{x}) = \frac{\partial \tilde{\mathbf{f}}}{\partial \mathbf{x}}\mathbf{g}^{\mathrm{I0}} \tag{4.31}$$

$$L_{\mathbf{g}^{\mathrm{P}}}\tilde{\mathbf{f}}(\mathbf{x}) = \frac{\partial \tilde{\mathbf{f}}}{\partial \mathbf{x}}\mathbf{g}^{\mathrm{P}} \tag{4.32}$$

If the matrix $L_{\mathbf{B}}\tilde{\mathbf{f}}(\mathbf{x})$ is invertible (which is typically true, as will be shown in the following examples), the index of the DAE system (4.27) is two (i.e., a solution for $\mathbf{z}$ is obtained directly from Equation (4.29)), and in this case the dimension of the underlying ODE system describing the slow dynamics is 1.

In this case, an explicit ODE representation (state-space realization) of the DAE system (4.27) can be obtained be employing a coordinate change of the form suggested in Section 2.3:

$$\begin{bmatrix} \zeta \\ \eta \end{bmatrix} = \mathbf{T}(\mathbf{x}) = \begin{bmatrix} \phi(\mathbf{x}) \\ \tilde{\mathbf{f}}(\mathbf{x}, \mathbf{u}^{\mathrm{l}}) \end{bmatrix} \tag{4.33}$$

In these new coordinates, the model of the slow dynamics becomes

$$\begin{aligned} \frac{d\zeta}{d\tau} &= \frac{\partial \phi}{\partial \mathbf{x}}\mathbf{B}(\mathbf{x})\mathbf{z}|_{\mathbf{x}=\mathbf{T}^{-1}(\zeta)} + \frac{\partial \phi}{\partial \mathbf{x}}\mathbf{g}^{\mathrm{I0}}(\mathbf{x})|_{\mathbf{x}=\mathbf{T}^{-1}(\zeta)} + \frac{\partial \phi}{\partial \mathbf{x}}\mathbf{g}^{\mathrm{P}}(\mathbf{x})u_{\mathrm{P}}|_{\mathbf{x}=\mathbf{T}^{-1}(\zeta)} \\ \eta &\equiv \mathbf{0} \end{aligned} \tag{4.34}$$

Remark 4.1. *According to the developments in Section 2.3 (see also Theorem A.1 in Appendix A), it is possible to choose the function* $\phi(\mathbf{x})$ *so that* $(\partial\phi/\partial\mathbf{x})\mathbf{B}(\mathbf{x}) = 0$. *In this case, the variable* ζ *evolves independently of the variables* $\mathbf{z}$, *and represents a true "slow" variable in the system (whereas the original state variables exhibit both fast and slow dynamics). Its transient evolution is given by*

$$\begin{aligned} \frac{d\zeta}{d\tau} &= \frac{\partial \phi}{\partial \mathbf{x}}\mathbf{g}^{\mathrm{I0}}(\mathbf{x})|_{\mathbf{x}=\mathbf{T}^{-1}(\zeta)} + \frac{\partial \phi}{\partial \mathbf{x}}\mathbf{g}^{\mathrm{P}}(\mathbf{x})u_{\mathrm{P}}|_{\mathbf{x}=\mathbf{T}^{-1}(\zeta)} \\ \eta &\equiv \mathbf{0} \end{aligned} \tag{4.35}$$

Remark 4.2. *Equation (4.34) depends on the impurity input and output flow rates (respectively, via* $\mathbf{g}^{\mathrm{I0}}(\mathbf{x})$ *and* $\mathbf{g}^{\mathrm{P}}(\mathbf{x})u_{\mathrm{P}}$*). This further confirms that the slow dynamics are associated with the impurity inventory in the process. It is also quite intuitive (as we will demonstrate in the examples below) that a coordinate change of the type mentioned in Remark 4.1 could entail the use of the total impurity holdup as a slow variable.*

These findings are in agreement with empirical knowledge, which has, for a long time, associated the presence of (inert) impurities in a process with long response times, often spanning many hours or even days (Luyben 2000).

4.5 Motivating examples (continued)

Let us continue with the analysis of the motivating examples introduced in Section 4.2, and present the derivation of reduced-order models for the fast and slow dynamics of the two process systems, according to the theoretical framework developed above.

4.5.1 Processes with light impurities

We begin with a description of the fast dynamics of the stiff model in Equation (4.5). This is readily obtained in the form of Equation (4.20) by considering the dynamic model (4.5) in the limit as $\epsilon \to 0$:

$$
\begin{aligned}
\dot{M}_{\mathrm{R}} &= F_0 + R - F \\
\dot{y}_{\mathrm{A,R}} &= \frac{1}{M_{\mathrm{R}}}[F_0(1 - y_{\mathrm{A,R}}) + R(y_{\mathrm{A}} - y_{\mathrm{A,R}})] - r_{\mathrm{A}} \\
\dot{y}_{\mathrm{I,R}} &= \frac{1}{M_{\mathrm{R}}}[-F_0 y_{\mathrm{I,R}} + R(y_{\mathrm{I}} - y_{\mathrm{I,R}})] \\
\dot{M}_{\mathrm{V}} &= F - R - (N_{\mathrm{A}} + N_{\mathrm{B}}) \\
\dot{y}_{\mathrm{A}} &= \frac{1}{M_{\mathrm{V}}}[F(y_{\mathrm{A,R}} - y_{\mathrm{A}}) - N_{\mathrm{A}} + y_{\mathrm{A}}(N_{\mathrm{A}} + N_{\mathrm{B}})] \\
\dot{y}_{\mathrm{I}} &= \frac{1}{M_{\mathrm{V}}}[F(y_{\mathrm{I,R}} - y_{\mathrm{I}}) + y_{\mathrm{I}}(N_{\mathrm{A}} + N_{\mathrm{B}})] \\
\dot{N}_{\mathrm{L}} &= (N_{\mathrm{A}} + N_{\mathrm{B}}) - L \\
\dot{x}_{\mathrm{A}} &= \frac{1}{M_{\mathrm{L}}}[N_{\mathrm{A}} - x_{\mathrm{A}}(N_{\mathrm{A}} + N_{\mathrm{B}})] \\
\dot{x}_{\mathrm{I}} &= -\frac{1}{M_{\mathrm{L}}}[x_{\mathrm{I}}(N_{\mathrm{A}} + N_{\mathrm{B}})]
\end{aligned}
\tag{4.36}
$$

Equation (4.36) is a non-stiff model that approximates the dynamics of the reactor–condenser system in Figure 4.1 in the original (fast) time scale t.

Note that the differential equations in Equation (4.36) are not independent, or, equivalently, the corresponding quasi-steady-state constraints are not linearly independent. Specifically, the third constraint can be expressed as a linear combination of the others, i.e., there exist only eight linearly independent constraints that can be written in the form of Equation (4.22), with

$$
\mathbf{B}(\mathbf{x}) = \begin{bmatrix}
1 & 0 & 0 & 0 & 0 & 0 & 0 & 0 \\
0 & 1/M_{\mathrm{R}} & 0 & 0 & 0 & 0 & 0 & 0 \\
-y_{\mathrm{I,R}}/M_{\mathrm{R}} & 0 & -y_{\mathrm{I}}/M_{\mathrm{R}} & 0 & -1/M_{\mathrm{R}} & 0 & 0 & 0 \\
0 & 0 & 1 & 0 & 0 & 0 & 0 & 0 \\
0 & 0 & 0 & 1/M_{\mathrm{V}} & 0 & 0 & 0 & 0 \\
0 & 0 & 0 & 0 & 1/M_{\mathrm{V}} & 0 & 0 & 0 \\
0 & 0 & 0 & 0 & 0 & 1 & 0 & 0 \\
0 & 0 & 0 & 0 & 0 & 0 & 1/M_{\mathrm{L}} & 0 \\
0 & 0 & 0 & 0 & 0 & 0 & 0 & 1/M_{\mathrm{L}}
\end{bmatrix}
\tag{4.37}
$$

and

$$\tilde{\mathbf{f}}(\mathbf{x}, \mathbf{u}^{\mathrm{l}}) = \begin{bmatrix} F_0 + R - F \\ F_0(1 - y_{\mathrm{A,R}}) + R(y_{\mathrm{A}} - y_{\mathrm{A,R}}) - r_{\mathrm{A}} M_{\mathrm{R}} \\ F - R - (N_{\mathrm{A}} + N_{\mathrm{B}}) \\ F(y_{\mathrm{A,R}} - y_{\mathrm{A}}) - N_{\mathrm{A}} + y_{\mathrm{A}}(N_{\mathrm{A}} + N_{\mathrm{B}}) \\ F(y_{\mathrm{I,R}} - y_{\mathrm{I}}) + y_{\mathrm{I}}(N_{\mathrm{A}} + N_{\mathrm{B}}) \\ (N_{\mathrm{A}} + N_{\mathrm{B}}) - L \\ N_{\mathrm{A}} - x_{\mathrm{A}}(N_{\mathrm{A}} + N_{\mathrm{B}}) \\ x_{\mathrm{I}}(N_{\mathrm{A}} + N_{\mathrm{B}}) \end{bmatrix} \tag{4.38}$$

This is consistent with the fact that these constraints correspond to the limit as the purge flow rate *and* the inflow of the impurity become zero. In this limit, the number of moles of the impurity leaving the reactor is identical to that leaving the condenser, hence the redundant constraint. Note also that, in the fast time scale, only the flow rates F, R, and L affect the dynamics and can be used for addressing control objectives such as stabilization of holdups, production rate, and product quality. The purge flow rate has, of course, no effect on the dynamics in this fast time scale.

Turning now to the slow dynamics, we define the slow time scale $\tau = t\epsilon$ and consider the limit $\epsilon \to 0$, obtaining a description of the slow dynamics of the form (4.27):

$$\begin{aligned}
\frac{dM_{\mathrm{R}}}{d\tau} &= \lim_{\epsilon\to 0} \frac{1}{\epsilon}(F_0 + R - F) \\
\frac{dy_{\mathrm{A,R}}}{d\tau} &= \lim_{\epsilon\to 0} \frac{1}{\epsilon M_{\mathrm{R}}}[F_0(1 - y_{\mathrm{A,R}}) + R(y_{\mathrm{A}} - y_{\mathrm{A,R}}) - r_{\mathrm{A}} M_{\mathrm{R}}] - \frac{1}{M_{\mathrm{R}}} F_0 \beta_1 \\
\frac{dy_{\mathrm{I,R}}}{d\tau} &= \lim_{\epsilon\to 0} \frac{1}{\epsilon M_{\mathrm{R}}}[-F_0 y_{\mathrm{I,R}} + R(y_{\mathrm{I}} - y_{\mathrm{I,R}})] + \frac{1}{M_{\mathrm{R}}} F_0 \beta_1 \\
\frac{dM_{\mathrm{V}}}{d\tau} &= \lim_{\epsilon\to 0} \frac{1}{\epsilon}[F - R - (N_{\mathrm{A}} + N_{\mathrm{B}})] - F_{0,\mathrm{s}} \frac{P}{P_{\mathrm{s}}} \\
\frac{dy_{\mathrm{A}}}{d\tau} &= \lim_{\epsilon\to 0} \frac{1}{\epsilon M_{\mathrm{V}}}[F(y_{\mathrm{A,R}} - y_{\mathrm{A}}) - N_{\mathrm{A}} + y_{\mathrm{A}}(N_{\mathrm{A}} + N_{\mathrm{B}})] \\
\frac{dy_{\mathrm{I}}}{d\tau} &= \lim_{\epsilon\to 0} \frac{1}{\epsilon M_{\mathrm{V}}}[F(y_{\mathrm{I,R}} - y_{\mathrm{I}}) + y_{\mathrm{I}}(N_{\mathrm{A}} + N_{\mathrm{B}})] \\
\frac{dM_{\mathrm{L}}}{d\tau} &= \lim_{\epsilon\to 0} \frac{1}{\epsilon}[N_{\mathrm{A}} + N_{\mathrm{B}} - L] \\
\frac{dx_{\mathrm{A}}}{d\tau} &= \lim_{\epsilon\to 0} \frac{1}{\epsilon M_{\mathrm{L}}}[N_{\mathrm{A}} - x_{\mathrm{A}}(N_{\mathrm{A}} + N_{\mathrm{B}})] \\
\frac{dx_{\mathrm{I}}}{d\tau} &= \lim_{\epsilon\to 0} \frac{-1}{\epsilon M_{\mathrm{L}}}[x_{\mathrm{I}}(N_{\mathrm{A}} + N_{\mathrm{B}})]
\end{aligned} \tag{4.39}$$

subject to the quasi-steady-state constraints obtained by setting the terms in Equation (4.38) equal to zero.

It can be shown that on setting the reactor effluent flow rate F and the product flow rate L with the proportional control laws

$$\begin{aligned} F &= F_{\mathrm{s}}(1 - k_{\mathrm{p},1}(M_{\mathrm{R,sp}} - M_{\mathrm{R}})) \\ L &= L_{\mathrm{s}}(1 - k_{\mathrm{p},2}(M_{\mathrm{L,sp}} - M_{\mathrm{L}})) \end{aligned} \tag{4.40}$$

where the index s denotes nominal values and the index sp setpoints, the matrix $L_{\mathbf{B}}\tilde{\mathbf{f}}(\mathbf{x})$ is invertible, and hence a coordinate change of the type (4.33) exists. Note that the control laws in Equation (4.40) correspond to the stabilization of the reactor and condenser liquid holdups.

In order to obtain an ODE description of the slow dynamics, the total impurity holdup in the recycle loop, i.e.,

$$\phi(x) = M_{\mathrm{R}}y_{\mathrm{I,R}} + M_{\mathrm{V}}y_{\mathrm{I}} \tag{4.41}$$

represents a meaningful choice of the function $\phi(x)$ in the coordinate change (4.33). This coordinate change then yields

$$\frac{d\zeta}{d\tau} = F_{\mathrm{O}}\beta_1 - F_{\mathrm{O,s}}\frac{P}{P_{\mathrm{s}}}y_{\mathrm{I}}(\zeta) \tag{4.42}$$

or, equivalently,

$$\frac{d\zeta}{d\tau} = F_{\mathrm{O}}\beta_1 - F_{\mathrm{O,s}}y_{\mathrm{I}}(\zeta)u_{\mathrm{p}} \tag{4.43}$$

where $y_{\mathrm{I}}(\zeta)$ is computed from the steady-state constraints obtaining by setting $\tilde{\mathbf{f}}(\mathbf{x}, \mathbf{u}^{\mathrm{l}})$ in Equation (4.38) equal to zero and inverting the coordinate transformation (4.33), with ζ defined as above. Equation (4.43) represents a 1D non-stiff description of the slow dynamics of the process in Figure 4.1. The single slow mode of the system is therefore associated with the total holdup of the inert impurity, which, as anticipated, is a "true slow variable" of the system.

4.5.2 Processes with heavy impurities

Following a similar procedure to the one employed above, it is easy to verify that we obtain a model that approximates the fast dynamics of the system in Figure 4.2, in the form of Equation (4.20). Also, it can be verified that only $2N+8$ of the $2N+9$ steady-state constraints that correspond to the fast dynamics are independent. After controlling the reactor holdup M_{R}, the distillate holdup M_{D}, and the reboiler holdup M_{B} with proportional controllers using respectively F, D, and B as manipulated inputs, the matrix $L_{\mathbf{B}}\tilde{\mathbf{f}}(\mathbf{x})$ is nonsingular, and hence the coordinate change

$$\zeta = M_{\mathrm{R}}x_{\mathrm{I,R}} + M_{\mathrm{f}}x_{\mathrm{I,f}} + M_{\mathrm{B}}x_{\mathrm{I,B}} + \sum_{i=1}^{N} M_i x_{\mathrm{I},i} \tag{4.44}$$

which corresponds to the total holdup of the impurity in the recycle loop, exists. After applying (4.44), we obtain the following 1D description of the slow dynamics of the system:

$$\frac{d\zeta}{d\tau} = \beta_1 F_0 - F_{0,\mathrm{s}} \frac{P}{P_\mathrm{s}} x_{\mathrm{I,B}}(\zeta) \tag{4.45}$$

As in the previous example we considered, the total holdup of the impurity represents the "true slow variable" in the 1D slow dynamics of the system.

4.6 Further applications

4.6.1 Processes with slow secondary reactions

A similar analysis can be carried out in the case of a process system in which the "impurities" are generated in the reactor, rather than introduced into the feed stream. Let us consider the process system in Figure 4.4, which has a structure very similar to that of the reactor–condenser system presented in Section 4.2.

The difference consists in the fact that *two* first-order reactions, A → B and A → I, take place in the reactor, with reaction rate constants k_1 and k_2, respectively. B is the desired product and is separated in the condenser, while the undesired light byproduct I (which is assumed to be generated in small quantities) does not separate and a purge stream P is used for its removal. Carrying over the notation and modeling assumptions of Section 4.2, the model of the process in Figure 4.4 can be written as

$$\begin{aligned}
\dot{M}_R &= F_0 + R - F \\
\dot{y}_{\mathrm{A,R}} &= \frac{1}{M_\mathrm{R}}[F_0(y_{\mathrm{A}0} - y_{\mathrm{A,R}}) + R(y_\mathrm{A} - y_{\mathrm{A,R}}) - k_1 M_\mathrm{R} y_{\mathrm{A,R}} - k_2 M_\mathrm{R} y_{\mathrm{A,R}}] \\
\dot{y}_{\mathrm{I,R}} &= \frac{1}{M_\mathrm{R}}[-F_0 y_{\mathrm{I,R}} + R(y_\mathrm{I} - y_{\mathrm{I,R}}) + k_2 M_\mathrm{R} y_{\mathrm{A,R}}] \\
\dot{M}_\mathrm{V} &= F - R - N - P
\end{aligned}$$

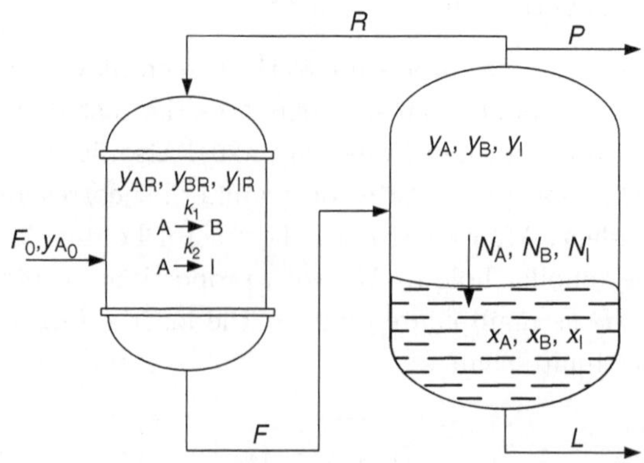

Figure 4.4 Process system in which a secondary reaction takes place, yielding small quantities of an unwanted byproduct.

$$
\begin{aligned}
\dot{y}_{\mathrm{A}} &= \frac{1}{M_{\mathrm{V}}}[F(y_{\mathrm{A,R}} - y_{\mathrm{A}}) - N_{\mathrm{A}} + y_{\mathrm{A}}N] \\
\dot{y}_{\mathrm{I}} &= \frac{1}{M_{\mathrm{V}}}[F(y_{\mathrm{I,R}} - y_{\mathrm{I}}) - N_{\mathrm{I}} + y_{\mathrm{I}}N] \\
\dot{M}_{\mathrm{L}} &= N - L \\
\dot{x}_{\mathrm{A}} &= \frac{1}{M_{\mathrm{L}}}[N_{\mathrm{A}} - x_{\mathrm{A}}N] \\
\dot{x}_{\mathrm{I}} &= \frac{1}{M_{\mathrm{L}}}[N_{\mathrm{I}} - x_{\mathrm{I}}N]
\end{aligned} \tag{4.46}
$$

In this case, the form of Assumptions 4.1 and 4.3 remains the same as in the second section, while Assumption 4.2 implies that the rate constant of the reaction which leads to the formation of the impurity is very small, or $k_2 M_{\mathrm{R,s}}/F_{0,\mathrm{s}} = \beta_1\varepsilon$, with β_1 being an $\mathcal{O}(1)$ quantity.

Notice that here Assumption 4.2 is expressed as a ratio of the characteristic time for the chemical reaction and the characteristic time for convection, being thus equivalent to considering that the second reaction has a low Damköhler number.

With the aforementioned assumptions, the dynamic model of the system takes the form

$$
\begin{aligned}
\dot{M}_{\mathrm{R}} &= F_0 + R - F \\
\dot{y}_{\mathrm{AR}} &= \frac{1}{M_{\mathrm{R}}}\Big[F_0(y_{\mathrm{A0}} - y_{\mathrm{A,R}}) + R(y_{\mathrm{A}} - y_{\mathrm{A,R}}) \\
&\qquad - k_1 M_{\mathrm{R}} y_{\mathrm{A,R}} - \varepsilon\beta_1 \frac{F_{0,\mathrm{s}}}{M_{\mathrm{R,s}}} M_{\mathrm{R}} y_{\mathrm{A,R}}\Big] \\
\dot{y}_{\mathrm{IR}} &= \frac{1}{M_{\mathrm{R}}}\Big[-F_0 y_{\mathrm{I,R}} + R(y_{\mathrm{I}} - y_{\mathrm{I,R}}) + \varepsilon\beta_1 \frac{F_{0,\mathrm{s}}}{M_{\mathrm{R,s}}} M_{\mathrm{R}} y_{\mathrm{A,R}}\Big] \\
\dot{M}_{\mathrm{V}} &= F - R - (N_{\mathrm{A}} + N_{\mathrm{B}}) - \beta_2\epsilon^2 y_{\mathrm{I}} + \beta_2\beta_3\epsilon x_{\mathrm{I}} - \epsilon F_{0,\mathrm{s}}\frac{P}{P_{\mathrm{s}}} \\
\dot{y}_{\mathrm{A}} &= \frac{1}{M_{\mathrm{V}}}[F(y_{\mathrm{A,R}} - y_{\mathrm{A}}) - N_{\mathrm{A}} + y_{\mathrm{A}}(N_{\mathrm{A}} + N_{\mathrm{B}}) \\
&\qquad + y_{\mathrm{A}}(\beta_2\epsilon^2 y_{\mathrm{I}} - \beta_2\beta_3\epsilon x_{\mathrm{I}})] \\
\dot{y}_{\mathrm{I}} &= \frac{1}{M_{\mathrm{V}}}[F(y_{\mathrm{I,R}} - y_{\mathrm{I}}) - (\beta_2\epsilon^2 y_{\mathrm{I}} - \beta_2\beta_3\epsilon x_{\mathrm{I}}) \\
&\qquad + y_{\mathrm{I}}(N_{\mathrm{A}} + N_{\mathrm{B}}) + y_{\mathrm{I}}(\beta_2\epsilon^2 y_{\mathrm{I}} - \beta_2\beta_3\epsilon x_{\mathrm{I}})] \\
\dot{M}_{\mathrm{L}} &= (N_{\mathrm{A}} + N_{\mathrm{B}}) + \beta_2\epsilon^2 y_{\mathrm{I}} - \beta_2\beta_3\epsilon x_{\mathrm{I}} - L \\
\dot{x}_{\mathrm{A}} &= \frac{1}{M_{\mathrm{L}}}[N_{\mathrm{A}} - x_{\mathrm{A}}(N_{\mathrm{A}} + N_{\mathrm{B}}) - x_{\mathrm{A}}(\beta_2\epsilon^2 y_{\mathrm{I}} - \beta_2\beta_3\epsilon x_{\mathrm{I}})] \\
\dot{x}_{\mathrm{I}} &= \frac{1}{M_{\mathrm{L}}}[\beta_2\epsilon^2 y_{\mathrm{I}} - \beta_2\beta_3\epsilon x_{\mathrm{I}} - x_{\mathrm{I}}(N_{\mathrm{A}} + N_{\mathrm{B}}) - x_{\mathrm{I}}(\beta_2\epsilon^2 y_{\mathrm{I}} - \beta_2\beta_3\epsilon x_{\mathrm{I}})]
\end{aligned} \tag{4.47}
$$

which is in the form of Equation (4.18).

After applying the model reduction proposed above, considering again the total holdup of the impurity as a 1D function required by the coordinate change (4.33) we obtain the following state-space realization of the slow dynamics of the system in Figure 4.4:

$$\frac{d\zeta}{d\tau} = \frac{F_{0,s}}{M_{R,s}} \beta_1 M_R(\zeta) y_{A,R}(\zeta) - F_{0,s} \frac{P}{P_s} y_I(\zeta) \tag{4.48}$$

with $M_R(\zeta)$, $y_{A,R}(\zeta)$, and $y_I(\zeta)$ being computed as in the case of the reactor condenser system with feed impurities.

4.6.2 An analogy with systems with large recycle

In the analyses presented above, the mole fractions of the impurity in the units of the recycle loop are $\mathcal{O}(1)$, and hence the amount of impurity that is recycled in the system is much larger than the impurity throughput of the process. The presence of a single slow mode associated with the impurity is therefore in complete agreement with the analysis of systems with large recycle developed in Chapter 3, which predicts a slow model of dimension equal to the number of components for which the recycle flow rate is much larger than the throughput. This perspective is reflected in the following example.

Example 4.1. Cryogenic air separation is currently the most economical means for producing oxygen, nitrogen, and argon on a large scale. A typical air-separation unit (ASU) consists of three heat-integrated columns (Figure 4.5). Air is compressed in the main air compressor (MAC) and impurities such as water and hydrocarbons are removed in the prepurifier (PP). The air stream

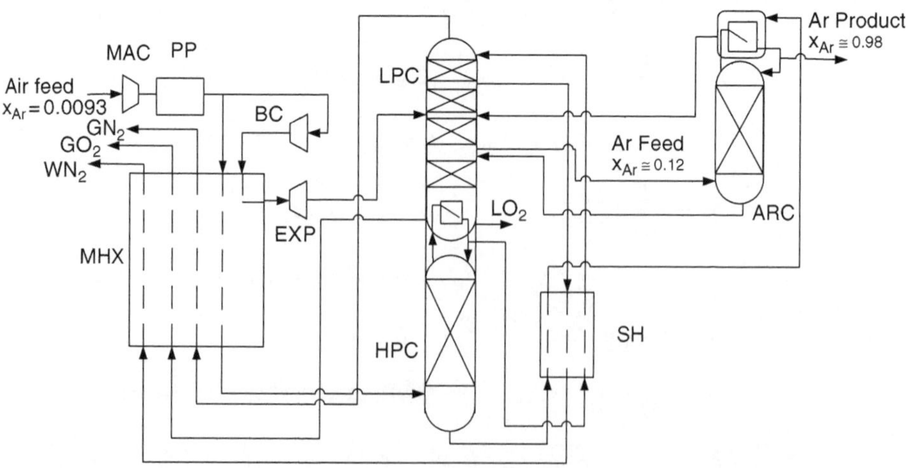

Figure 4.5 Cryogenic process for the production of oxygen, nitrogen, and argon.

is then divided into two substreams; the first substream (main air stream) is cooled in the main heat exchanger (MHX) against warming cryogenic products, while the second substream (expansion air) undergoes additional compression before being fed to the MHX. The cooled expansion air is then fed to a turbine expander, where it is further cooled to provide additional refrigeration to the system.

On exiting the MHX, the main air stream enters the high-pressure column (HPC), where it is separated into nearly pure nitrogen (at the top of the HPC) and an oxygen-enriched stream at the HPC bottom. The condensing stream at the top of the HPC is used to boil the near-pure oxygen liquid at the bottom of the low-pressure column (LPC). The nitrogen-rich HPC top stream also provides reflux to the LPC, while the oxygen-enriched HPC bottoms stream is used as a condensing utility for the argon column (ARC) before being fed to the LPC. Nearly pure nitrogen is collected at the top of the LPC and a waste-nitrogen stream WN_2 is drawn at an intermediate stage close to the top of the LPC to provide additional refrigeration in the MHX.

The air stream fed to the process contains a small amount of argon, which accumulates (reaching between 12% and 14% Ar) on an intermediate stage close to the bottom of the LPC; the argon column gets its feed as an intermediate vapor stream drawn at this point. A small argon product stream is removed at the ARC top, and most of the ARC feed is returned to the LPC (i.e., the ARC operates at a very high reflux ratio).

Argon is a valuable product on many global markets and modern ASUs are designed to recover more than 90% of the argon in the air feed as a relatively high-purity (98% or higher, with the remainder being oxygen) argon product stream. From steady-state considerations, the flow rate of the argon stream is about 0.8% of the flow rate of the air feed stream.

In the sense of the framework developed in this chapter, the argon present in the air feed can be construed as a "feed impurity," while the argon product stream can be regarded as a "purge" stream. From this point of view, ASUs are process systems with recycle and purge, or, alternatively, processes in which argon is recycled at a high rate compared with the throughput.

This observation allows us a novel insight into ASU dynamics and control: we can expect that the dynamics of the ASU exhibit a slow component, related to the presence of the argon in the feed stream and the associated argon-recovery system. Specifically, changes in the controlled variables of the process will be reflected in the composition of the argon product stream after a long period of time, a fact that has been confirmed by practical observations (Vinson 2006). Furthermore, maximizing argon recovery is typically a key operating objective in ASUs (Vinson 2006), and, considering the developments which we present later in this chapter, the control of the purity of the argon product stream should be undertaken over a longer time horizon, potentially using a separate controller designed on the basis of the model of the slow dynamics.

4.6.3 Processes with multiple impurities

The framework developed above can be extended to the case in which multiple impurities are present in the system. It is straightforward to show that, in such circumstances, the dimension of the state-space realization of the model of the slow dynamics (Equation (4.34)) will be equal to the number of components whose rate of input to the process (either as feed impurities or as reaction byproducts) is small.

4.7 Control implications

The analysis presented in Section 4.4 indicates that, impurity levels notwithstanding, the rate at which the overall process evolves is not significantly different (i.e., faster or slower) than the dynamics of the individual units. On the other hand, we can expect the impurity levels in the process to respond very slowly to changes in the manipulated variables. This suggests that the control of integrated processes in which impurities are present should be approached using two layers of control action.

- Control objectives related to the operation of the process units and the process itself (production rate, product quality, unit-level, and total inventory) should be addressed in the fast time scale. For instance, when a multi-loop linear control strategy is considered, the reset time for the controllers should be of the order of magnitude of the time constants of the individual process units.
- The control of the impurity levels in the process should be undertaken in the slow time scale, and any control strategy should account for the long time horizon that the respective variables evolve in. One could, for example, employ a description of the slow dynamics (Equation (4.35)) for synthesizing a model-based controller.

  It is important to note that, in typical practical situations in which cost constraints play an important role, impurity-concentration measurements are available for only a few units (and, more often than not, just for a single unit). Thus, a model of the evolution of the total impurity inventory (such as those developed in the examples above, i.e., Equations (4.43) and (4.45)) is not well suited for controller design. Rather, an appropriate coordinate change of the type in Equation (4.33) should be used to obtain a model of the evolution of the measured concentration variable in the slow time scale. An example of this approach is presented in the case study following this section.

Our analysis also provides clues regarding the manipulated inputs that are available for use in each of the control layers. Recall that the flow rates $\mathbf{u}^l$ of the material streams not connected with the process impurity input and output are present only in the model of the fast dynamics (4.20). $\mathbf{u}^l$ thus represent the inputs of choice for addressing all the control objectives not related to impurity levels.

Conversely, impurity levels should be controlled by varying the flow rate u_p of the purge stream. u_p is clearly the *only* manipulated input available in the model of the slow dynamics (4.35). The rate at which impurity is input to or generated in the process (as captured by the term $(\partial\phi/\partial\mathbf{x})\mathbf{g}^{I0}(\mathbf{x})|_{\mathbf{x}=\mathbf{T}^{-1}(\zeta)}$) cannot be set by a process operator and constitutes a (typically unmeasured) process *disturbance* that the control system must deal with.

4.8 Case study: control of a reactor–condenser process

4.8.1 Process description

A process designed to generate product B from a raw-material stream containing the reactant A consists of a reactor in which the reaction A $\rightarrow$ B takes place, followed by a condenser, where product B is separated in liquid form and the unreacted A is recycled to the reactor in vapor form, as in Figure 4.1. The feed stream contains a small quantity of noncondensing impurity I, which is eliminated by purging a small portion of the recycle stream. The impurity has an inhibitive effect on the reaction, which is reflected in the rate expression:

$$r_A = \frac{k_1 y_{A,R}}{1 + y_{I,R}} \tag{4.49}$$

The process parameters and the nominal values of the state variables are given in Table 4.1 (the subscript c indicates process parameters for the condenser). The model of the process is identical (with the exception of the reaction rate expression) to the one developed in Section 4.2.1. The data in Table 4.1 also follow the notation introduced in Section 4.2.1.

The operating objective for this system is to control the mole fraction of the product B in the liquid product stream at $x_{B,sp} = 0.819$, in the presence of disturbances in the inlet composition and changes in the production rate.

Table 4.1. Nominal process parameters (adapted from (Baldea *et al.* 2006))

F_0	100.00 mol/min	T_c	279.00 K	M_R	2411.90 mol
R	100.10 mol/min	P_c	2.83 MPa	M_V	1225.60 mol
F	200.10 mol/min	$\mathcal{A}$	200.00 m^2/m^3	M_L	14940.00 mol
L	97.10 mol/min	$\mathcal{K}_A$	342.00 mol m^{-2} min^{-1}	y_A	0.255
P	3.90 mol/min	$\mathcal{K}_B$	360.00 mol m^{-2} min^{-1}	y_I	0.511
$y_{A,0}$	0.98	$\mathcal{K}_I$	1.8×10^{-5} mol m^{-2} min^{-1}	$y_{A,R}$	0.219
$y_{I,0}$	0.02	$P_A^S(T_c)$	4.00 MPa	$y_{I,R}$	0.266
$k_{p,1}$	0.01 min^{-1}	$P_B^S(T_c)$	0.80 MPa	x_A	0.181
$k_{p,2}$	0.01 min^{-1}	$P_I^S(T_c)$	90.00 MPa	x_I	1.90×10^{-5}
V_c	2.00 m^3	ρ_L	15.00 kmol/m^3		

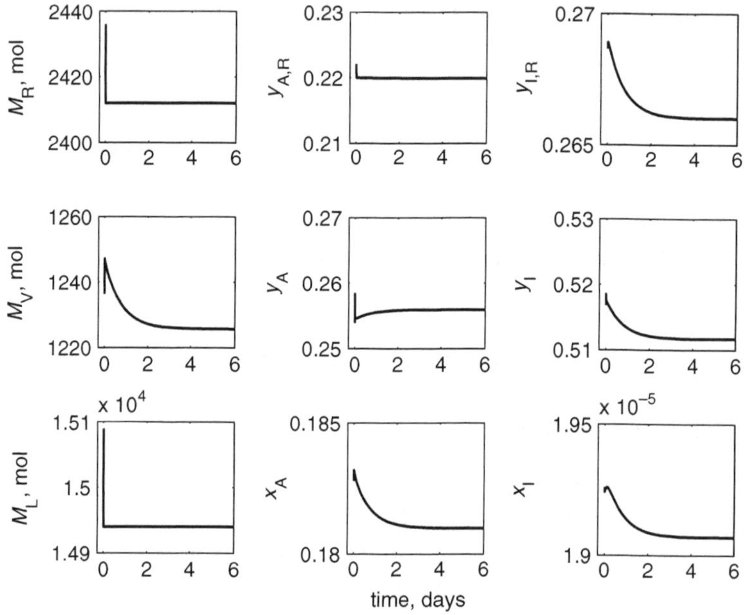

Figure 4.6 Time responses of all state variables.

4.8.2 System analysis

We will first concentrate on studying the process dynamics, so let us consider a numerical experiment that consists of starting a dynamic simulation of the process from initial conditions that are slightly perturbed from the nominal, steady-state values of the state variables. Although material holdups are stabilized using the proportional controllers in Equation (4.40), in view of the process-level operating objective stated above, this can be considered an "open-loop" simulation.

The process response is presented in Figure 4.6. Observe that *all* the state variables exhibit a fast transient, followed by a slow approach to steady state, which is indicative of the two-time-scale behavior of the system, and is consistent with our observation that processes with impurities and purge are modeled by systems of ODEs that are in a *nonstandard* singularly perturbed form.

Figure 4.7 shows the evolution of the total impurity holdup for the same simulation; note that this variable exhibits dynamics only in the slow time scale, which is – again – consistent with our previous findings.

4.8.3 Controller design

According to the results presented in Section 4.7, the control of the product purity x_B should be addressed in the fast time scale, together with inventory/holdup control. After implementing the controllers (4.40), which use F and L as

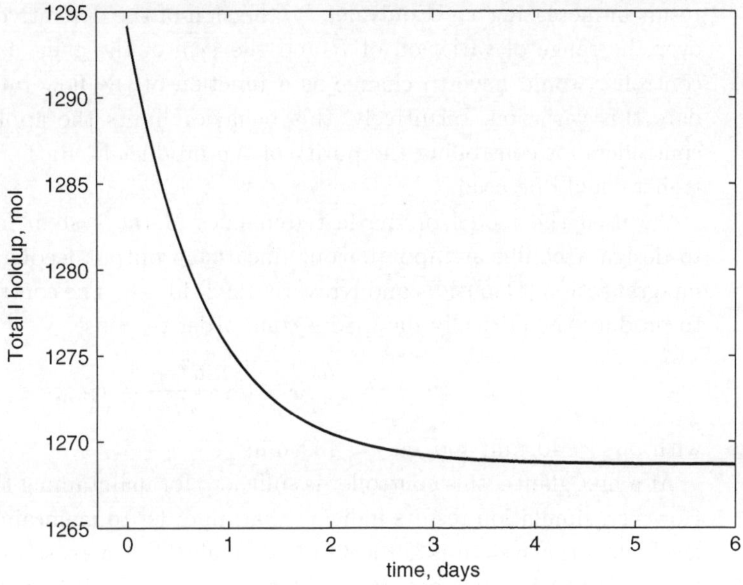

Figure 4.7 Evolution of the total impurity holdup.

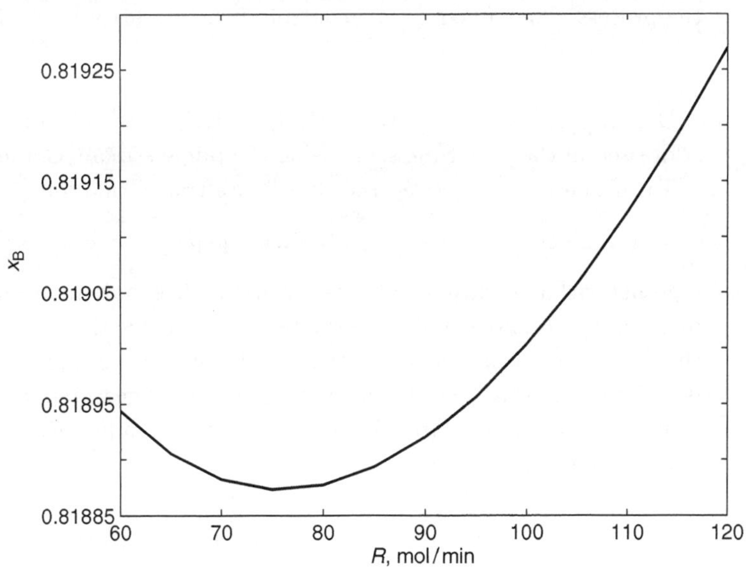

Figure 4.8 Input multiplicity of product purity loop.

manipulated inputs, the remaining available manipulated input for controlling the product purity is the recycle flow rate R.

A plot of x_B vs. R at steady state (Figure 4.8) reveals an input multiplicity: at low values of the recycle flow rate, an increase in R will yield a decrease in the purity of the product. If R is, however, increased further, it will eventually result in increasing x_B. Equivalently, the sign of the derivative $\partial x_B/\partial R$ changes over the range of variation of R and the *sign* of the gain of any *linear* purity controller would have to change as a function of the flow rate R to accommodate this variation. Intuitively, this behavior limits the applicability of linear controllers for controlling the purity of the product B. Rather, a nonlinear controller should be used.

We used the model of the fast dynamics of the system in Equation (4.36) to design a nonlinear input–output linearizing output feedback controller with integral action (Daoutidis and Kravaris 1992) for x_B. The controller was designed to produce the critically damped second-order response

$$x_B + \beta_{B,1}\frac{dx_B}{dt} + \beta_{B,2}\frac{d^2x_B}{dt^2} = x_{B,sp} \tag{4.50}$$

with $\beta_{B,1} = 40$ min and $\beta_{B,2} = 400$ min^2.

At a first glance, this controller is sufficient for maintaining the product purity. However, simulation results indicate that, in order to maintain x_B at the desired level when the system is subjected to a small (5%) increase in the mole fraction $y_{I,0}$, the recycle flow rate R would need to rise to 501.3 mol/min (a fivefold increase from the nominal value). Thus, due to its inhibitive effect on the reaction rate, the accumulation of the impurity I is highly detrimental to the operation of the process. Consequently, the control of the impurity levels in the reactor is of critical importance and directly linked to the main objective of product-purity control.

Drawing again on our theoretical analysis, the control of $y_{I,R}$ should be addressed in the slow time scale using the purge stream as a manipulated input. To this end, we employed the coordinate transformation (4.33) with

$$\phi(\mathbf{x}) = y_{I,R} \tag{4.51}$$

together with the quasi-steady-state constraints stemming from setting the terms in Equation (4.38) equal to zero, to obtain a description of the evolution of the reactor impurity mole fraction in the slow time scale. We then used this description as the basis for synthesizing a nonlinear input–output linearizing controller that manipulates the purge flow rate to induce the following first-order response for $y_{I,R}$:

$$y_{I,R} + \beta_Y\frac{dy_{I,R}}{dt} = v \tag{4.52}$$

with $\beta_Y = 500$ min and integral action imposed on the $v - y_{I,R}$ dynamics for offset-free tracking.

4.8.4 Simulation results and discussion

Figures 4.9–4.11 present the closed-loop response for a 10% rise in the system throughput (imposed by changing the feed flow rate F_0). Figures 4.12–4.14 show the closed-loop response of the reactor–condenser system in the case of an

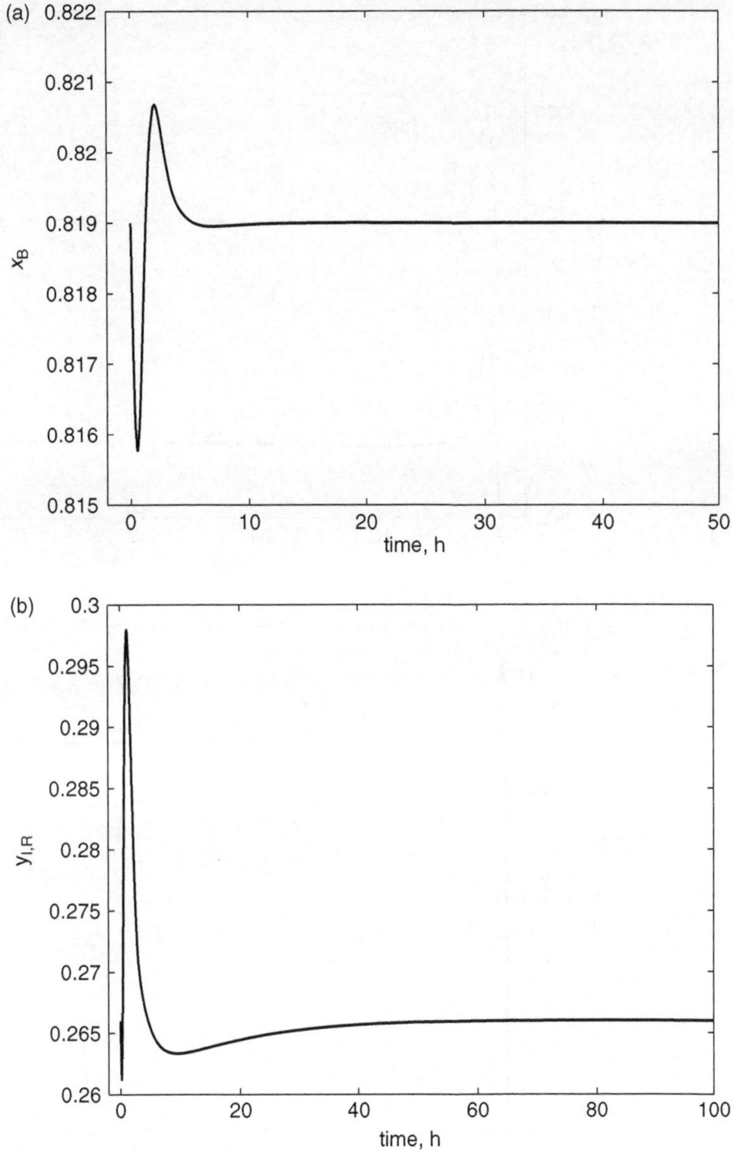

Figure 4.9 Evolution of the process composition variables for a 10% increase in the production rate at $t = 0$. (a) Product purity and (b) reactor impurity level.

unmeasured increase in the inlet mole fraction of the impurity y_{I0}, from $y_{I0} = 0.02$ to $y_{I0} = 0.025$. The proposed nonlinear control structure exhibits excellent performance in both cases, showing small changes in the product purity.

The closed-loop behavior in the presence of the same disturbances but considering, additionally, a 10% modeling error in the mass-transfer coefficient of the product, $\mathcal{K}_B$, and a 5% modeling error for the reaction rate constant k_1

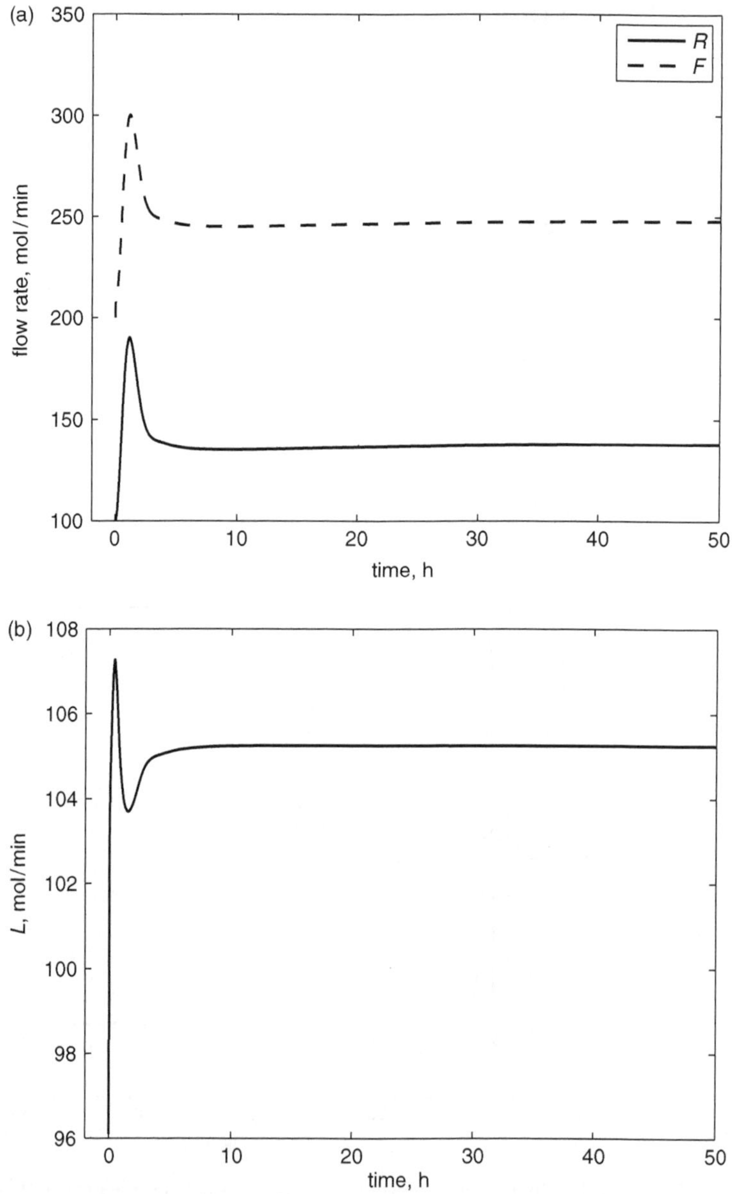

Figure 4.10 Evolution of the process stream flow rates for a 10% increase in the production rate at $t = 0$. (a) Effluent and recycle flow rates, and (b) product flow rate.

is presented, respectively, in Figures 4.15–4.20. The proposed nonlinear control structure also exhibits excellent performance in the presence of unmeasured disturbances, even when model mismatch is considered. The disturbances are rejected with small changes in the recycle flow rate and minimal effect on the product purity.

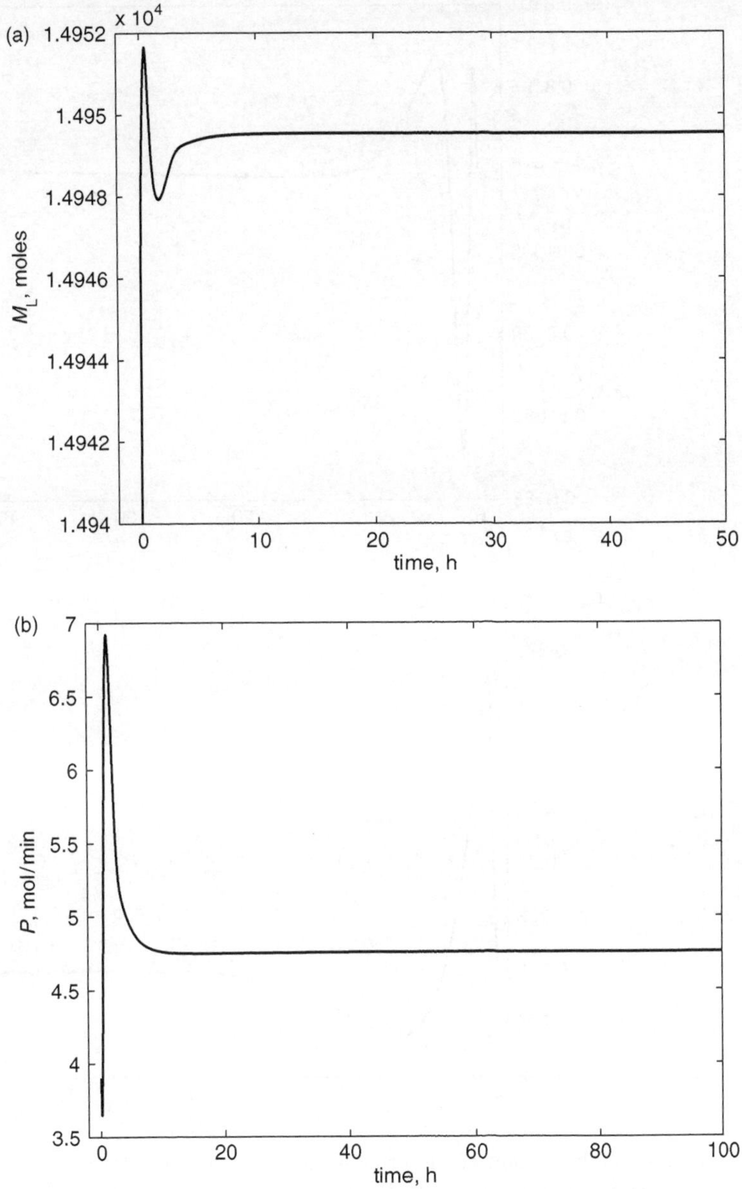

Figure 4.11 Evolution of (a) the condenser liquid holdup and (b) the purge flow rate for a 10% increase in the production rate at $t = 0$.

Notice that, according to the developments above, a change in the inlet impurity fraction is a disturbance that strongly impacts the slow dynamics of the process. This is apparent in the simulation scenarios presented in Figures 4.12–4.14 and 4.18–4.20: the time required to reach steady state after an increase in y_{I0} is clearly longer than the response time for an increase in the feed flow rate F_0 (Figures 4.9–4.11 and 4.15–4.17).

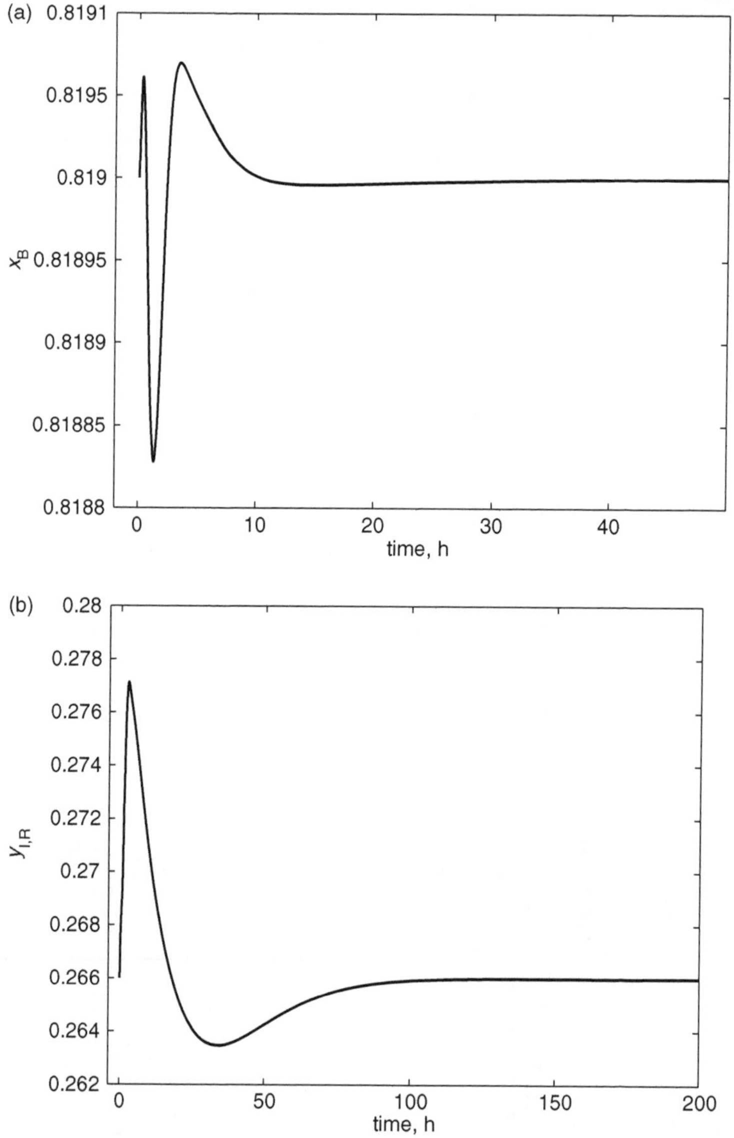

Figure 4.12 Evolution of the process composition variables for a 25% unmeasured increase in the inlet impurity levels y_{I0} occurring at $t = 0$. (a) Product purity and (b) reactor impurity level.

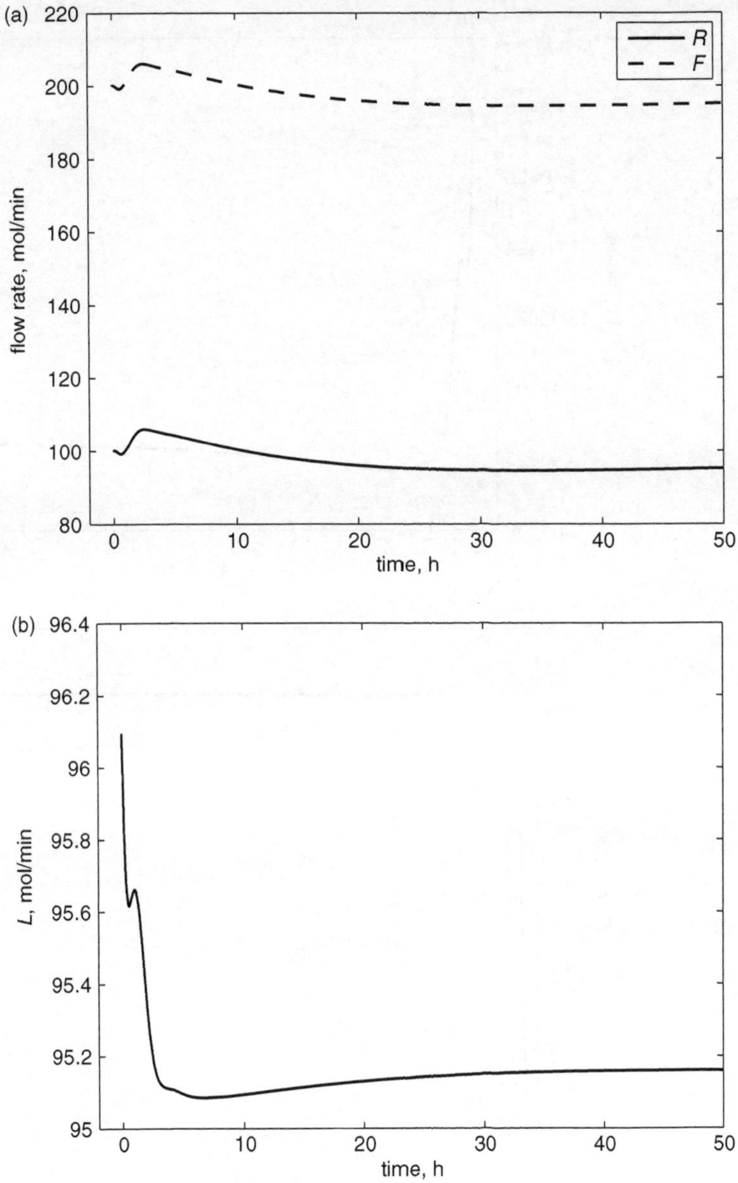

Figure 4.13 Evolution of the process stream flow rates for a 25% unmeasured increase in the inlet impurity levels y_{I0} occurring at $t = 0$. (a) Effluent and recycle flow rates, and (b) product flow rate.

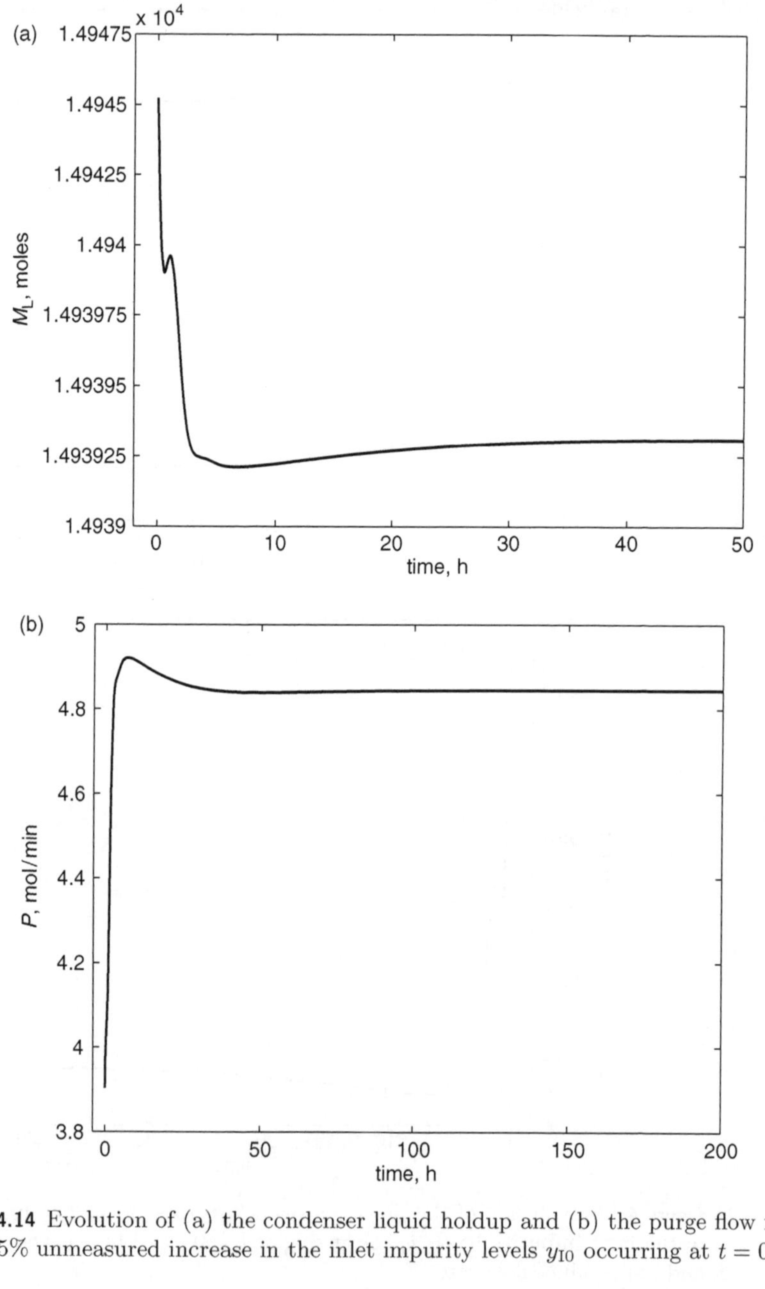

Figure 4.14 Evolution of (a) the condenser liquid holdup and (b) the purge flow rate for a 25% unmeasured increase in the inlet impurity levels y_{I0} occurring at $t = 0$.

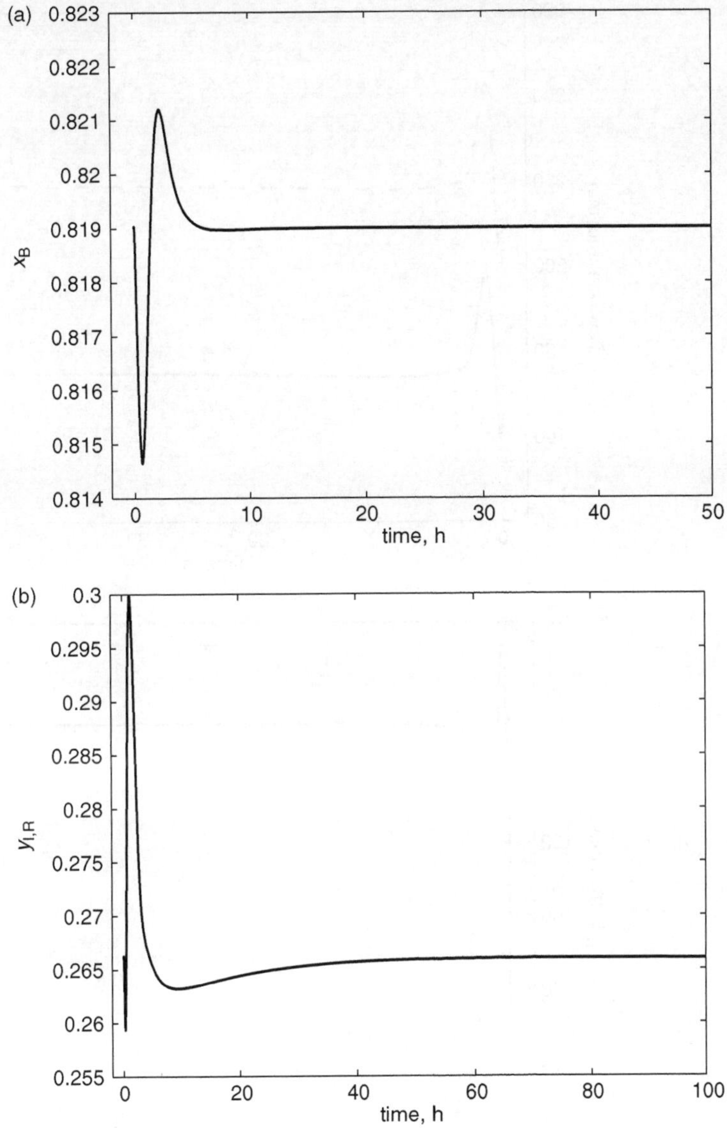

Figure 4.15 Evolution of the process composition variables for a 10% increase in the production rate at $t = 0$, under plant–model parameter mismatch. (a) Product purity and (b) reactor impurity level.

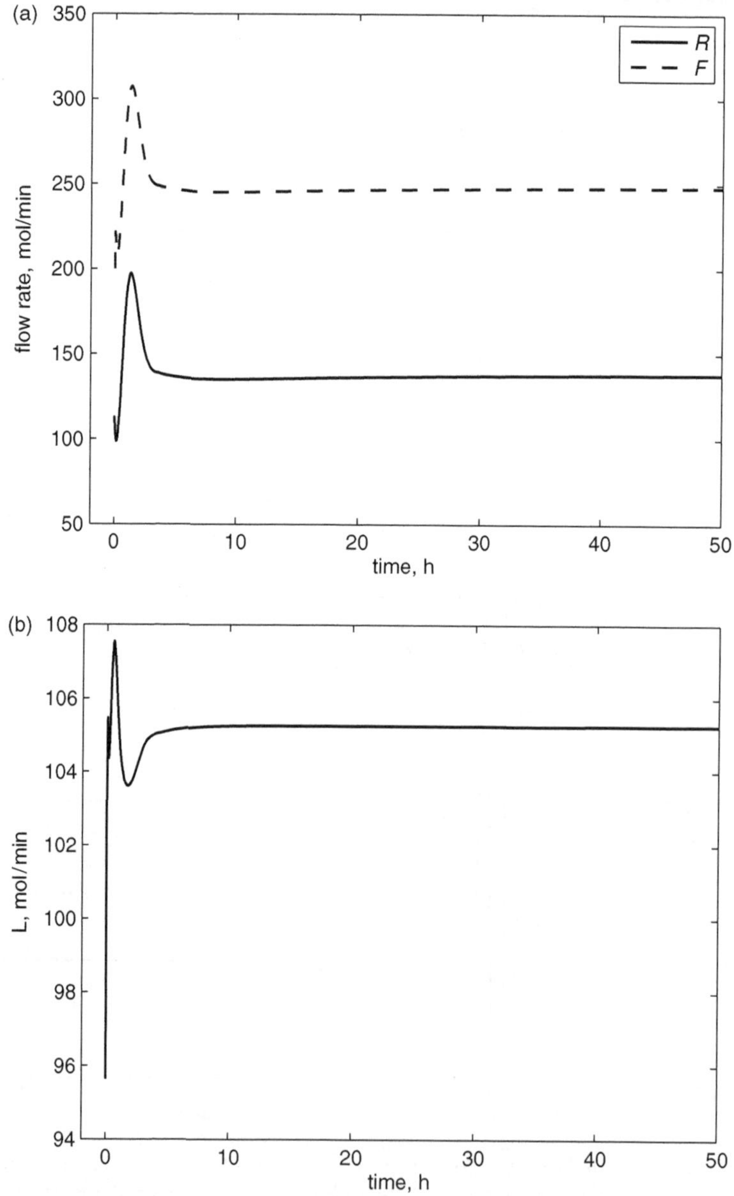

Figure 4.16 Evolution of the process stream flow rates for a 10% increase in the production rate at $t = 0$, under plant–model parameter mismatch. (a) Effluent and recycle flow rates, and (b) product flow rate.

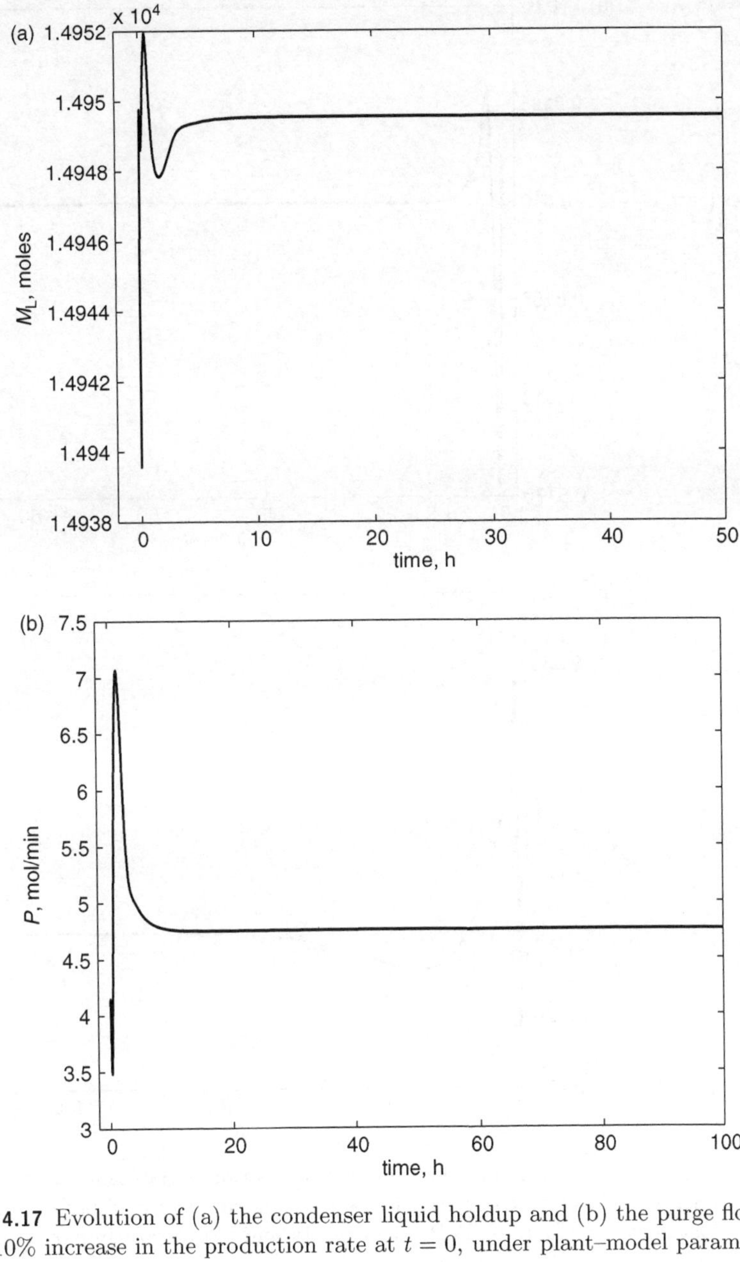

Figure 4.17 Evolution of (a) the condenser liquid holdup and (b) the purge flow rate for a 10% increase in the production rate at $t = 0$, under plant–model parameter mismatch.

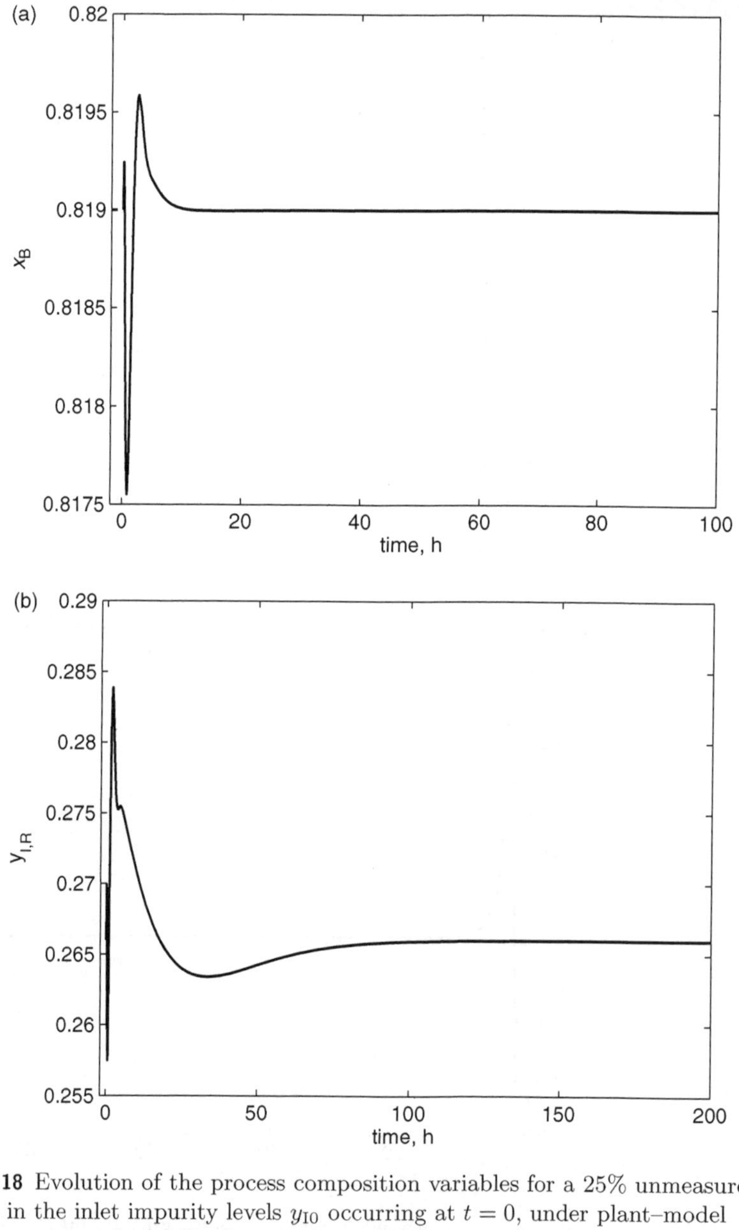

Figure 4.18 Evolution of the process composition variables for a 25% unmeasured increase in the inlet impurity levels y_{I0} occurring at $t = 0$, under plant–model parameter mismatch. (a) Product purity and (b) reactor impurity level.

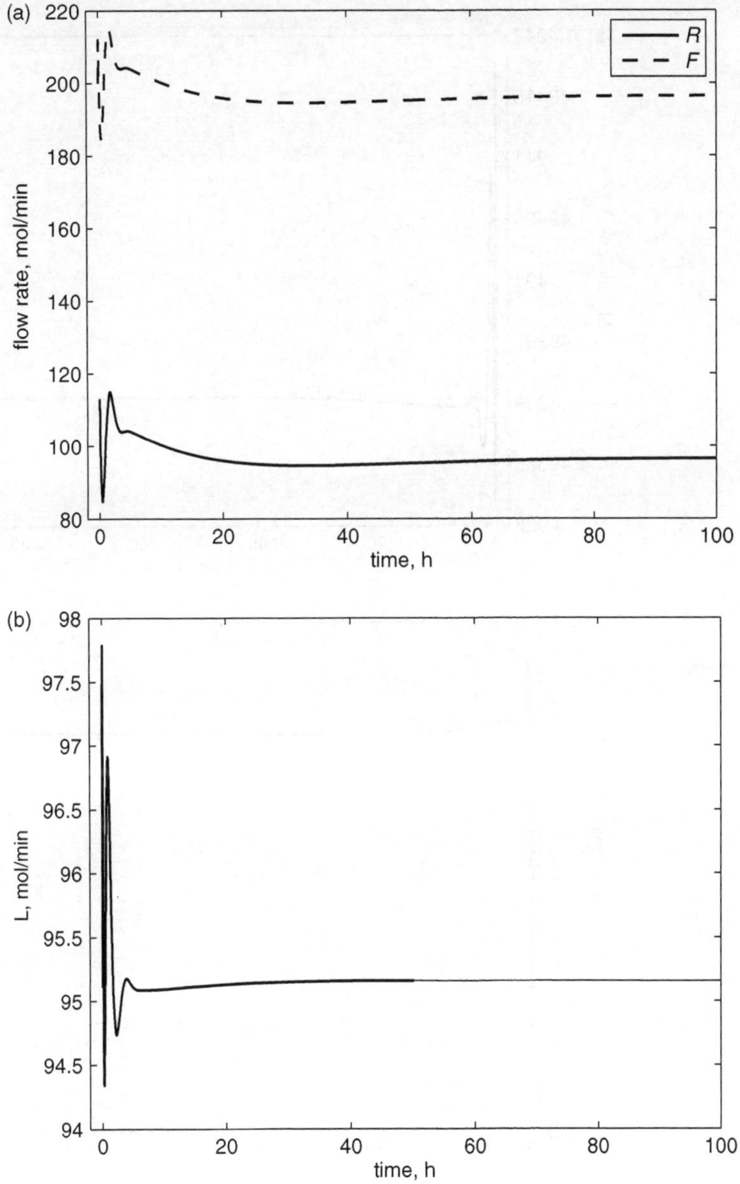

Figure 4.19 Evolution of the process stream flow rates for a 25% unmeasured increase in the inlet impurity levels y_{I0} occurring at $t = 0$, under plant–model parameter mismatch. (a) Effluent and recycle flow rates, and (b) product flow rate.

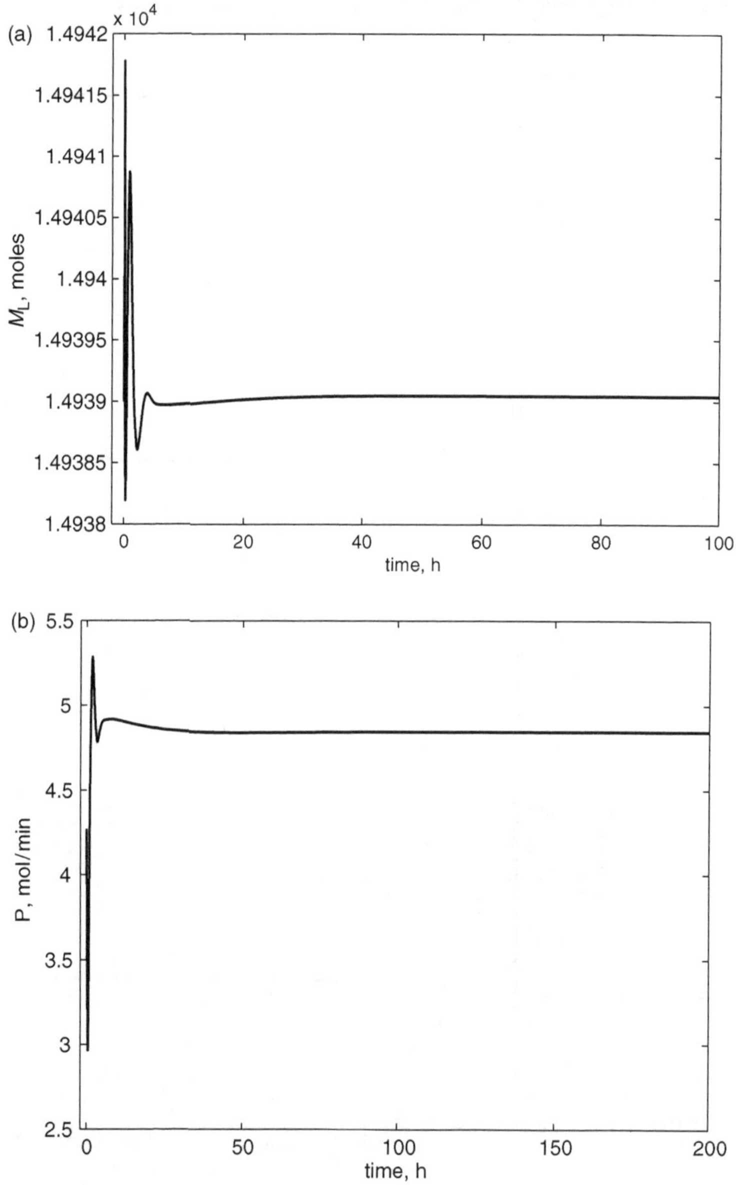

Figure 4.20 Evolution of (a) the condenser liquid holdup and (b) the purge flow rate for a 25% unmeasured increase in the inlet impurity levels y_{I0} occurring at $t = 0$, under plant–model parameter mismatch.

4.9 Synopsis

In this chapter, we have considered the effect of the presence of difficult-to-separate impurities (either present in the feed or generated as a reaction by product) and of a purge stream used to remove them on the dynamics of integrated processes. We have shown that the impurity levels in the process evolve over a much longer time scale than the dynamics of the individual units (and of the process itself) and that, consequently, processes with impurities and purge have a two-time-scale dynamic behavior.

We then proposed a method for deriving reduced-order, non-stiff models for the behavior of such systems in each time scale. In the general case, the slow dynamics of the systems was shown to be 1D and directly associated with the total impurity holdup.

Finally, we analyzed the control implications of the presence of impurities in a process, concluding that the control of impurity levels must be addressed over an extended time horizon using the flow rate of the purge stream as a manipulated input. To close the impurity-levels loop, one should resort either to an appropriately tuned linear controller (e.g., a PI controller with long reset time) or to a (nonlinear) model-based controller that uses (an inverse of) the reduced-order model of the slow dynamics – as developed in this chapter – to compute the necessary control action.

5 Dynamics and control of generalized integrated process systems

5.1 Introduction

The purpose of this chapter is to create a general framework that captures the dynamic effect of the simultaneous presence of *both* significant material recycle streams *and* purge streams. We will use this result presented in (Baldea and Daoutidis 2007) to rationalize at the theoretical level the development and use of a hierarchical process control structure, consisting of several interconnected control and optimization layers.

5.2 System description and modeling

In order to generalize the discussion of the process structures discussed in Chapters 3 and 4, let us consider the class of processes in Figure 5.1. This generic structure consists of $S+1$ units and a single recycle loop that includes units $1,\ldots,S$. One (multi-stage) separator is present and we denote by S the part of the separator that is included in the recycle loop, and by $S+1$ the part that is

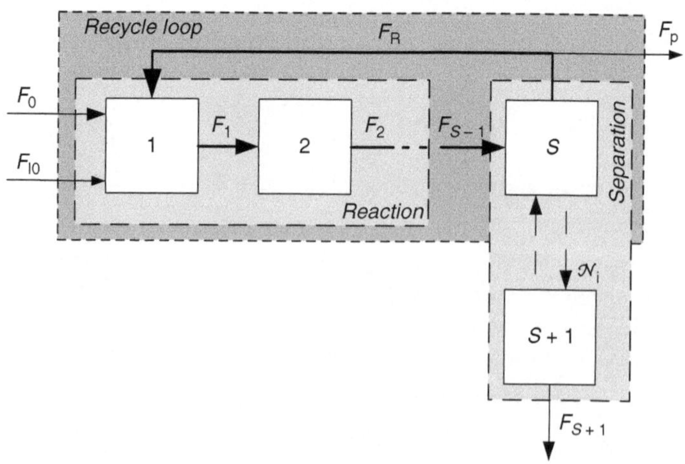

Figure 5.1 A generic reaction–separation process system with material recycle and purge.

outside the loop. One of the output streams of the separator is recycled, with the flow rate F_{R} of the recycle stream being much larger than the process throughput F_0. The other output stream leaves the process as a product stream. There are in total $\mathcal{C}$ components in the process, including an impurity I which enters the process at a small rate F_{I0}. We assume that I does not separate readily in the separator and that a purge stream is used to prevent its accumulation in the recycle loop. In order to derive a model for this generic process, let F_0 denote the feed flow rate to the first unit, F_{I0} the rate at which the impurity is input to the process, F_j, $j = 1, \ldots, S+1$, the outlet flow rate from the jth unit, F_{R} the recycle flow rate, and F_{P} the purge flow rate. Also, let $\mathcal{N}_i$, $i = 1, \ldots, \mathcal{C}-1$, and $\mathcal{N}_{\mathrm{I}}$ denote the net rates at which the ith component and the impurity, respectively, are separated from the recycle loop.

Considering that the individual process units are modeled as lumped-parameter systems, the mathematical model that describes the overall and component material balances of the process has the generic form

$$\begin{aligned} \dot{\mathbf{x}} = \mathbf{f}(\mathbf{x}) + \sum_{j=1}^{S-1} \mathbf{g}_j(\mathbf{x})F_j + \mathbf{g}_{\mathrm{R}}(\mathbf{x})F_{\mathrm{R}} + \sum_{j=0,S+1} \mathbf{g}_j(\mathbf{x})F_j + \sum_{i=1}^{\mathcal{C}-1} \mathbf{g}_{c,i}(\mathbf{x})\mathcal{N}_i \\ + \mathbf{g}_{\mathrm{I0}}(\mathbf{x})F_{\mathrm{I0}} + \mathbf{g}_{\mathrm{I}}(\mathbf{x})\mathcal{N}_{\mathrm{I}} + \mathbf{g}_{\mathrm{P}}(\mathbf{x})F_{\mathrm{p}} \end{aligned} \tag{5.1}$$

where $\mathbf{f}(\mathbf{x})$, $\mathbf{g}_j(\mathbf{x})$, $\mathbf{g}_{c,i}(\mathbf{x})$, $\mathbf{g}_{\mathrm{R}}(\mathbf{x})$, $\mathbf{g}_{\mathrm{I0}}(\mathbf{x})$, $\mathbf{g}_{\mathrm{I}}(\mathbf{x})$, and $\mathbf{g}_{\mathrm{P}}(\mathbf{x})$ are appropriately defined n-dimensional vector functions. $\mathbf{x} \in \mathcal{X} \subset \mathbb{R}^n$ represents the state vector.

The premises that we rely upon in the analysis presented in this chapter draw on Assumptions 4.1–4.4 and the discussion in Section 3.2, and are summarized below.

Assumption 5.1. *There are no significant side-streams in the process. Consequently, the nominal flow rates of the streams in the recycle loop are of comparable magnitude, i.e.,*

$$\frac{F_{j,\mathrm{s}}}{F_{\mathrm{R,s}}} = k_j = \mathcal{O}(1), \; j = 1, \ldots, S-1 \tag{5.2}$$

where the index s *denotes steady-state values.*

Assumption 5.2. *The net rates at which the components* $1, \ldots, \mathcal{C}-1$ *are separated from the recycle loop, as well as the nominal flow rate of the product stream, are of comparable magnitude. Also, from mass-balance considerations, they are of the same magnitude as the flow rate of the process feed stream:*

$$\frac{\mathcal{N}_{i,\mathrm{s}}}{F_{0,\mathrm{s}}} = \mathcal{O}(1), \; i = 1, \ldots, \mathcal{C}-1 \tag{5.3}$$

$$\frac{F_{S+1,\mathrm{s}}}{F_{0,\mathrm{s}}} = \mathcal{O}(1) \tag{5.4}$$

Assumption 5.3. *The process conditions and constraints (e.g., low single-pass conversion) are such that, at steady state, the flow rate of the material recycle stream, $F_{\mathrm{R,s}}$, must be kept significantly larger than the process throughput $F_{0,\mathrm{s}}$. This is reflected in a large recycle number* **Rc***:*

$$\mathsf{Rc} = \frac{F_{\mathrm{R,s}}}{F_{0,\mathrm{s}}} \gg 1 \tag{5.5}$$

Assumption 5.4. *In order to minimize the loss of raw material and the release of potentially hazardous chemicals into the environment, the steady-state flow rate of the purge stream $F_{\mathrm{P,s}}$ is much smaller than the flow rate of the reactant feed stream $F_{0,\mathrm{s}}$. Equivalently, the purge number* **Pu** *of the process is small:*

$$\mathsf{Pu} = \frac{F_{\mathrm{P,s}}}{F_{0,\mathrm{s}}} \ll 1 \tag{5.6}$$

Also, the inlet flow rate $F_{\mathrm{I0,s}}$ of the impurity is comparable in magnitude to the purge flow rate:

$$\frac{F_{\mathrm{I0,s}}}{F_{0,\mathrm{s}}} = \beta_1 \mathsf{Pu} \tag{5.7}$$

where β_1 is an $\mathcal{O}(1)$ quantity.

Assumption 5.5. *The impurity is not readily separated from the rest of the components present in the process. Equivalently, the net rate of impurity removal from the recycle loop in the separation unit is significantly smaller than the rate at which the impurity is input to the process:*

$$\frac{\mathcal{N}_{\mathrm{I,s}}}{F_{0,\mathrm{s}}} = \beta_2 \mathsf{Pu}^2 \tag{5.8}$$

where β_2 is an $\mathcal{O}(1)$ quantity.

On defining $u_j = F_j/F_{j,\mathrm{s}}$, $j = 0, 1, \ldots, S+1$, R, P to be the scaled inputs that correspond to the flow rates $F_0, \ldots, F_j, \ldots, F_{\mathrm{R}}$ and F_{P}, and using the above assumptions, the model of the generic process becomes

$$\begin{aligned} \dot{\mathbf{x}} &= \mathbf{f}(\mathbf{x}) + F_{0,\mathrm{s}} \sum_{j=0,S+1} \mathbf{g}_j(\mathbf{x}) u_j + \sum_{i=1}^{\mathcal{C}-1} \mathbf{g}_{\mathrm{c},i}(\mathbf{x}) \mathcal{N}_i \\ &+ \mathsf{Pu}\, F_{0,\mathrm{s}} \left[\mathbf{g}_{\mathrm{I0}}(\mathbf{x}) \beta_1 + \mathsf{Pu}\, \mathbf{g}_{\mathrm{I}}(\mathbf{x}) \beta_2 \frac{\mathcal{N}_{\mathrm{I}}}{\mathcal{N}_{\mathrm{I,s}}} + \mathbf{g}_{\mathrm{P}}(\mathbf{x}) u_{\mathrm{P}} \right] \\ &+ \mathsf{Rc} \left[\sum_{j=1}^{S-1} k_j \mathbf{g}_j(\mathbf{x}) u_j + \mathbf{g}_{\mathrm{R}}(\mathbf{x}) u_{\mathrm{R}} \right] \end{aligned} \tag{5.9}$$

Equation (5.9) can be written in a more compact and mathematically convenient form as

$$\frac{d\mathbf{x}}{dt} = \underbrace{\bar{\mathbf{f}}(\mathbf{x}, \mathbf{u}^{\mathrm{s}})}_{\substack{\text{process inlet/outlet} \\ \text{chemical reactions}}} + \underbrace{\frac{1}{\varepsilon_1}\mathbf{G}^{\mathrm{l}}(\mathbf{x})\mathbf{u}^{\mathrm{l}}}_{\text{material recycling}}$$

$$+ \underbrace{\varepsilon_2\left[\mathbf{g}^{\mathrm{I0}}(\mathbf{x}) + \varepsilon_2\mathbf{g}^{\mathrm{I}}(\mathbf{x}) + \mathbf{g}^{\mathrm{P}}(\mathbf{x})u_{\mathrm{p}}\right]}_{\substack{\text{impurity inlet/outlet} \\ \text{impurity purging}}} \tag{5.10}$$

with $\mathbf{u}^{\mathrm{l}} \in \mathbb{R}^{m^{\mathrm{l}}}$ being the vector of scaled input variables corresponding to the "large" flow rates $F_1, \ldots, F_{S-1}, F_{\mathrm{R}}$, and $\mathbf{u}^{\mathrm{s}} \in \mathbb{R}^{m^{\mathrm{s}}}$ being the vector of scaled input variables corresponding to the "small" flow rates F_0 and F_{S+1}. $\bar{\mathbf{f}}(\mathbf{x}, \mathbf{u}^{\mathrm{s}})$, $\mathbf{g}^{\mathrm{I0}}(\mathbf{x})$, $\mathbf{g}^{\mathrm{I}}(\mathbf{x})$, and $\mathbf{g}^{\mathrm{P}}(\mathbf{x})$ are n-dimensional vector functions, and $\mathbf{G}^{\mathrm{l}}(\mathbf{x}) \in \mathbb{R}^{n \times m^{\mathrm{l}}}$. As established in the previous chapters, we used the notation $\varepsilon_1 = 1/\mathbf{Rc}$ and $\varepsilon_2 = \mathbf{Pu}$, and, since the mass-transfer rates $\mathcal{N}_i$, $i = 1, \ldots, \mathcal{C} - 1$, are generally defined as functions of the physical parameters of the system, the corresponding terms have been included in the expressions for $\bar{\mathbf{f}}(\mathbf{x})$ and $\mathbf{g}^{\mathrm{I}}(\mathbf{x})$.

Clearly, the general model (5.10) contains terms of very different magnitudes, corresponding, respectively, to the process input and output flows and to the chemical reactions, to the presence of the large material recycle stream, and to the presence of the impurity and the purge stream used for its removal. While (as we have argued in the previous chapters of the book) the presence of these terms is purely a *steady-state*, design feature of the process, it is intuitive that their impact on the process *dynamics* will also be very different.

From a mathematical point of view, we can see that Equation (5.10) is in a (nonstandard) singularly perturbed form. This suggests that the integrated processes under consideration will feature a dynamic behavior with at least two distinct time scales. Drawing on the developments in Chapters 2, 3, and 4, the following section demonstrates that these systems evolve in effect over *three* distinct time scales and proposes a method for deriving reduced-order, non-stiff models for the dynamics in each time scale.

5.3 Time-scale decomposition and nonlinear model reduction

5.3.1 Fast dynamics at the unit level

Let us define the new time variable $\tau = 1/\varepsilon_1$, which is of the order of magnitude of the residence time in an individual process unit. In this fast ("stretched") time scale, the model of Equation (5.10) becomes

$$\frac{d\mathbf{x}}{d\tau} = \varepsilon_1\{\bar{\mathbf{f}}(\mathbf{x}, \mathbf{u}^{\mathrm{s}}) + \varepsilon_2[\mathbf{g}^{\mathrm{I0}}(\mathbf{x}) + \varepsilon_2\mathbf{g}^{\mathrm{I}}(\mathbf{x}) + \mathbf{g}^{\mathrm{P}}(\mathbf{x})u_{\mathrm{p}}]\} + \mathbf{G}^{\mathrm{l}}(\mathbf{x})\mathbf{u}^{\mathrm{l}} \tag{5.11}$$

Following the procedure devised in Chapter 3, we consider the limit $\varepsilon_1 \to 0$, corresponding to the recycle flow rate becoming infinite. This results in the following description of the fast dynamics of the system:

$$\frac{d\mathbf{x}}{d\tau} = \mathbf{G}^{\mathrm{l}}(\mathbf{x})\mathbf{u}^{\mathrm{l}} \tag{5.12}$$

Equation (5.12) effectively corresponds to the dynamics of the individual process units that are part of the recycle loop. The description of the fast dynamics (5.12) involves only the large flow rates $\mathbf{u}^{\mathrm{l}}$ of the recycle-loop streams, and does not involve the small feed/product flow rates $\mathbf{u}^{\mathrm{s}}$ or the purge flow rate u_{P}. As shown in Chapter 3, it is easy to verify that the large flow rates $\mathbf{u}^{\mathrm{l}}$ of the internal streams do not affect the *total holdup* of any of the components $1, \dots, \mathcal{C} - 1$ (which is influenced only by the small flow rates $\mathbf{u}^{\mathrm{s}}$), or the total holdup of I (which is influenced exclusively by the inflow $F_{\mathrm{I}0}$, the transfer rate $\mathcal{N}_{\mathrm{I}}$ in the separator, and the purge stream u_{P}). By way of consequence, the differential equations in (5.12) are not independent. Equivalently, the quasi-steady-state condition $\mathbf{0} = \mathbf{G}^{\mathrm{l}}(\mathbf{x})\mathbf{u}^{\mathrm{l}}$ corresponding to the dynamical system (5.12) does not specify a set of isolated equilibrium points, but, rather, a low-dimensional equilibrium manifold.

As in the previous chapters, we will assume that the linearly independent quasi-steady-state conditions in $\mathbf{G}^{\mathrm{l}}(\mathbf{x})\mathbf{u}^{\mathrm{l}}$ can be isolated, i.e., the vector function $\mathbf{G}^{\mathrm{l}}$ can be reformulated as

$$\mathbf{G}^{\mathrm{l}}(\mathbf{x}) = \mathbf{B}(\mathbf{x})\tilde{\mathbf{G}}^{\mathrm{l}}(\mathbf{x}) \tag{5.13}$$

with $\mathbf{B}(\mathbf{x}) \in \mathbb{R}^{n\times(n-\mathcal{C}-m)}$ being a full column rank matrix and the matrix $\tilde{\mathbf{G}}^{\mathrm{l}}(\mathbf{x}) \in \mathbb{R}^{(n-\mathcal{C}-m)\times m^{\mathrm{l}}}$, with $m < n$ being the number of states associated with unit(s) $S+1$, having linearly independent rows. This suggests that the dimension of the equilibrium manifold of the fast dynamics of the units within the recycle loop has an upper bound in $\mathcal{C} + m$.

5.3.2 Process-level dynamics

In order to obtain a description of the dynamics after the fast transient, we first recognize that the equations describing the process evolution in the fast time scale can be replaced, in the time scale t, by the corresponding quasi-steady-state constraints. These constraints are obtained by multiplying Equation (5.10) by ε_1 and considering the limit $\varepsilon_1 \to 0$. Taking into account (5.13), the constraints that must be satisfied in the slow(er) time scale(s) are

$$\mathbf{0} = \tilde{\mathbf{G}}^{\mathrm{l}}(\mathbf{x})\mathbf{u}^{\mathrm{l}} \tag{5.14}$$

In this limit, the terms $(\tilde{\mathbf{G}}^{\mathrm{l}}(\mathbf{x})\mathbf{u}^{\mathrm{l}})/\varepsilon_1$ become indeterminate. On defining $\mathbf{z} = \lim_{\varepsilon\to 0}(\tilde{\mathbf{G}}^{\mathrm{l}}(\mathbf{x})\mathbf{u}^{\mathrm{l}})/\varepsilon_1$, $\mathbf{z} \in \mathbb{R}^{n-\mathcal{C}-m}$, as these finite but unknown terms, the process model after the fast "boundary-layer" dynamics takes the form

$$\begin{aligned}\frac{d\mathbf{x}}{dt} &= \bar{\mathbf{f}}(\mathbf{x},\mathbf{u}^{\mathrm{s}}) + \mathbf{B}(\mathbf{x})\mathbf{z} + \varepsilon_2[\mathbf{g}^{\mathrm{I0}}(\mathbf{x}) + \varepsilon_2\mathbf{g}^{\mathrm{I}}(\mathbf{x}) + \mathbf{g}^{\mathrm{P}}(\mathbf{x})u_{\mathrm{p}}] \\ 0 &= \tilde{\mathbf{G}}^{\mathrm{l}}(\mathbf{x})\mathbf{u}^{\mathrm{l}}\end{aligned} \tag{5.15}$$

Once the flow rates $\mathbf{u}^{\mathrm{l}}$ have been specified by appropriate control laws, it is possible to differentiate the constraints in Equation (5.15) to obtain (after differentiating a sufficient number of times) a solution for the algebraic variables $\mathbf{z}$. One differentiation in time will yield

$$\begin{aligned}\mathbf{z} = &-[L_{\mathbf{B}}(\tilde{\mathbf{G}}(\mathbf{x})\mathbf{u}^{\mathrm{l}}(\mathbf{x}))]^{-1}\{L_{\bar{\mathbf{f}}}(\tilde{\mathbf{G}}(\mathbf{x})\mathbf{u}^{\mathrm{l}}(\mathbf{x})) \\ &+ \varepsilon_2[L_{\mathbf{g}^{\mathrm{I0}}}(\tilde{\mathbf{G}}(\mathbf{x})\mathbf{u}^{\mathrm{l}}(\mathbf{x})) + \varepsilon_2 L_{\mathbf{g}^{\mathrm{I}}}(\tilde{\mathbf{G}}(\mathbf{x})\mathbf{u}^{\mathrm{l}}(\mathbf{x})) + L_{\mathbf{g}^{\mathrm{P}}}(\tilde{\mathbf{G}}(\mathbf{x})\mathbf{u}^{\mathrm{l}}(\mathbf{x}))u_{\mathrm{p}}]\}\end{aligned} \tag{5.16}$$

provided that the matrix $L_{\mathbf{B}}(\tilde{\mathbf{G}}(\mathbf{x})\mathbf{u}^{\mathrm{l}}(\mathbf{x}))$ is invertible (which is typically true). In this case, the index of the DAE system (5.15) is two and the dimension of the underlying ODE system describing the dynamics after the fast boundary layer is $\mathcal{C} + m$. An ODE description of this dynamics can be obtained by substituting $\mathbf{z}$ into Equation (5.15), to obtain

$$\begin{aligned}\frac{d\mathbf{x}}{dt} &= \tilde{\mathbf{f}}(\mathbf{x},\mathbf{u}^{\mathrm{s}}) + \varepsilon_2[\tilde{\mathbf{g}}^{\mathrm{I0}}(\mathbf{x}) + \varepsilon_2\tilde{\mathbf{g}}^{\mathrm{I}}(\mathbf{x}) + \tilde{\mathbf{g}}^{\mathrm{P}}(\mathbf{x})u_{\mathrm{p}}] \\ 0 &= \tilde{\mathbf{G}}^{\mathrm{l}}(\mathbf{x})\mathbf{u}^{\mathrm{l}}(\mathbf{x})\end{aligned} \tag{5.17}$$

with

$$\begin{aligned}\tilde{\mathbf{f}}(\mathbf{x},\mathbf{u}^{\mathrm{s}}) &= \bar{\mathbf{f}}(\mathbf{x},\mathbf{u}^{\mathrm{s}}) - [L_{\mathbf{B}}(\tilde{\mathbf{G}}(\mathbf{x})\mathbf{u}^{\mathrm{l}})]^{-1}L_{\bar{\mathbf{f}}}(\tilde{\mathbf{G}}(\mathbf{x})\mathbf{u}^{\mathrm{l}}) \\ \tilde{\mathbf{g}}^{\mathrm{I0}} &= \mathbf{g}^{\mathrm{I0}}(\mathbf{x}) - [L_{\mathbf{B}}(\tilde{\mathbf{G}}(\mathbf{x})\mathbf{u}^{\mathrm{l}})]^{-1}L_{\mathbf{g}^{\mathrm{I0}}}(\tilde{\mathbf{G}}(\mathbf{x})\mathbf{u}^{\mathrm{l}}) \\ \tilde{\mathbf{g}}^{\mathrm{I}} &= \mathbf{g}^{\mathrm{I}}(\mathbf{x}) - [L_{\mathbf{B}}(\tilde{\mathbf{G}}(\mathbf{x})\mathbf{u}^{\mathrm{l}})]^{-1}L_{\mathbf{g}^{\mathrm{I}}}(\tilde{\mathbf{G}}(\mathbf{x})\mathbf{u}^{\mathrm{l}}) \\ \tilde{\mathbf{g}}^{\mathrm{P}} &= \mathbf{g}^{\mathrm{P}}(\mathbf{x}) - [L_{\mathbf{B}}(\tilde{\mathbf{G}}(\mathbf{x})\mathbf{u}^{\mathrm{l}})]^{-1}L_{\mathbf{g}^{\mathrm{P}}}(\tilde{\mathbf{G}}(\mathbf{x})\mathbf{u}^{\mathrm{l}})u_{\mathrm{p}}\end{aligned}$$

A minimal-order ODE representation of the system (5.17) can be subsequently obtained by employing a coordinate change of the form

$$\begin{bmatrix}\zeta \\ \eta\end{bmatrix} = \mathbf{T}_1(\mathbf{x}) = \begin{bmatrix}\phi(\mathbf{x}) \\ \tilde{\mathbf{G}}^{\mathrm{l}}(\mathbf{x})\mathbf{u}^{\mathrm{l}}(\mathbf{x})\end{bmatrix} \tag{5.18}$$

Specifically, the dynamics after the fast boundary layer will be described by

$$\begin{aligned}\frac{d\zeta}{dt} &= \frac{d\phi}{d\mathbf{x}}\tilde{\mathbf{f}}(\mathbf{x},\mathbf{u}^{\mathrm{s}})|_{\mathbf{x}=\mathbf{T}_1^{-1}(\zeta)} \\ &+ \varepsilon_2\frac{d\phi}{d\mathbf{x}}\left[\tilde{\mathbf{g}}^{\mathrm{I0}}(\mathbf{x})|_{\mathbf{x}=\mathbf{T}_1^{-1}(\zeta)} + \varepsilon_2\tilde{\mathbf{g}}^{\mathrm{I}}(\mathbf{x})|_{\mathbf{x}=\mathbf{T}_1^{-1}(\zeta)} + \tilde{\mathbf{g}}^{\mathrm{P}}(\mathbf{x})|_{\mathbf{x}=\mathbf{T}_1^{-1}(\zeta)}u_{\mathrm{p}}\right] \\ \eta &\equiv 0\end{aligned} \tag{5.19}$$

For convenience of notation, we define

$$
\begin{aligned}
\frac{d\phi}{d\mathbf{x}}\tilde{\mathbf{f}}(\mathbf{x}, \mathbf{u}^{\mathrm{s}})|_{\mathbf{x}=\mathbf{T}_1^{-1}(\boldsymbol{\zeta})} &= \hat{\mathbf{f}}(\boldsymbol{\zeta}, \mathbf{u}^{\mathrm{s}}) \\
\frac{d\phi}{d\mathbf{x}}\tilde{\mathbf{g}}^{\mathrm{I0}}(\mathbf{x})|_{\mathbf{x}=\mathbf{T}_1^{-1}(\zeta)} &= \hat{\mathbf{g}}^{\mathrm{I0}}(\zeta) \\
\frac{d\phi}{d\mathbf{x}}\tilde{\mathbf{g}}^{\mathrm{I}}(\mathbf{x})|_{\mathbf{x}=\mathbf{T}_1^{-1}(\zeta)} &= \hat{\mathbf{g}}^{\mathrm{I}}(\zeta) \\
\frac{d\phi}{d\mathbf{x}}\tilde{\mathbf{g}}^{\mathrm{P}}(\mathbf{x})|_{\mathbf{x}=\mathbf{T}_1^{-1}(\zeta)} &= \hat{\mathbf{g}}^{\mathrm{P}}(\zeta)
\end{aligned}
$$

with which Equations (5.19) become

$$
\begin{aligned}
\frac{d\boldsymbol{\zeta}}{dt} &= \hat{\mathbf{f}}(\boldsymbol{\zeta}, \mathbf{u}^{\mathrm{s}}) + \varepsilon_2[\hat{\mathbf{g}}^{\mathrm{I0}}(\zeta) + \varepsilon_2\hat{\mathbf{g}}^{\mathrm{I}}(\zeta) + \hat{\mathbf{g}}^{\mathrm{P}}(\zeta)u_{\mathrm{p}}] \\
\eta &\equiv \mathbf{0}
\end{aligned}
\tag{5.20}
$$

Equations (5.20) capture the core dynamics of the process, present due to the large recycle stream. However, this model still contains both $\mathcal{O}(1)$ and $\mathcal{O}(\varepsilon_2)$ terms and is, therefore, stiff. The time evolution of the process after the fast boundary layer thus itself has the potential to feature two time scales.

5.3.3 Slow dynamics of the impurity levels

We proceed with the model reduction using the method developed in Chapter 4, by considering the limiting case of the purge flow rate and the impurity feed being set to zero, i.e., $\varepsilon_2 \to 0$. In this limit, we obtain a description of the dynamics in the intermediate time scale, that is, the time scale immediately succeeding the fast boundary layer:

$$
\frac{d\boldsymbol{\zeta}}{dt} = \hat{\mathbf{f}}(\boldsymbol{\zeta}, \mathbf{u}^{\mathrm{s}}) \tag{5.21}
$$

The description of the intermediate dynamics in Equation (5.21) involves only the flow rates $\mathbf{u}^{\mathrm{s}}$. However, it was demonstrated that these flow rates do not affect the total holdup of the impurity in the recycle loop, since the total holdup of I is influenced only by the inflow of impurity, by its transfer rate in the separator, and by the purge stream, which, as can be seen from Equation (5.21), have no influence on the dynamics in this intermediate time scale. Consequently, one of the differential equations describing the intermediate dynamics is redundant, and Equations (5.21) are not independent. Correspondingly, the steady-state conditions

$$
\mathbf{0} = \hat{\mathbf{f}}(\boldsymbol{\zeta}, \mathbf{u}^{\mathrm{s}}) \tag{5.22}
$$

specify a 1D sub-manifold in which a slower dynamics will evolve. Following the approach taken above, we assume that we can rewrite the vector function $\hat{\mathbf{f}}(\boldsymbol{\zeta}, \mathbf{u}^s)$ as

$$\hat{\mathbf{f}}(\boldsymbol{\zeta}, \mathbf{u}^s) = \hat{\mathbf{B}}(\mathbf{x})\check{\mathbf{f}}(\boldsymbol{\zeta}, \mathbf{u}^s) \tag{5.23}$$

where the matrix $\hat{\mathbf{B}}(\mathbf{x}) \in \mathbb{R}^{(\mathcal{C}+m)\times(\mathcal{C}+m-1)}$ has full column rank, and the vector $\check{\mathbf{f}}(\boldsymbol{\zeta}, \mathbf{u}^s) \in \mathbb{R}^{(\mathcal{C}+m-1)}$ has linearly independent rows.

Next, in order to obtain a description of the slow dynamics, we define the slow, compressed, time scale $\theta = \varepsilon_2 t$, in which the model of the process becomes

$$\varepsilon_2 \frac{d\boldsymbol{\zeta}}{d\theta} = \hat{\mathbf{f}}(\boldsymbol{\zeta}, \mathbf{u}^s) + \varepsilon_2[\hat{\mathbf{g}}^{\mathrm{I0}}(\zeta) + \varepsilon_2 \hat{\mathbf{g}}^{\mathrm{I}}(\zeta) + \hat{\mathbf{g}}^{\mathrm{P}}(\zeta)u_{\mathrm{p}}] \tag{5.24}$$

and we consider the limit $\varepsilon_2 \to 0$, in which the constraints $\mathbf{0} = \hat{\mathbf{f}}(\boldsymbol{\zeta}, \mathbf{u}^s)$, or, equivalently, the linearly independent constraints $\mathbf{0} = \check{\mathbf{f}}(\boldsymbol{\zeta}, \mathbf{u}^s)$ are obtained. These constraints must be satisfied in the slow time scale.

On dividing Equation (5.24) by ε_2, and considering the same limiting case under the constraints above, we obtain a description of the slow dynamics of the system. Note that, in this limit, the term $\check{\mathbf{f}}(\boldsymbol{\zeta}, \mathbf{u}^s)/\varepsilon_2$ becomes indeterminate. On defining $\hat{\mathbf{z}} = \lim_{\varepsilon_2 \to 0} \check{\mathbf{f}}(\boldsymbol{\zeta}, \mathbf{u}^s)/\varepsilon_2$, $\hat{\mathbf{z}} \in \mathbb{R}^{\mathcal{C}+m-1}$, the slow dynamics of process (5.10) takes the form

$$\begin{aligned} \frac{d\mathbf{x}}{d\tau} &= \hat{\mathbf{g}}^{\mathrm{I0}}(\boldsymbol{\zeta}) + \hat{\mathbf{g}}^{\mathrm{P}}(\boldsymbol{\zeta})u_{\mathrm{p}} + \hat{\mathbf{B}}(\boldsymbol{\zeta})\hat{\mathbf{z}} \\ 0 &= \check{\mathbf{f}}(\boldsymbol{\zeta}, \mathbf{u}^s) \end{aligned} \tag{5.25}$$

In the DAE system (5.25), the variables $\hat{\mathbf{z}} \in \mathbb{R}^{\mathcal{C}+m-1}$ are implicitly fixed by the algebraic constraints, rather than specified explicitly, and thus the index of the system is again nontrivial (i.e., higher than 1). Also, note that, as in the previous reduction step, the index of (5.25) is well-defined only if the flow rates $\mathbf{u}^s$ are specified as a function of the state variables (in this case, expressed in the new coordinates $\boldsymbol{\zeta}$), i.e., $\mathbf{u}^s = \mathbf{u}^s(\boldsymbol{\zeta})$. Specifying these flow rates via feedback control laws allows $\hat{\mathbf{z}}$ to be determined through differentiation of the algebraic constraints in Equation (5.25). Differentiating these constraints once yields

$$\hat{\mathbf{z}} = [L_{\hat{\mathbf{B}}}\check{\mathbf{f}}(\boldsymbol{\zeta}, \mathbf{u}^s(\boldsymbol{\zeta}))]^{-1}\{L_{\hat{\mathbf{B}}}\hat{\mathbf{g}}^{\mathrm{I0}}(\boldsymbol{\zeta}) + L_{\hat{\mathbf{B}}}\hat{\mathbf{g}}^{\mathrm{P}}(\boldsymbol{\zeta})u_{\mathrm{p}}\} \tag{5.26}$$

under the assumption that the matrix $L_{\hat{\mathbf{B}}}\check{\mathbf{f}}(\boldsymbol{\zeta}, \mathbf{u}^s(\boldsymbol{\zeta}))$ is invertible. In this case, the DAE model describing the slow dynamics (5.25) is of index 2, and the underlying dimension of the ODE system describing the evolution of the process in the slow time scale is 1. An explicit state-space realization of the slow dynamics can then be obtained via a coordinate change of the form

$$\begin{bmatrix} \gamma \\ \boldsymbol{\delta} \end{bmatrix} = \mathbf{T}_2(\boldsymbol{\zeta}) = \begin{bmatrix} \psi(\boldsymbol{\zeta}) \\ \check{\mathbf{f}}(\boldsymbol{\zeta}, \mathbf{u}^s) \end{bmatrix} \tag{5.27}$$

and takes the form

$$\frac{d\gamma}{d\theta} = \frac{d\psi}{d\zeta}\left[\hat{\mathbf{g}}^{\mathrm{I0}}(\zeta) + \hat{\mathbf{g}}^{\mathrm{P}}(\zeta)u_{\mathrm{p}} + \hat{\mathbf{B}}(\zeta)\hat{\mathbf{z}}\right]\Big|_{\zeta=\mathbf{T}_2^{-1}(\gamma)} \qquad (5.28)$$
$$\boldsymbol{\delta} \equiv \mathbf{0}$$

This is the dynamics associated with the small amount of feed impurity removed by the small purge stream.

5.4 Hierarchical controller design

The results above indicate clearly that the presence of flow rates of different magnitudes (a steady-state design feature of many process systems) impacts upon the dynamic behavior of the process. In what follows, we further our analysis by emphasizing the implications of steady-state design on the selection of control structures and the synthesis of well-conditioned controllers for the class of processes considered.

Each of the reduced-order models derived for the fast, intermediate, and slow dynamics (Equations (5.12), (5.21), and (5.28)) involves only one group of manipulated inputs, namely the large internal flow rates $\mathbf{u}^{\mathrm{l}}$, the small flow rates $\mathbf{u}^{\mathrm{s}}$, and the purge flow rate u_{p}, respectively. Thus, control objectives in each of the time scales can be addressed by using the manipulated inputs that are available and act upon the dynamics in the respective time scale, starting from the fastest.

5.4.1 Distributed control at the unit level

The fast dynamics of the process (5.12), corresponding to the evolution of the individual process units in the recycle loop, are influenced only by the large flow rates $\mathbf{u}^{\mathrm{l}}$ of the internal (recycle loop) streams. Thus, the flow rates of these streams are available for addressing regulatory control objectives at the unit level, such as liquid-level/holdup control, as well as for the rejection of fast disturbances. Similar control objectives for the units outside the recycle loop are to be addressed using the flow rates $\mathbf{u}^{\mathrm{s}}$ of the corresponding material streams. Typically, the above control objectives are fulfilled using simple linear controllers, possibly with integral action, depending on the stringency of the control objectives.

5.4.2 Supervisory control at the process level

The flow rates $\mathbf{u}^{\mathrm{s}}$ of the streams *outside* the recycle loop appear as the manipulated inputs available for controlling the "overall," process dynamics (5.21) in the intermediate time scale. Control objectives at the process level include the product purity, the stabilization of the total material holdup, and setting the production rate.

Very often (especially because they serve also for the regulatory control for the units outside the recycle loop), the number of available manipulated inputs $\mathbf{u}^s$ is exceeded by the number of process-level control objectives. In this case, it is possible to use the setpoints $\mathbf{y}^l_{sp}$ of the controllers in the fast time scale as manipulated inputs in the intermediate time scale, which leads to cascaded control configurations.[1] Such configurations are beneficial from the point of view of achieving a tighter coordination between the distributed and supervisory control levels. However, in this case, the constrained state space of the DAE description of the dynamics after the fast dynamics (5.15) becomes control-dependent (i.e., $\tilde{\mathbf{G}}^l(\mathbf{x})\mathbf{u}^l = \tilde{\mathbf{G}}^l(\mathbf{x})\mathbf{u}^l\left(\mathbf{x}, \mathbf{y}^l_{sp}\right)$,) and the derivation of a corresponding ODE representation of the type (5.17) – which is more challenging from a technical point of view – has to be addressed using, e.g., the method proposed in Contou-Carrère *et al.* (2004) and discussed in the previous chapter.

5.4.3 Control of impurity levels

The concentration of impurities (present in the feed) in the process evolves over a very slow horizon (days or, possibly, weeks). Moreover, the presence of impurities in the feed stream, together with significant material recycling, can lead to the accumulation of impurities in the recycle loop, with detrimental effects on the operation of the process and on its economics (Baldea *et al.* 2006). Therefore, as was shown in Chapter 4, the control of the impurity levels in the process is an important operational objective, and, according to the analysis presented above, it should be addressed in the slow time scale, using the flow rate of the purge stream, u_P, as a manipulated input.

5.4.4 Real-time optimization

A fourth, optimization, tier can be naturally integrated into the control structure delineated above. In particular, the models of the overall process behavior in the intermediate (5.21) and slow (5.28) time scales are non-stiff, and can be used to formulate well-conditioned optimization problems for computing the setpoints of the corresponding controllers, respectively, $\mathbf{y}^s_{sp}$ and y^P_{sp}.

The resulting hierarchical control structure is represented schematically in Figure 5.2. Note that, while controller design proceeds in a "bottom-up" manner, starting from the fastest time scale, during the operation of the process there will exist a tight "top-down" interconnection via control cascades between the supervisory and regulatory layers.

[1] The use of a "non-square" controller (e.g., an MPC), such that the number of manipulated inputs is lower than the number of controlled variables, is certainly possible. While this approach eschews the use of cascaded configurations, it is intuitively detrimental to closed-loop performance due to the reduced number of manipulated variables.

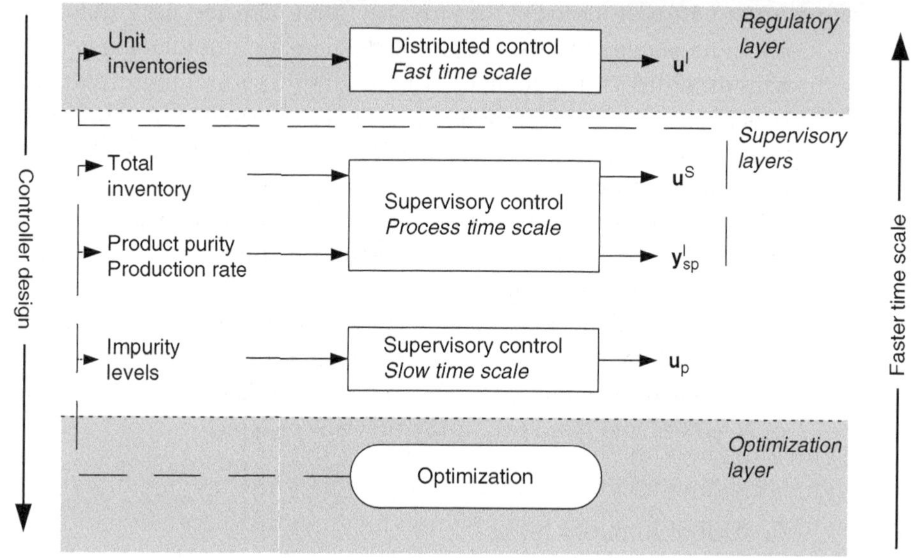

Figure 5.2 Hierarchical control structure.

5.5 Case study: dynamics and control of a reactor–separator process core

5.5.1 Process description

In what follows, we consider the process of Figure 5.3, consisting of a gas-phase reactor and a condenser that are part of a recycle loop. This process is similar to the one discussed in Chapter 4 in that the feed stream contains the reactant A (of mole fraction $y_{A,0}$) as well as a small quantity $y_{I,0}$ of an inert, volatile impurity I. However, in the present case, the slow first-order reaction A → B results in a low single-pass conversion for the given equipment size. Consequently, the reaction mass contains an appreciable quantity of unreacted reactant A and

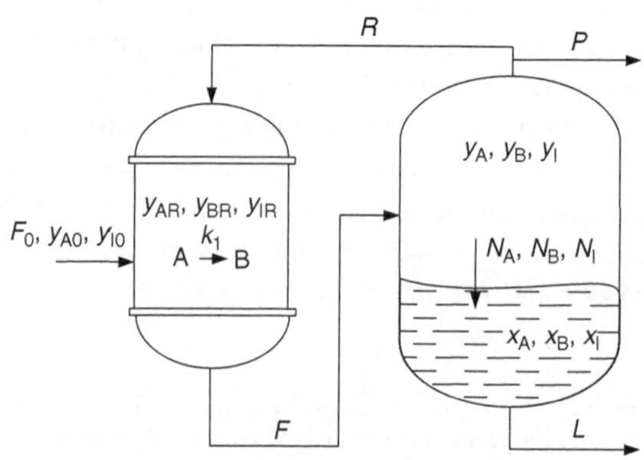

Figure 5.3 A reactor–single-stage-separator process.

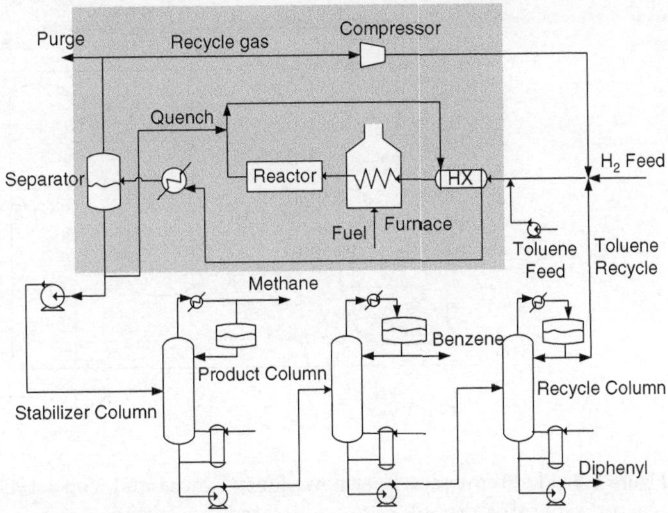

Figure 5.4 The toluene hydrodealkylation (HDA) plant (Douglas 1988). The reactor–separator process core is highlighted.

a significant portion of it must be recycled to the reactor. In order to meet the purity requirements, the recycle flow rate R needs to be much larger than the feed flow rate F_0. As in our previous encounter with this system, the feed impurity is removed via a purge stream of small flow rate P. The objectives for this process are stable operation and ensuring that the concentration of B in the liquid product stream is within specifications. A third objective concerns preventing accumulation of the inert impurity **I** in the recycle loop.

While apparently simple, such reactor–single-stage-separator processes with high recycle rates and purge streams are omnipresent in the chemical industry, lying at the core of complex integrated plants. This is also reflected in the fact that the most frequently cited challenge processes proposed in the literature, i.e., the toluene hydrodealkylation plant (HDA) (Figure 5.4), the Tennessee Eastman process (Figure 5.5), and the vinyl acetate monomer plant (Figure 5.6), feature such reactor–separator cores. Thus, a rigorous analysis of the class of processes considered in this case study is an important step in tackling the dynamic analysis and control of integrated chemical plants.

Assuming isothermal operation of the reactor and condenser, the dynamic model of the process has the form

$$\begin{aligned}
\dot{M}_{\mathrm{R}} &= F_0 + R - F \\
\dot{y}_{\mathrm{A,R}} &= (1/M_{\mathrm{R}})\left[F_0(y_{\mathrm{A},0} - y_{\mathrm{A,R}}) + R(y_{\mathrm{A}} - y_{\mathrm{A,R}}) - k_1 M_{\mathrm{R}} y_{\mathrm{A,R}}\right] \\
\dot{y}_{\mathrm{I,R}} &= (1/M_{\mathrm{R}})\left[F_0(y_{\mathrm{I},0} - y_{\mathrm{I,R}}) + R(y_{\mathrm{I}} - y_{\mathrm{I,R}})\right] \\
\dot{M}_{\mathrm{V}} &= F - R - N - P \\
\dot{y}_{\mathrm{A}} &= (1/M_{\mathrm{V}})\left[F(y_{\mathrm{A,R}} - y_{\mathrm{A}}) - N_{\mathrm{A}} + y_{\mathrm{A}} N\right]
\end{aligned} \tag{5.29}$$

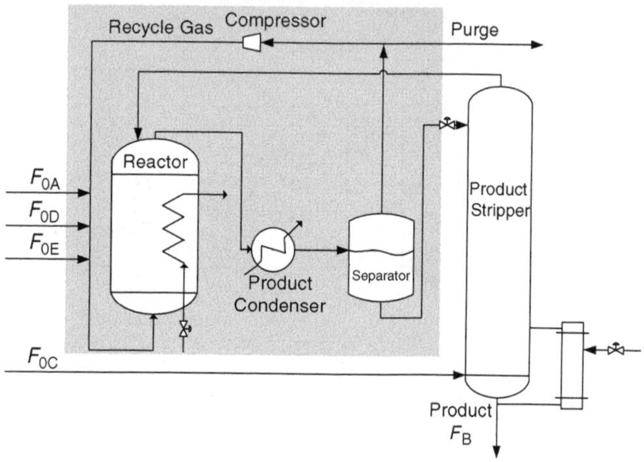

Figure 5.5 The Tennessee Eastman plant (Downs and Vogel 1993), with the reactor–separator process core highlighted.

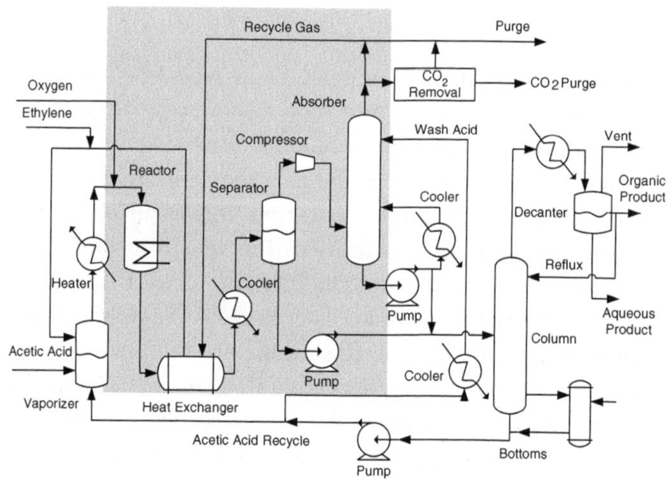

Figure 5.6 The vinyl acetate monomer plant (Luyben and Tyreus 1998). The reactor–separator process core is highlighted.

$$\dot{y}_{\mathrm{I}} = (1/M_{\mathrm{V}})\left[F(y_{\mathrm{I,R}} - y_{\mathrm{I}}) - N_{\mathrm{I}} + y_{\mathrm{I}}N\right]$$
$$\dot{M}_{\mathrm{L}} = N - L$$
$$\dot{x}_{\mathrm{A}} = (1/M_{\mathrm{L}})\left[N_{\mathrm{A}} - x_{\mathrm{A}}N\right]$$
$$\dot{x}_{\mathrm{I}} = (1/M_{\mathrm{L}})\left[N_{\mathrm{I}} - x_{\mathrm{I}}N\right]$$

where M_{R}, M_{V} and M_{L} denote the molar holdups in the reactor and in the condenser vapor and liquid phases, respectively. The interphase mole transfer

Table 5.1. Nominal values of the state variables and parameter values of the reactor–separator process core in Figure 5.3

F_0	100.00 mol/min	T_c	298.00 K	M_R	4500.00 mol
R	445.48 mol/min	P_c	0.327 MPa	M_V	621.70 mol
F	545.48 mol/min	$\mathcal{A}$	200.00 m^{-1}	M_L	4500.00 mol
L	95.00 mol/min	$\mathcal{K}_A$	140 mol m^{-2} min^{-1}	y_A	0.429
P	5.00 mol/min	$\mathcal{K}_B$	145 mol m^{-2} min^{-1}	y_I	0.399
$y_{A,0}$	0.98	$\mathcal{K}_I$	5×10^{-5} mol m^{-2} min^{-1}	$y_{A,R}$	0.375
$y_{I,0}$	0.02	$P_A^S(T_c)$	1.17 MPa	$y_{I,R}$	0.330
k_{uR}	0.1 mol^{-1}	$P_B^S(T_c)$	0.06 MPa	x_A	0.12
k_{uF}	0.1 mol^{-1}	$P_I^S(T_c)$	56.7 MPa	x_I	1.25×10^{-5}
V_c	5.00 m^3	ρ_L	15.00 kmol/m^3	k_1	0.05 min^{-1}

rates in the condenser are governed by rate expressions of the form $N_j = \mathcal{K}_j\mathcal{A}(y_j - (\mathcal{P}_j^s/\mathcal{P})x_j)(M_L/\rho_L)$, $i \in$ A, B, I, where $\mathcal{K}_j$ represents the mass-transfer coefficient, y_j the mole fraction in the gas phase, x_j the mole fraction in the liquid phase, $\mathcal{P}_j^s$ the saturation vapor pressure of the component j, and $\mathcal{P}$ the pressure in the condenser. N represents the total mass-transfer rate in the condenser, $N = N_A + N_B + N_I$.

The parameters of the model are summarized in Table 5.1, together with the nominal values of the state variables (the subscript c denotes the parameters of the condenser).

5.5.2 System analysis

Inspecting Equation (5.29), we notice that three of the state variables (namely, M_R, M_V, and M_L) are material holdups, which act as integrators and render the system open-loop unstable. Our initial focus will therefore be a "pseudo-open loop" analysis consisting of simulating the model in Equation (5.29) after the holdup of the reactor, and the vapor and liquid holdup in the condenser, have been stabilized. This task is accomplished by defining the reactor effluent, recycle, and liquid-product flow rates as functions of M_R, M_V, and M_L via appropriate control laws (specifically, via the proportional controllers (5.42) and (5.48), as discussed later in this section). With this primary control structure in place, we carried out a simulation using initial conditions that were slightly perturbed from the steady-state values in Table 5.1.

The responses of all the state variables (Figure 5.7) exhibit an initial fast transient, followed by a slower dynamics. The states approach their nominal steady-state values after a period of time that exceeds 48 h (*nota bene*, two days!), indicating that a very slow component is also present in the process dynamics. The analysis in the following section will use the framework developed earlier in the chapter to provide a theoretical explanation for these findings.

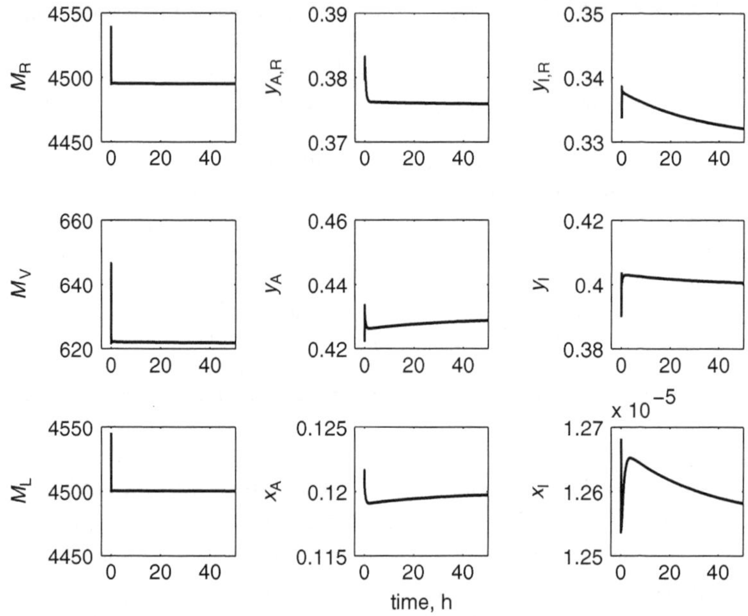

Figure 5.7 Evolution of the state variables of the "pseudo-open-loop" system starting from initial conditions slightly perturbed from the steady-state values.

5.5.3 Reduced-order modeling

In order to elucidate the dynamic behavior of the process and to devise representations of the dynamics in each time scale, let us define the recycle number of the process, **Rc**, as the ratio of the steady-state values of the feed and recycle flow rates:

$$\mathbf{Rc} = \frac{R_s}{F_{0s}} \tag{5.30}$$

and the purge number

$$\mathbf{Pu} = \frac{P_s}{F_{0s}} \tag{5.31}$$

as the ratio of the purge and feed flow rates at steady state.

According to the developments above and with the data in Table 5.1, we have $y_{\mathrm{I},0} = \alpha_{\mathrm{I}}\mathbf{Pu}$, with $\alpha_{\mathrm{I}} = \mathcal{O}(1)$. Upon further inspection of the data in Table 5.1, we also notice that the vapor pressure of the impurity is very high compared with the vapor pressures of components A and B. Given this, we can write

$$\frac{\mathcal{P}_{\mathrm{I}}^{\mathrm{S}}}{\mathcal{P}} = \frac{\alpha_2}{\mathbf{Pu}} \tag{5.32}$$

Corroborating this with the observation that the mass transfer coefficient of I is very low compared with those of A and B (and therefore $\mathcal{K}_{\mathrm{I}} = \alpha_1 \mathbf{Pu}^2$, with $\alpha_1 = O(1)$) we can infer that only a negligible amount of I will leave the recycle loop through the liquid stream at the bottom of the condenser.

We also define the scaled flow rates $u_{\mathrm{R}} = R/R_{\mathrm{s}}$, $u_{\mathrm{F}} = F/F_{\mathrm{s}}$, and $u_{\mathrm{p}} = P/P_{\mathrm{s}}$, and the ratio $k = F_{\mathrm{s}}/R_{\mathrm{s}}$, which, according to the data in Table 5.1, is of $\mathcal{O}(1)$. Using these definitions and observations (and the notation $\varepsilon_1 = 1/\mathbf{Rc}$ and $\varepsilon_2 = \mathbf{Pu}$), the model in Equation (5.29) becomes

$$
\begin{aligned}
\dot{M}_{\mathrm{R}} &= F_0 + (1/\varepsilon_1)F_{0\mathrm{s}}(u_{\mathrm{R}} - k u_{\mathrm{F}}) \\
\dot{y}_{\mathrm{A,R}} &= (1/M_{\mathrm{R}})[F_0(1 - \alpha_{\mathrm{I}}\varepsilon_2 - y_{\mathrm{A,R}}) - k_1 M_{\mathrm{R}} y_{\mathrm{A,R}} \\
&\qquad + (1/\varepsilon_1)F_{0\mathrm{s}} u_{\mathrm{R}}(y_{\mathrm{A}} - y_{\mathrm{A,R}})] \\
\dot{y}_{\mathrm{I,R}} &= (1/M_{\mathrm{R}})[F_0(\alpha_{\mathrm{I}}\varepsilon_2 - y_{\mathrm{I,R}}) + (1/\varepsilon_1)F_{0\mathrm{s}} u_{\mathrm{R}}(y_{\mathrm{I}} - y_{\mathrm{I,R}})] \\
\dot{M}_{\mathrm{V}} &= -N_{\mathrm{A}} - N_{\mathrm{B}} - (\alpha_1\varepsilon_2^2 y_{\mathrm{I}} + \alpha_1\alpha_2\varepsilon_2 x_{\mathrm{I}})M_{\mathrm{L}}/\varrho_{\mathrm{L}} \\
&\qquad + (1/\varepsilon_1)F_{0\mathrm{s}}(k u_{\mathrm{F}} - u_{\mathrm{R}}) - \varepsilon_2 F_{0\mathrm{s}} u_{\mathrm{p}} \\
\dot{y}_{\mathrm{A}} &= (1/M_{\mathrm{V}})[(1/\varepsilon_1)F_{0\mathrm{s}} k u_{\mathrm{F}}(y_{\mathrm{A,R}} - y_{\mathrm{A}}) - N_{\mathrm{A}} + y_{\mathrm{A}}(N_{\mathrm{A}} + N_{\mathrm{B}}) \\
&\qquad + y_{\mathrm{A}}(\alpha_1\varepsilon_2^2 y_{\mathrm{I}} + \alpha_1\alpha_2\varepsilon_2 x_{\mathrm{I}})M_{\mathrm{L}}/\varrho_{\mathrm{L}}] \\
\dot{y}_{\mathrm{I}} &= (1/M_{\mathrm{V}})[(1/\varepsilon_1)F_{0\mathrm{s}} k u_{\mathrm{F}}(y_{\mathrm{I,R}} - y_{\mathrm{I}}) - (\alpha_1\varepsilon_2^2 y_{\mathrm{I}} + \alpha_1\alpha_2\varepsilon_2 x_{\mathrm{I}})M_{\mathrm{L}}/\varrho_{\mathrm{L}} \\
&\qquad + y_{\mathrm{I}}(N_{\mathrm{A}} + N_{\mathrm{B}}) + y_{\mathrm{I}}(\alpha_1\varepsilon_2^2 y_{\mathrm{I}} + \alpha_1\alpha_2\varepsilon_2 x_{\mathrm{I}})M_{\mathrm{L}}/\varrho_{\mathrm{L}}] \\
\dot{M}_{\mathrm{L}} &= N_{\mathrm{A}} + N_{\mathrm{B}} + (\alpha_1\varepsilon_2^2 y_{\mathrm{I}} + \alpha_1\alpha_2\varepsilon_2 x_{\mathrm{I}})M_{\mathrm{L}}/\varrho_{\mathrm{L}} - L \\
\dot{x}_{\mathrm{A}} &= (1/M_{\mathrm{L}})[N_{\mathrm{A}} - x_{\mathrm{A}}(N_{\mathrm{A}} + N_{\mathrm{B}}) - x_{\mathrm{A}}(\alpha_1\varepsilon_2^2 y_{\mathrm{I}} + \alpha_1\alpha_2\varepsilon_2 x_{\mathrm{I}})M_{\mathrm{L}}/\varrho_{\mathrm{L}}] \\
\dot{x}_{\mathrm{I}} &= (1/M_{\mathrm{L}})[(\alpha_1\varepsilon_2^2 y_{\mathrm{I}} + \alpha_1\alpha_2\varepsilon_2 x_{\mathrm{I}})M_{\mathrm{L}}/\varrho_{\mathrm{L}} \\
&\qquad - x_{\mathrm{I}}(N_{\mathrm{A}} + N_{\mathrm{B}}) - x_{\mathrm{I}}(\alpha_1\varepsilon_2^2 y_{\mathrm{I}} + \alpha_1\alpha_2\varepsilon_2 x_{\mathrm{I}})M_{\mathrm{L}}/\varrho_{\mathrm{L}}]
\end{aligned}
\tag{5.33}
$$

which is in the generic form of Equation (5.10), with $\mathbf{x} = [M_{\mathrm{R}}\ y_{\mathrm{A,R}}\ y_{\mathrm{I,R}}\ M_{\mathrm{V}}\ y_{\mathrm{A}}\ y_{\mathrm{I}}\ M_{\mathrm{L}}\ x_{\mathrm{A}}\ x_{\mathrm{I}}]^{\mathrm{T}}$, $\mathbf{u}^{\mathrm{l}} = [u_{\mathrm{R}}\ u_{\mathrm{F}}]^{\mathrm{T}}$, $\mathbf{u}^{\mathrm{s}} = [F_0\ L]^{\mathrm{T}}$, $n = 9$, and $m = 3$, and

$$
\mathbf{G}^{\mathrm{l}}(\mathbf{x}) = F_{0\mathrm{s}} \begin{bmatrix}
1 & -k \\
(1/M_{\mathrm{R}})(y_{\mathrm{A}} - y_{\mathrm{A,R}}) & 0 \\
(1/M_{\mathrm{R}})(y_{\mathrm{I}} - y_{\mathrm{I,R}}) & 0 \\
-1 & k \\
0 & (1/M_{\mathrm{V}})k(y_{\mathrm{A,R}} - y_{\mathrm{A}}) \\
0 & (1/M_{\mathrm{V}})k(y_{\mathrm{I,R}} - y_{\mathrm{I}}) \\
0 & 0 \\
0 & 0 \\
0 & 0
\end{bmatrix}
\tag{5.34}
$$

$$\bar{\mathbf{f}}(\mathbf{x}, \mathbf{u}^{\mathrm{s}}) = \begin{bmatrix} F_0 \\ (1/M_{\mathrm{R}})(F_0(1 - y_{\mathrm{A,R}}) - k_1 M_{\mathrm{R}} y_{\mathrm{A,R}}) \\ (1/M_{\mathrm{R}})(F_0(-y_{\mathrm{I,R}})) \\ -N_{\mathrm{A}} - N_{\mathrm{B}} \\ (1/M_{\mathrm{V}})(-N_{\mathrm{A}} + y_{\mathrm{A}}(N_{\mathrm{A}} + N_{\mathrm{B}})) \\ (1/M_{\mathrm{V}}) y_{\mathrm{I}}(N_{\mathrm{A}} + N_{\mathrm{B}}) \\ N_{\mathrm{A}} + N_{\mathrm{B}} - L \\ (1/M_{\mathrm{L}})(N_{\mathrm{A}} - x_{\mathrm{A}}(N_{\mathrm{A}} + N_{\mathrm{B}})) \\ (1/M_{\mathrm{L}})(-x_{\mathrm{I}}(N_{\mathrm{A}} + N_{\mathrm{B}})) \end{bmatrix} \tag{5.35}$$

$$\mathbf{g}^{\mathrm{I0}}(\mathbf{x}) = \begin{bmatrix} 0 \\ -(1/M_{\mathrm{R}}) F_0 \alpha_{\mathrm{I}} \\ (1/M_{\mathrm{R}}) F_0 \alpha_{\mathrm{I}} \\ -\alpha_1 \alpha_2 x_{\mathrm{I}} M_{\mathrm{L}} / \varrho_{\mathrm{L}} \\ (1/M_{\mathrm{V}}) y_{\mathrm{A}} \alpha_1 \alpha_2 x_{\mathrm{I}} M_{\mathrm{L}} / \varrho_{\mathrm{L}} \\ (1/M_{\mathrm{V}})[-\alpha_1 \alpha_2 x_{\mathrm{I}} + y_{\mathrm{I}} \alpha_1 \alpha_2 x_{\mathrm{I}}] M_{\mathrm{L}} / \varrho_{\mathrm{L}} \\ \alpha_1 \alpha_2 x_{\mathrm{I}} M_{\mathrm{L}} / \varrho_{\mathrm{L}} \\ -(1/M_{\mathrm{L}}) x_{\mathrm{A}} (\alpha_1 \alpha_2 x_{\mathrm{I}}) M_{\mathrm{L}} / \varrho_{\mathrm{L}} \\ (1/M_{\mathrm{L}})[\alpha_1 \alpha_2 x_{\mathrm{I}} - x_{\mathrm{I}}^2 \alpha_1 \alpha_2] M_{\mathrm{L}} / \varrho_{\mathrm{L}} \end{bmatrix} \tag{5.36}$$

$$\mathbf{g}^{\mathrm{I}} = \begin{bmatrix} 0 \\ 0 \\ 0 \\ -\alpha_1 y_{\mathrm{I}} M_{\mathrm{L}} / \varrho_{\mathrm{L}} \\ (1/M_{\mathrm{V}}) y_{\mathrm{A}} \alpha_1 y_{\mathrm{I}} M_{\mathrm{L}} / \varrho_{\mathrm{L}} \\ (1/M_{\mathrm{V}})[-\alpha_1 y_{\mathrm{I}} + \alpha_1 y_{\mathrm{I}}^2] M_{\mathrm{L}} / \varrho_{\mathrm{L}} \\ \alpha_1 y_{\mathrm{I}} M_{\mathrm{L}} / \varrho_{\mathrm{L}} \\ -(1/M_{\mathrm{L}}) x_{\mathrm{A}} \alpha_1 y_{\mathrm{I}} M_{\mathrm{L}} / \varrho_{\mathrm{L}} \\ (1/M_{\mathrm{L}})[\alpha_1 y_{\mathrm{I}} - x_{\mathrm{I}} \alpha_1 y_{\mathrm{I}}] M_{\mathrm{L}} / \varrho_{\mathrm{L}} \end{bmatrix} \tag{5.37}$$

$$\mathbf{g}^{\mathrm{P}} = [0\ 0\ 0\ F_{0\mathrm{s}}\ 0\ 0\ 0\ 0\ 0]^{\mathrm{T}} \tag{5.38}$$

We proceed with the derivation of approximate models for the process dynamics in each time scale, beginning with the fastest. To this end, we define the fast, stretched time scale $\tau_1 = t/\varepsilon_1$, in which the process model takes the form of Equation (5.11), and, in the limit $\varepsilon_1 \to 0$, corresponding to an infinitely large recycle flow rate, we obtain a description of the process dynamics in the fast time scale that is in the form of Equation (5.12):

$$
\begin{aligned}
\frac{dM_{\mathrm{R}}}{d\tau} &= F_{0\mathrm{s}}(u_{\mathrm{R}} - ku_{\mathrm{F}}) \\
\frac{dy_{\mathrm{A,R}}}{d\tau} &= \frac{F_{0\mathrm{s}}}{M_{\mathrm{R}}} u_{\mathrm{R}}(y_{\mathrm{A}} - y_{\mathrm{A,R}}) \\
\frac{dy_{\mathrm{I,R}}}{d\tau} &= \frac{F_{0\mathrm{s}}}{M_{\mathrm{R}}} u_{\mathrm{R}}(y_{\mathrm{I}} - y_{\mathrm{I,R}}) \\
\frac{dM_{\mathrm{V}}}{d\tau} &= F_{0\mathrm{s}}(ku_{\mathrm{F}} - u_{\mathrm{R}}) \\
\frac{dy_{\mathrm{A}}}{d\tau} &= \frac{F_{0\mathrm{s}}}{M_{\mathrm{V}}} ku_{\mathrm{F}}(y_{\mathrm{A,R}} - y_{\mathrm{A}}) \\
\frac{dy_{\mathrm{I}}}{d\tau} &= \frac{F_{0\mathrm{s}}}{M_{\mathrm{V}}} ku_{\mathrm{F}}(y_{\mathrm{I,R}} - y_{\mathrm{I}}) \\
\frac{dM_{\mathrm{L}}}{d\tau} &= 0 \\
\frac{dx_{\mathrm{A}}}{d\tau} &= 0 \\
\frac{dx_{\mathrm{I}}}{d\tau} &= 0
\end{aligned}
\tag{5.39}
$$

Clearly, not all the nontrivial algebraic equations that correspond to the equilibrium of the fastest dynamics in Equation (5.39) are linearly independent. Specifically, the last three equations can be expressed as functions of the first three, reformulating $\mathbf{G}^{\mathrm{l}}$ as in Equation (5.13), with

$$
\tilde{\mathbf{G}}^{\mathrm{l}}(\mathbf{x}) = \begin{bmatrix} 1 & -k \\ (y_{\mathrm{A}} - y_{\mathrm{A,R}}) & 0 \\ (y_{\mathrm{I}} - y_{\mathrm{I,R}}) & 0 \end{bmatrix}
\tag{5.40}
$$

and

$$
\mathbf{B}(\mathbf{x}) = F_{0\mathrm{s}} \begin{bmatrix} 1 & 0 & 0 \\ 0 & 1/M_{\mathrm{R}} & 0 \\ 0 & 0 & 1/M_{\mathrm{R}} \\ -1 & 0 & 0 \\ -(1/M_{\mathrm{V}})(y_{\mathrm{A,R}} - y_{\mathrm{A}}) & -1/M_{\mathrm{V}} & 0 \\ -(1/M_{\mathrm{V}})(y_{\mathrm{I,R}} - y_{\mathrm{I}}) & 0 & -1/M_{\mathrm{V}} \\ 0 & 0 & 0 \\ 0 & 0 & 0 \\ 0 & 0 & 0 \end{bmatrix}
$$

Consequently, the steady-state condition associated with the fast dynamics specifies a six-dimensional equilibrium manifold in which a slower dynamics evolves.

Turning to the slower dynamics, and considering the limit $\varepsilon_1 \to 0$ in the original time scale t, the linearly independent constraints of Equation (5.14), i.e.,

$$\begin{aligned} 0 &= u_{\mathrm{R}} - k u_{\mathrm{F}} \\ 0 &= u_{\mathrm{R}}(y_{\mathrm{A}} - y_{\mathrm{A,R}}) \\ 0 &= u_{\mathrm{R}}(y_{\mathrm{I}} - y_{\mathrm{I,R}}) \end{aligned} \tag{5.41}$$

are obtained. Observe that the terms $\lim_{\varepsilon \to 0} \tilde{\mathbf{G}}^{\mathrm{l}}(\mathbf{x})\mathbf{u}^{\mathrm{l}}/\varepsilon_1$ are indeterminate, yet finite. On defining these unknown limit terms as $\mathbf{z}_1$, the model of the process takes the form of Equation (5.15), which represents a DAE description of the core process dynamics after the fast transients.

By setting the $\mathbf{u}^{\mathrm{l}}$ with the proportional laws

$$\mathbf{u}^{\mathrm{l}} = \begin{bmatrix} u_{\mathrm{R}} \\ u_{\mathrm{F}} \end{bmatrix} = \begin{bmatrix} 1 - k_{u\mathrm{R}}(M_{\mathrm{V,sp}} - M_{\mathrm{V}}) \\ 1 - k_{u\mathrm{F}}(M_{\mathrm{R,sp}} - M_{\mathrm{R}}) \end{bmatrix} \tag{5.42}$$

the algebraic variables $\mathbf{z}_1$ can be computed after one differentiation of the algebraic constraints (5.41), thus the index of this DAE system is exactly two.

One coordinate change that satisfies the conditions in (5.18) involves the total material holdup of the recycle loop and the holdups of the individual components in the recycle loop, i.e.,

$$\begin{aligned} \zeta_1 &= M_{\mathrm{R}} + M_{\mathrm{V}} \\ \zeta_2 &= M_{\mathrm{R}} y_{\mathrm{A,R}} + M_{\mathrm{V}} y_{\mathrm{A}} \\ \zeta_3 &= M_{\mathrm{R}} y_{\mathrm{I,R}} + M_{\mathrm{V}} y_{\mathrm{I}} \\ \zeta_4 &= M_{\mathrm{L}} \\ \zeta_5 &= x_{\mathrm{A}} \\ \zeta_6 &= x_{\mathrm{I}} \\ \eta_1 &= u_{\mathrm{R}}(x) + k u_{\mathrm{F}}(x) \\ \eta_2 &= u_{\mathrm{R}}(x)(y_{\mathrm{A}} - y_{\mathrm{A,R}}) \\ \eta_3 &= u_{\mathrm{R}}(x)(y_{\mathrm{I}} - y_{\mathrm{I,R}}) \end{aligned} \tag{5.43}$$

In these coordinates, the model of the process becomes

$$\begin{aligned} \dot{\zeta}_1 &= F_0 - N_{\mathrm{A}} - N_{\mathrm{B}} - \varepsilon_2 \tilde{N}_{\mathrm{I}} \varepsilon_2 F_{0\mathrm{s}} u_{\mathrm{p}} \\ \dot{\zeta}_2 &= -\frac{1}{\zeta_1 (k k_{u\mathrm{F}} + k_{u\mathrm{R}})} (\varepsilon_2 F_{0\mathrm{s}} k k_{u\mathrm{F}} \zeta_2 u_{\mathrm{p}} + k_1 \zeta_2 k_{u\mathrm{F}} M_{\mathrm{R,sp}} \\ &\quad - k F_0 y_{\mathrm{A},0} \zeta_1 k_{u\mathrm{F}} + k k_{u\mathrm{F}} N_{\mathrm{A}} \zeta_1 \\ &\quad - k k_1 \zeta_2 + k_1 \zeta_2 \zeta_1 k_{u\mathrm{R}} - F_0 y_{\mathrm{A}0} \zeta_1 k_{u\mathrm{R}} \\ &\quad + k_{u\mathrm{R}} N_{\mathrm{A}} \zeta_1 + k_1 \zeta_2 - k_1 \zeta_2 k_{u\mathrm{R}} M_{\mathrm{V,sp}} \\ &\quad + \zeta_2 k_{u\mathrm{R}} \varepsilon_2 u_{\mathrm{p}} F_{0\mathrm{s}}) \end{aligned} \tag{5.44}$$

$$
\begin{aligned}
\dot{\zeta}_3 &= (\varepsilon_2/\zeta_1)(-\zeta_3 u_{\rm p} F_{0{\rm s}} - \tilde{N}_{\rm I}\zeta_1 + F_0\alpha_{\rm I}\zeta_1)\\
\dot{\zeta}_4 &= N_{\rm A} + N_{\rm B} + \varepsilon_2\tilde{N}_{\rm I} - L\\
\dot{\zeta}_5 &= N_{\rm A} - \zeta_5(N_{\rm A} + N_{\rm B} + \varepsilon_2\tilde{N}_{\rm I})\\
\dot{\zeta}_6 &= \varepsilon_2\tilde{N}_{\rm I} - \zeta_6(N_{\rm A} + N_{\rm B} + \varepsilon_2\tilde{N}_{\rm I})
\end{aligned}
$$

with

$$
\tilde{N}_{\rm I} = (\alpha_1\varepsilon_2\zeta_3/\zeta_1 + \alpha_1\alpha_2\zeta_6)\zeta_4/\rho_{\rm L} \tag{5.45}
$$

Notice that the above model is still stiff, due to the presence of the parameter ε_2. Considering the limit $\varepsilon_2 \to 0$, corresponding to the absence of the inert component from the feed and a zero purge flow rate, we obtain the following description of the intermediate (process-level) dynamics:

$$
\begin{aligned}
\dot{\zeta}_1 &= F_0 - N_{\rm A} - N_{\rm B}\\
\dot{\zeta}_2 &= -(k_1\zeta_2 k_{u{\rm F}} M_{\rm R,sp} - kF_0 y_{\rm A,0}\zeta_1 k_{u{\rm F}} + kk_{u{\rm F}} N_{\rm A}\zeta_1 - kk_1\zeta_2\\
&\quad + k_1\zeta_2\zeta_1 k_{u{\rm R}} - F_0 y_{\rm A,0}\zeta_1 k_{u{\rm R}} + k_{u{\rm R}} N_{\rm A}\zeta_1 + k_1\zeta_2 - k_1\zeta_2 k_{u{\rm R}} M_{\rm V,sp}\\
&\quad + \zeta_2 k_{u{\rm R}}\varepsilon_2 u_{\rm p} F_{0{\rm s}})/(\zeta_1(kk_{u{\rm F}} + k_{u{\rm R}}))\\
\dot{\zeta}_3 &= 0\\
\dot{\zeta}_4 &= N_{\rm A} + N_{\rm B} - L\\
\dot{\zeta}_5 &= N_{\rm A} - \zeta_5(N_{\rm A} + N_{\rm B})\\
\dot{\zeta}_6 &= -\zeta_6(N_{\rm A} + N_{\rm B})
\end{aligned} \tag{5.46}
$$

It is evident from the equations above that the variable ζ_3 (the total impurity holdup) does not evolve in the time scale t, thus being a true slow variable.

On defining the slow, compressed, time scale $\theta = \varepsilon_2 t$ and considering the limit $\varepsilon_2 \to 0$, we obtain an expression for the slowest dynamics of the process, due to the presence of the inert impurity I. This has the form

$$
\frac{{\rm d}\zeta_3}{{\rm d}\theta} = (-\zeta_3 u_{\rm p} F_{0{\rm s}} + F_0\alpha_{\rm I}\zeta_1)/\zeta_1 \tag{5.47}
$$

under the quasi-steady-state constraints that arise from setting the right-hand sides of the equations in (5.46) equal to zero. Notice that the quasi-steady-state constraints corresponding to the intermediate dynamics are linearly independent in this case. Equivalently, in this case the conditions set forth in Theorem A.1 are fulfilled, and the coordinate change (5.43) results in an ODE representation of the dynamics after the fast boundary layer (Equation (5.44)) that is in a *standard* singularly perturbed form.

Remark 5.1. *The presence of three distinct time horizons is well captured by Figure 5.8, which provides a different perspective on the results of the "numerical experiment" discussed in Section 5.5.2. Namely, the reactor holdup (top plot) reaches steady state very quickly, and the* total holdup *of A (middle plot) reaches*

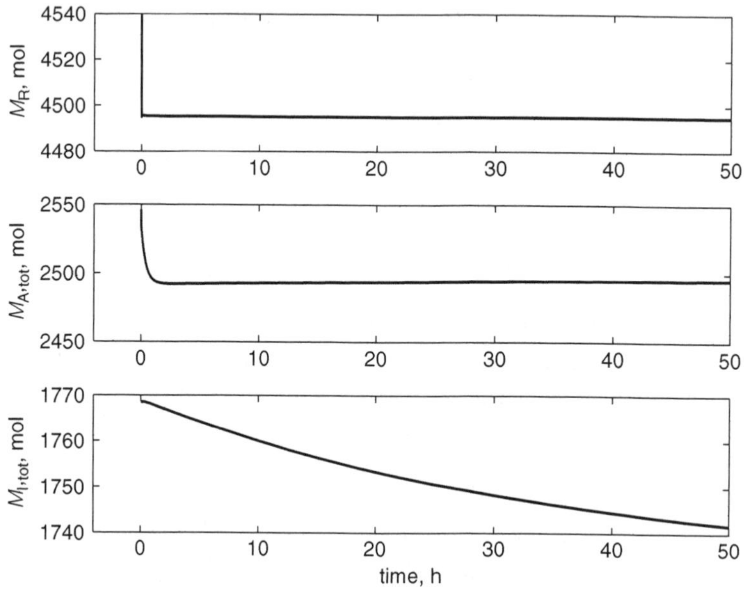

Figure 5.8 Three distinct time horizons in the dynamic response of the reactor–condenser process core. Top: fast, unit-level dynamics. Middle: the total holdup of component A has an intermediate response time. Bottom: the total holdup of impurity evolves in the slowest time scale. The plots depict simulation results with initial conditions slightly perturbed from their steady-state values.

steady state within a few hours, while the total holdup *of I (bottom plot) is approaching steady state after two days (i.e., the* total holdups *of component A and the impurity I evolve only over the intermediate and slow time horizons, respectively). These plots are in complete agreement with our theoretical results, which predict that the dynamics of the individual units (i.e., the reactor) are fast, that there exists a process-level dynamic component associated with the total material holdup and with the holdups of all the components (except the impurity), and that the slowest component of the process dynamics is related to the presence of the impurity I.*

5.5.4 Hierarchical control system design

The presence of three distinct time horizons in the process dynamics, as evinced by the analysis above, warrants the use of a hierarchical control structure that addresses the distributed (unit-level) and plant-wide control objectives separately.

The first (distributed) layer of the control structure proposed in Section 5.4 was implemented as described in Equation (5.42), i.e., by stabilizing the holdups of the units within the recycle loop (the reactor and the vapor phase of the condenser) with proportional control laws. The liquid holdup in the condenser (and, thereby, the total material holdup of the process) was also stabilized with a proportional controller:

$$L = L_s(1 - k_L(M_{L,sp} - M_L)) \tag{5.48}$$

with $k_L = 0.1\ \text{mol}^{-1}$.

For controller-design purposes, it is practical to derive a state-space realization of the dynamics after the fastest boundary layer (Equation (5.20)) using a coordinate change (5.18) in which the control objectives appear directly. Thus, rather than expressing the dynamics of the system in terms of total holdups, we used

$$\zeta = \begin{bmatrix} M_R \\ y_{A,R} \\ y_{I,R} \\ M_L \\ x_A \\ x_I \end{bmatrix} \tag{5.49}$$

$$\eta = \begin{bmatrix} u_R(x) + k u_F(x) \\ u_R(x)(y_A - y_{A,R}) \\ u_R(x)(y_I - y_{I,R}) \end{bmatrix} \tag{5.50}$$

Using the symbolic calculation engine available in Matlab®,[2] we obtained the following description of the intermediate dynamics of the reactor–condenser process:

$$\begin{aligned}
\dot{\zeta}_1 &= k_{uR}(-N_B - N_A + F_o)/(k k_{uF} + k_{uR}) \\
\dot{\zeta}_2 &= -k_{uR}(N_A + F_o\zeta_2 + k_1\zeta_1\zeta_2 - F_o - \zeta_2 N_B - \zeta_2 N_A)/A \\
\dot{\zeta}_3 &= -k_{uR}\zeta_3(-N_B - N_A + F_o)/A \\
\dot{\zeta}_4 &= N_A + N_B - L \\
\dot{\zeta}_5 &= -(-N_A + \zeta_5 N_A + \zeta_5 N_B)/\zeta_4 \\
\dot{\zeta}_6 &= -1/[\zeta_4\zeta_6(N_A + N_B)]
\end{aligned} \tag{5.51}$$

where

$$A = (-1 + k_{uR}M_{V,sp} + k - k k_{uF} M_{R,sp} + k k_{uF}\zeta_1 + \zeta_1 k_{uR}) \tag{5.52}$$

We then used the model (5.51) as the basis for synthesizing an input–output linearizing controller with integral action (Daoutidis and Kravaris 1992) for the product purity $x_B = 1 - \zeta_5 - \zeta_6$, using the condenser vapor holdup setpoint

[2] Matlab is a registered trademark of The Mathworks, Inc.

$M_{\mathrm{V,sp}}$ as a manipulated input[3] and requesting a second-order critically damped response:

$$x_{\mathrm{B}} + \beta_1 \frac{dx_{\mathrm{B}}}{dt} + \beta_2 \frac{d^2 x_{\mathrm{B}}}{dt^2} = x_{\mathrm{B,sp}} \tag{5.53}$$

with $\beta_1 = 40$ min and $\beta_2 = 400$ min^2.

The choice of manipulated input for the supervisory controller (the setpoint of a controller that belongs to the primary control structure used to stabilize the fast dynamics) is dictated by the low number (more precisely, one – the liquid product flow rate) of stream flow rates available as manipulated inputs in the intermediate time scale. The implementation of the resulting *cascaded* control structure is more elaborate from a technical point of view, since the equilibrium manifold of the fast dynamics (5.41) becomes control-dependent. We used the method proposed in Contou-Carrère *et al.* (2004) as discussed earlier in the book to overcome this difficulty.

Considering now the slowest (impurity) dynamics, we note that, under the coordinate change (5.49), only five of the six steady-state conditions (5.22) – corresponding to the model in Equation (5.51) – are linearly independent. Thus, we have

$$\hat{\mathbf{B}}(\mathbf{x}) = \begin{bmatrix} k_{u\mathrm{R}}/(kk_{u\mathrm{F}} + k_{u\mathrm{R}}) & 0 & 0 & 0 & 0 \\ 0 & -k_{u\mathrm{R}}/A & 0 & 0 & 0 \\ -k_{u\mathrm{R}}\zeta_3/A & 0 & 0 & 0 & 0 \\ 0 & 0 & 1 & 0 & 0 \\ 0 & 0 & 0 & -1/\zeta_4 & 0 \\ 0 & 0 & 0 & 0 & -1/\zeta_4 \end{bmatrix} \tag{5.54}$$

and

$$\check{\mathbf{f}} = \begin{bmatrix} -N_{\mathrm{B}} - N_{\mathrm{A}} + F_0 \\ N_{\mathrm{A}} + F_0\zeta_2 + k_1\zeta_1\zeta_2 - F_0 - \zeta_2 N_{\mathrm{B}} - \zeta_2 N_{\mathrm{A}} \\ N_{\mathrm{A}} + N_{\mathrm{B}} - L \\ -N_{\mathrm{A}} + \zeta_5 N_{\mathrm{A}} + \zeta_5 N_{\mathrm{B}} \\ \zeta_6(N_{\mathrm{A}} + N_{\mathrm{B}}) \end{bmatrix} \tag{5.55}$$

The above-mentioned steady-state conditions thus describe a 1D manifold in which the slow dynamics of the process evolve.

[3] Note that the setpoint of the vapor holdup in the reactor, $M_{\mathrm{R,sp}}$, is an equally valid choice of manipulated input.

Under the control law (5.48), the matrix $L_{\hat{\mathbf{B}}}\check{\mathbf{f}}(\boldsymbol{\zeta}, \mathbf{u}^{\mathrm{s}}(\boldsymbol{\zeta}))$ is invertible. Then, having

$$\hat{\mathbf{g}}^{\mathrm{P}} = \begin{bmatrix} (-F_{0\mathrm{s}} k_{u\mathrm{R}})/(k k_{u\mathrm{F}} + k_{u\mathrm{R}}) \\ 0 \\ 0 \\ 0 \\ 0 \\ 0 \end{bmatrix} \tag{5.56}$$

$$\hat{\mathbf{g}}^{\mathrm{I0}} = \begin{bmatrix} -k_{u\mathrm{R}}(\alpha_1\zeta_4\alpha_2\zeta_6)/[(k k_{u\mathrm{F}} + k_{u\mathrm{R}})/\rho_{\mathrm{L}}] \\ -k_{u\mathrm{R}}(F_0\alpha_{\mathrm{I}}\rho_{\mathrm{L}} - \zeta_4\alpha_1\alpha_2\zeta_6\zeta_2)/(\rho_{\mathrm{L}}/A) \\ k_{u\mathrm{R}} - \alpha_1\zeta_4\alpha_2\zeta_6 + F_0\alpha_{\mathrm{I}}\rho_{\mathrm{L}} + \zeta_4\alpha_1\alpha_2\zeta_6\zeta_2)/(\rho_{\mathrm{L}}/A) \\ \alpha_1\zeta_4\alpha_2\zeta_6/\rho_{\mathrm{L}} \\ -\zeta_5\alpha_1\zeta_4\alpha_2\zeta_6/(\zeta_4/\rho_{\mathrm{L}}) \\ -(-\alpha_1\zeta_4\alpha_2\zeta_6 + \alpha_1\zeta_4\alpha_2\zeta_6^2)/(\zeta_4/\rho_{\mathrm{L}}) \end{bmatrix} \tag{5.57}$$

$$\hat{\mathbf{g}}^{\mathrm{I}} = \begin{bmatrix} -k_{u\mathrm{R}}(\alpha_1\zeta_4\zeta_3)/[(k k_{u\mathrm{F}} + k_{u\mathrm{R}})/\rho_{\mathrm{L}}] \\ -k_{u\mathrm{R}}(-\zeta_2\alpha_1\zeta_3\zeta_4)/(\rho_{\mathrm{L}}/A) \\ k_{u\mathrm{R}}(\alpha_1\zeta_3^2\zeta_4 - \alpha_1\zeta_4\zeta_3)/(\rho_{\mathrm{L}}/A) \\ \alpha_1\zeta_4\zeta_3/\rho_{\mathrm{L}} \\ -\zeta_5\alpha_1\zeta_3/\rho_{\mathrm{L}} \\ -(-\alpha_1\zeta_3 + \alpha_1\zeta_3\zeta_6)/\rho_{\mathrm{L}} \end{bmatrix} \tag{5.58}$$

we derived via symbolic calculation (the details are omitted for brevity) a 1D, minimal-order representation of the slow dynamics (5.28) on the basis of which we synthesized an input–output-linearizing controller with integral action for the impurity mole fraction in the reactor, requesting a first-order response:

$$\zeta_3 + \gamma\frac{d\zeta_3}{d\theta} = \zeta_{3,\mathrm{sp}} \tag{5.59}$$

with $\gamma = 500$ min.

Table 5.2 summarizes the control principles outlined in this chapter as they apply to the reactor–condenser process.

Remark 5.2. *Industrial implementations of the distributed layer of this control structure would depend on sensor availability. Thus, the holdup of the reactor and the gas phase in the condenser would be stabilized by controlling the pressure in these vessels, while the liquid-level measurements would be used to estimate and control the liquid holdup. A potential process and instrumentation diagram (P&ID) for this process is presented in Figure 5.9.*

Table 5.2. Control structure for the reactor–separator process core

Time scale	Controlled output	Manipulated input	Controller
Fast	Reactor holdup M_R	Reactor effluent flow rate F	Proportional
	Condenser vapor holdup M_V	Recycle flow rate R	Proportional
Intermediate	Total material holdup	Liquid-product flow rate L	Proportional
	Product purity x_B	Condenser holdup setpoint $M_{V,sp}$	Nonlinear, model-based cascaded structure
Slow	Inert levels in the reactor, $y_{I,R}$	Purge flow rate P	Nonlinear, model-based

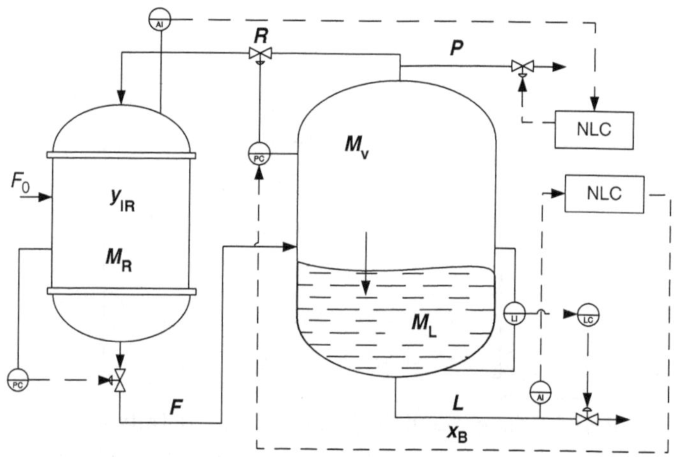

Figure 5.9 A potential process and instrumentation diagram (P&ID) for the reactor–condenser process core.

5.5.5 Simulation results and discussion

The theoretical concepts developed above were validated through numerical simulations. First, we considered the case of a 15% increase in the production rate (achieved by increasing the feed flow rate). A second case concerned the operation of the process under a significant (25%) rise in the inlet impurity concentration. Figures 5.10–5.12 and 5.13–5.15 present the closed-loop response for these two scenarios, evincing excellent controller performance. A second set of simulations considered the same operating scenarios, with the added challenge of a mismatch in two key parameters: the reaction rate k_1 and the mass transfer coefficient $\mathcal{K}_B$ of the product B in the controller model were assumed to be overestimated by 10% compared with their values in the process. The control performance also in these cases (Figures 5.16–5.18 and 5.19–5.21) is excellent.

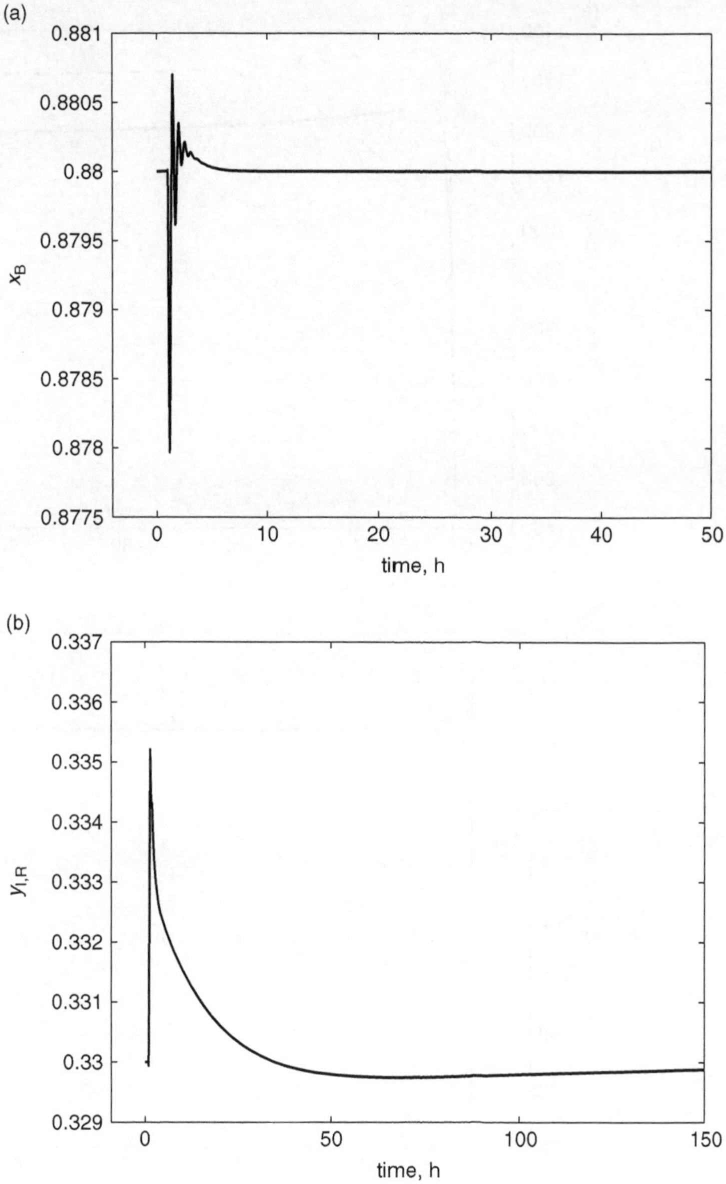

Figure 5.10 Evolution of the process composition variables for a 15% increase in the production rate at $t = 1\,\text{h}$. (a) Product purity and (b) reactor impurity level.

Remark 5.3. *Figures 5.10–5.21 show that the response of the mole fraction of the impurity I in the reactor is also slow in* **closed-loop** *operation. This is to be expected and was accounted for in the design of the impurity-mole-fraction controller by requesting a sufficiently long time constant for the closed-loop response (5.59).*

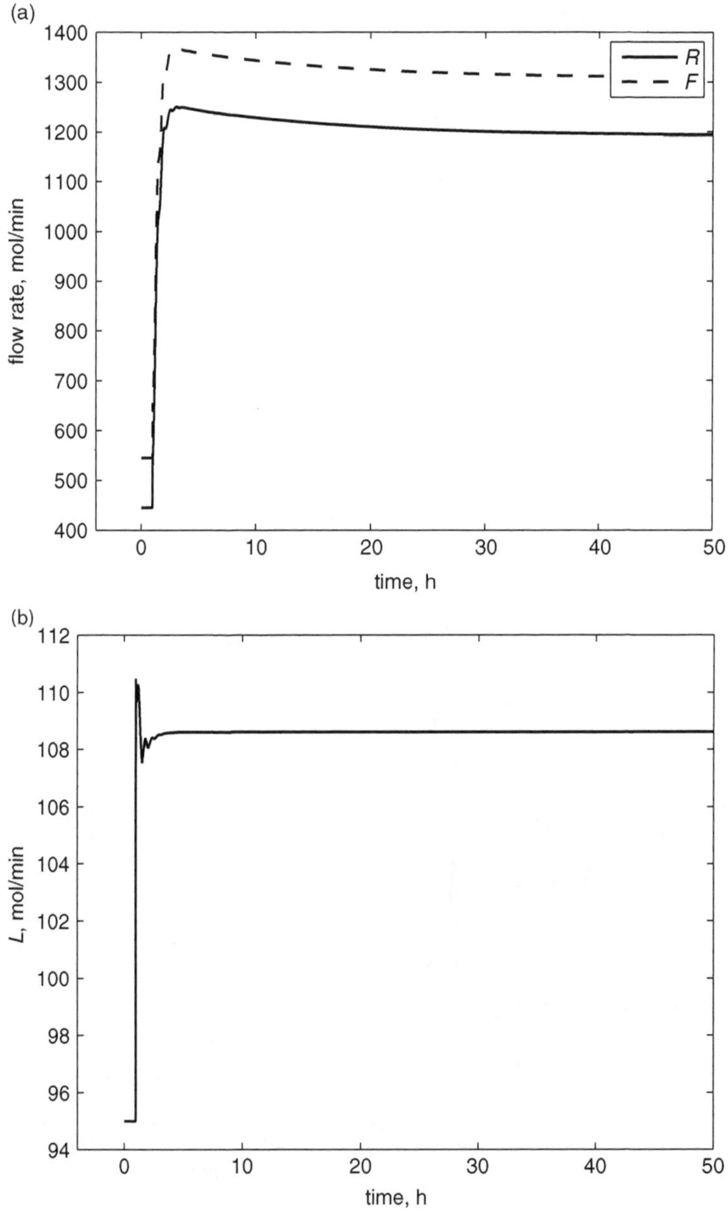

Figure 5.11 Evolution of the process stream flow rates for a 15% increase in the production rate at $t = 1\,\mathrm{h}$. (a) Effluent and recycle flow rates, and (b) product flow rate.

Remark 5.4. *The steady-state process parameters in Table 5.1 can be used to calculate the values of the recycle and purge numbers. For this process, we have* $\mathbf{Rc} = 4.45$ *and* $\mathbf{Pu} = 0.05$*, or, equivalently,* $\varepsilon_1 = 1/\mathbf{Rc} = 0.224$ *and* $\varepsilon_2 = \mathbf{Pu} = 0.05$*. It is noteworthy that its numerical value does not immediately warrant the "small-parameter" description for* ε_1 *(this is not an issue for* ε_2*). The*

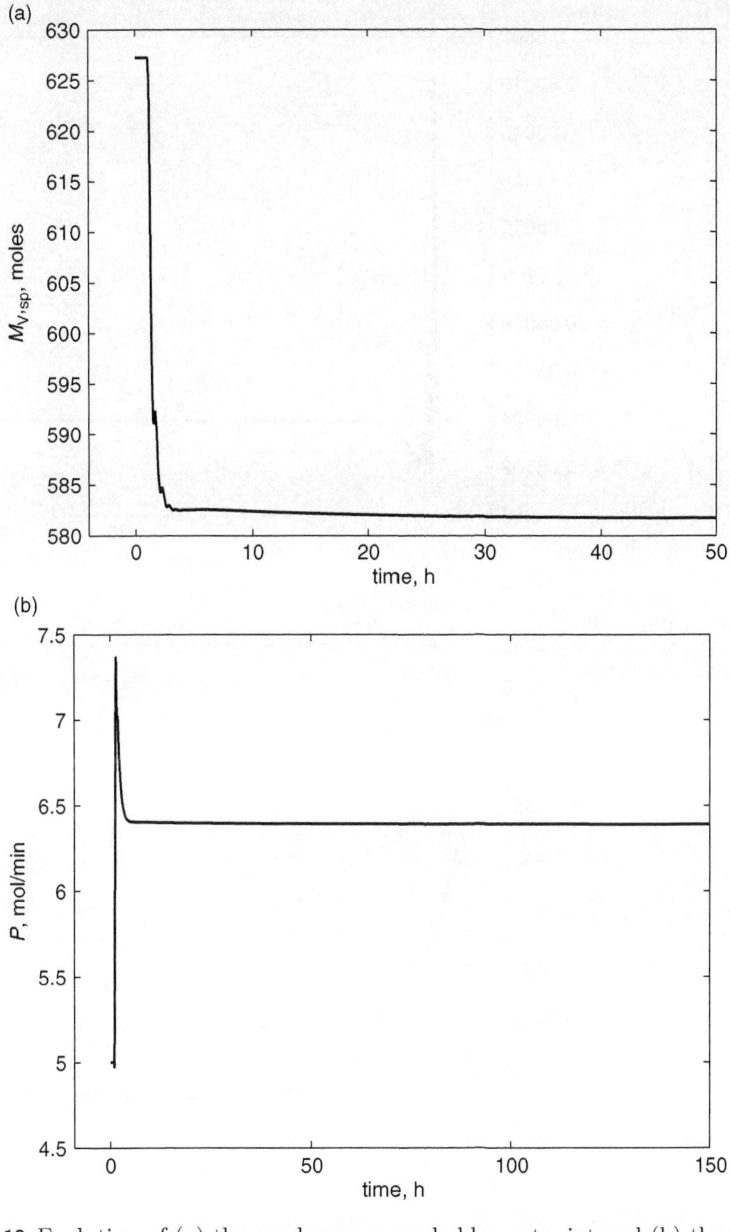

Figure 5.12 Evolution of (a) the condenser vapor holdup setpoint and (b) the purge flow rate for a 15% increase in the production rate at $t = 1\,\text{h}$.

reactor–condenser system does not therefore strictly fulfill the assumptions of our singular perturbation analysis. Yet, the model-based control structure designed using the reduced-order models of the dynamics in each time scale performs remarkably well. This case study thus provides further confirmation of the wide applicability and robustness of the model-reduction and controller design methods proposed in this chapter.

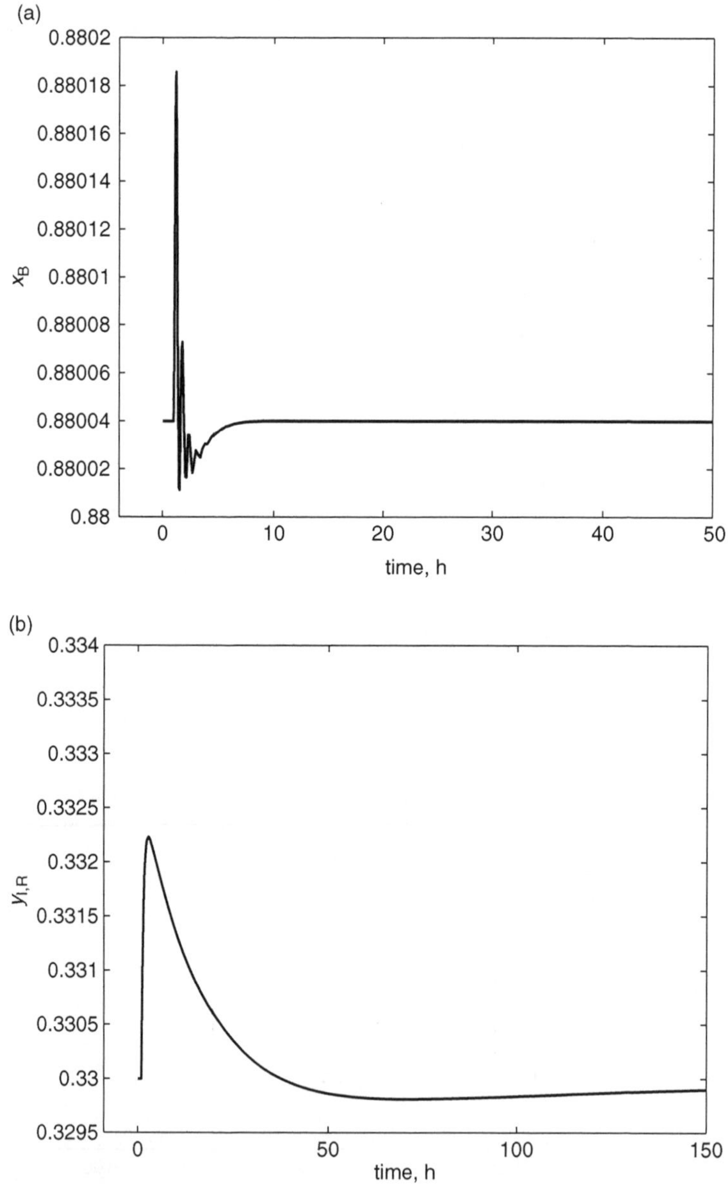

Figure 5.13 Evolution of the process composition variables for a 25% unmeasured increase in the inlet impurity levels y_{I0} occurring at $t = 1\,\text{h}$. (a) Product purity and (b) reactor impurity level.

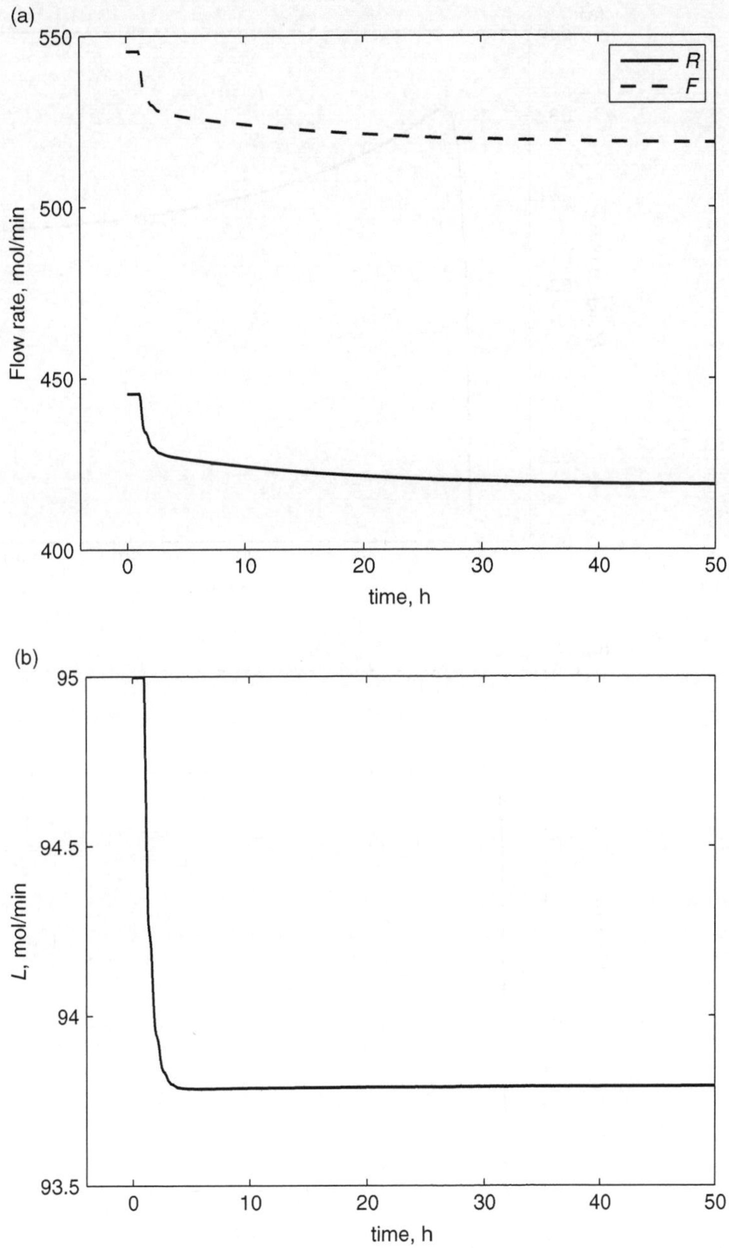

Figure 5.14 Evolution of the process stream flow rates for a 25% unmeasured increase in the inlet impurity levels y_{I0} occurring at $t = 1\,\mathrm{h}$. (a) Effluent and recycle flow rates, and (b) product flow rate.

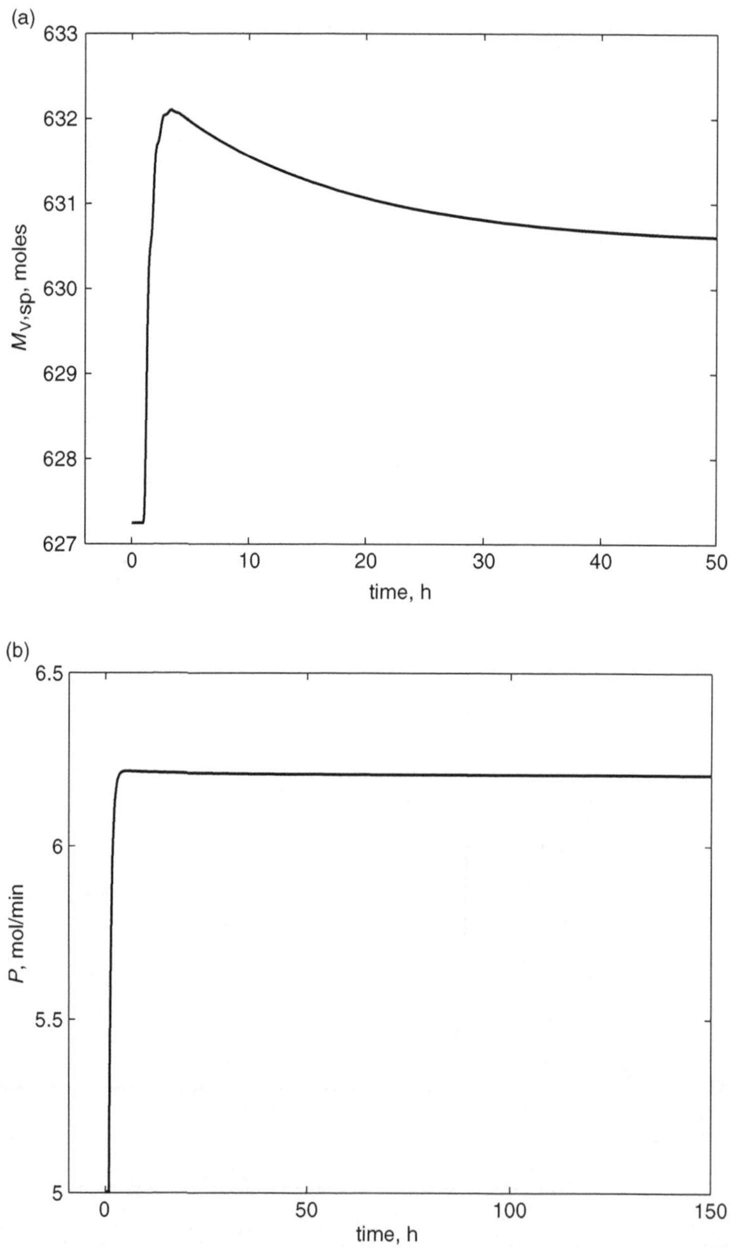

Figure 5.15 Evolution of (a) the condenser vapor holdup setpoint and (b) the purge flow rate for a 25% unmeasured increase in the inlet impurity levels y_{I0} occurring at $t = 1\,\mathrm{h}$.

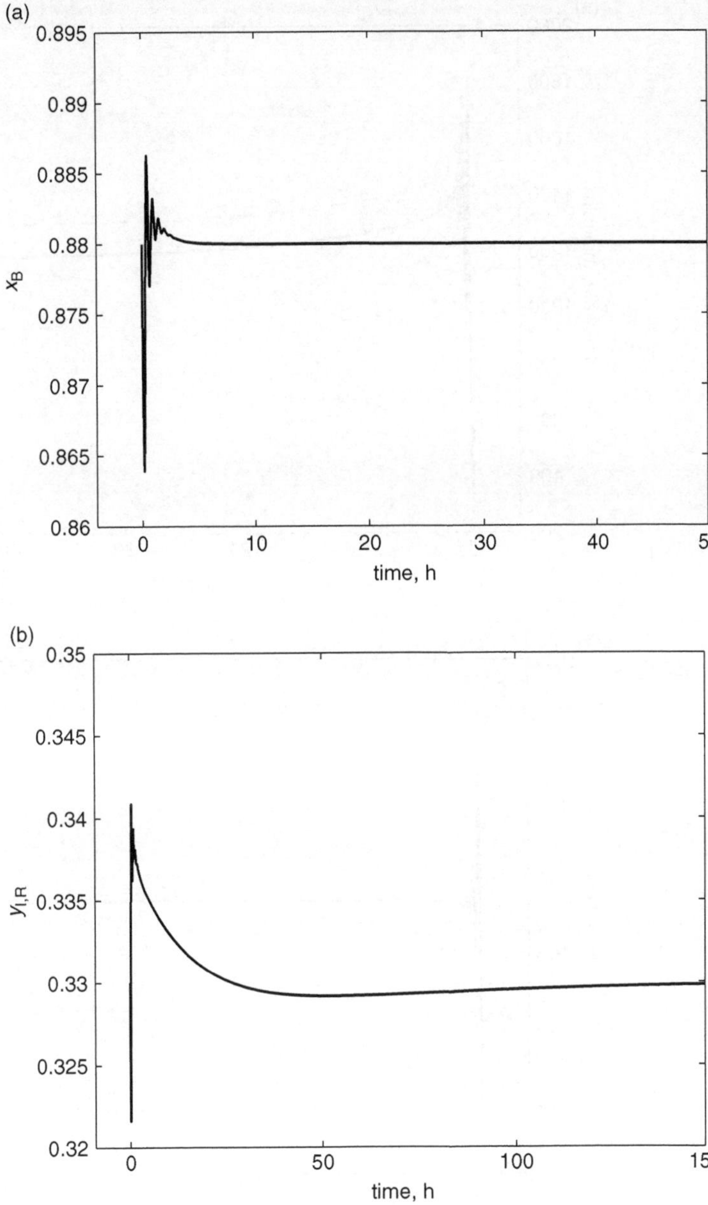

Figure 5.16 Closed-loop evolution of the process composition variables for a 15% increase in the production rate occurring at $t = 0$, under plant–model parameter mismatch. The reaction rate k_1 and the mass transfer coefficient $\mathcal{K}_B$ in the controller model are assumed to be overestimated by 10% compared with their values in the plant. (a) Product purity and (b) reactor impurity level.

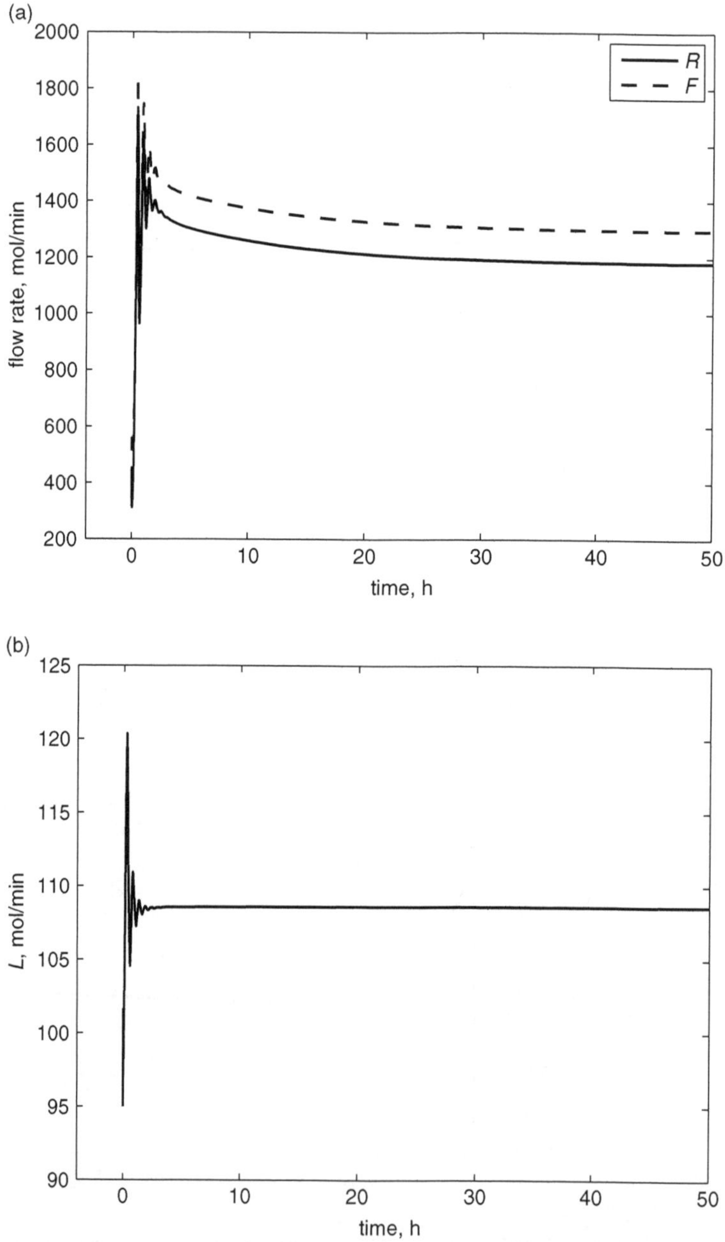

Figure 5.17 Closed-loop evolution of the process stream flow rates for a 15% increase in the production rate occurring at $t = 0$, under plant–model parameter mismatch. The reaction rate k_1 and the mass-transfer coefficient $\mathcal{K}_B$ in the controller model are assumed to be overestimated by 10% compared with their values in the plant. (a) Effluent and recycle flow rates, and (b) product flow rate.

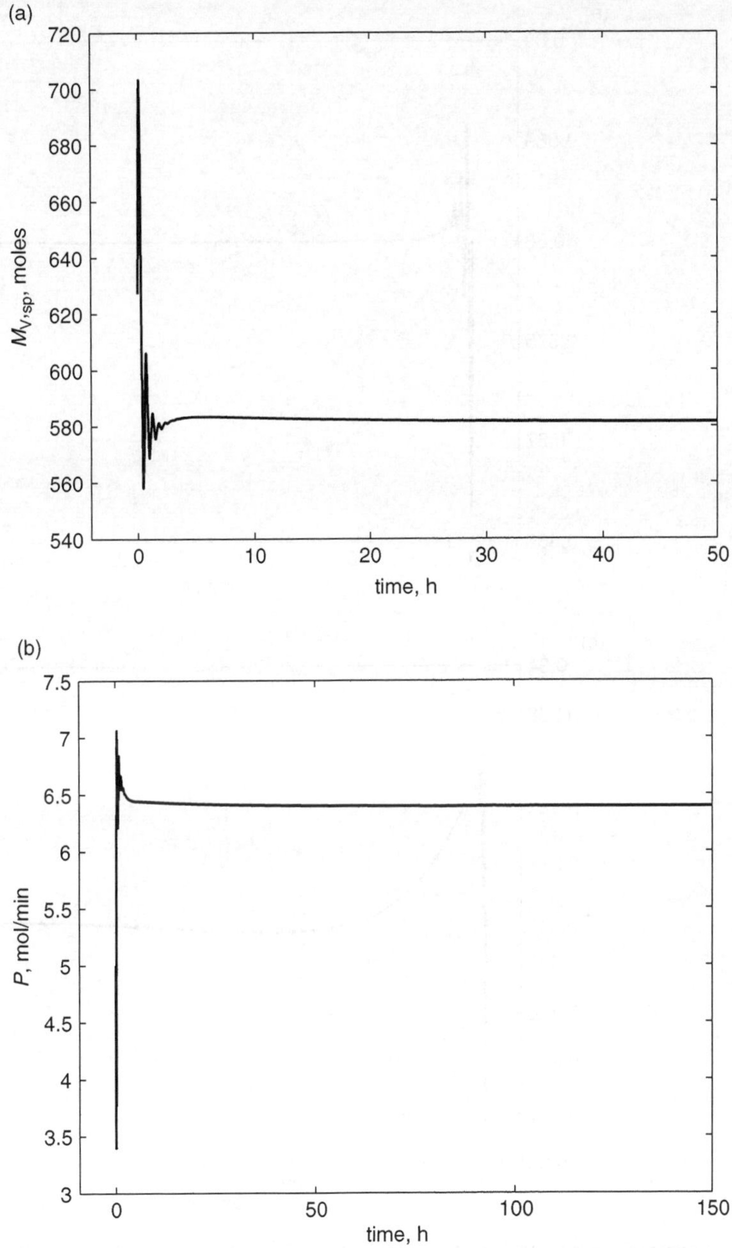

Figure 5.18 Closed-loop evolution of (a) the condenser vapor holdup setpoint and (b) the purge flow rate for a 15% increase in the production rate occurring at $t = 0$, under plant–model parameter mismatch. The reaction rate k_1 and the mass-transfer coefficient $\mathcal{K}_B$ in the controller model are assumed to be overestimated by 10% compared with their values in the plant.

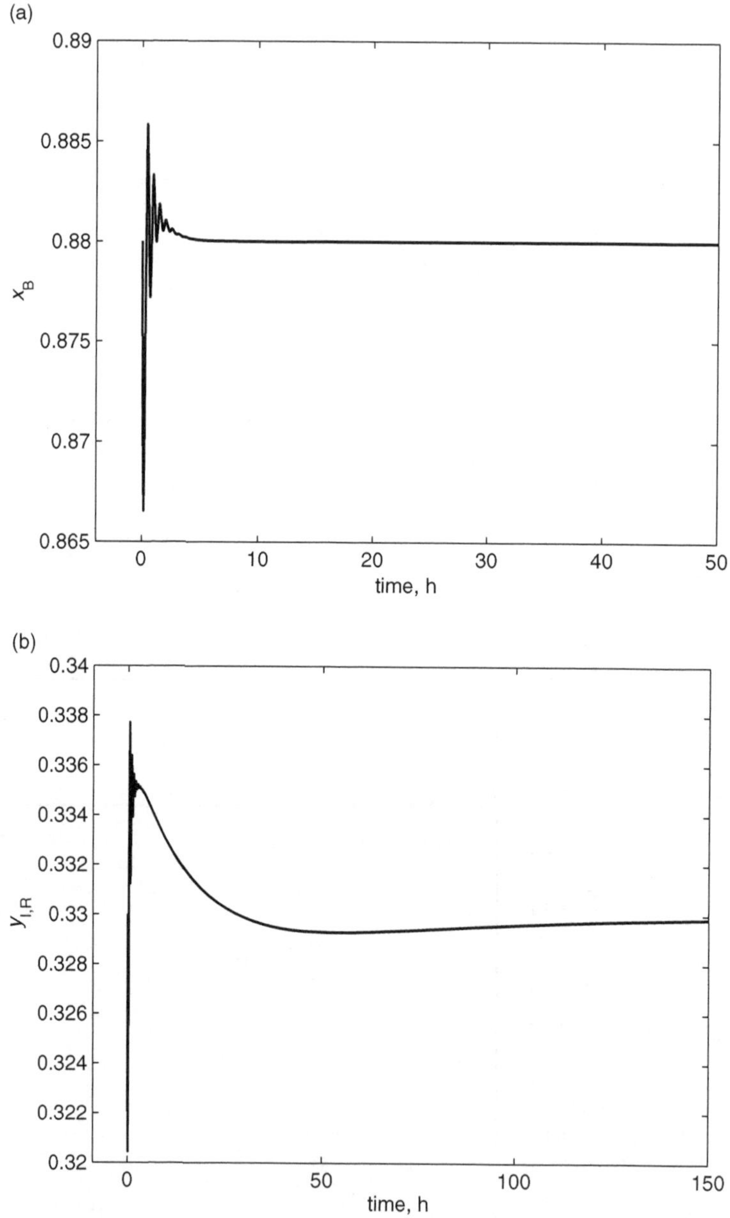

Figure 5.19 Evolution of the process composition variables for a 25% unmeasured increase in the inlet impurity levels y_{I0} occurring at $t = 0$, under plant–model parameter mismatch. The reaction rate k_1 and the mass-transfer coefficient $\mathcal{K}_B$ in the controller model are assumed to be overestimated by 10% compared with their values in the plant. (a) Product purity and (b) reactor impurity level.

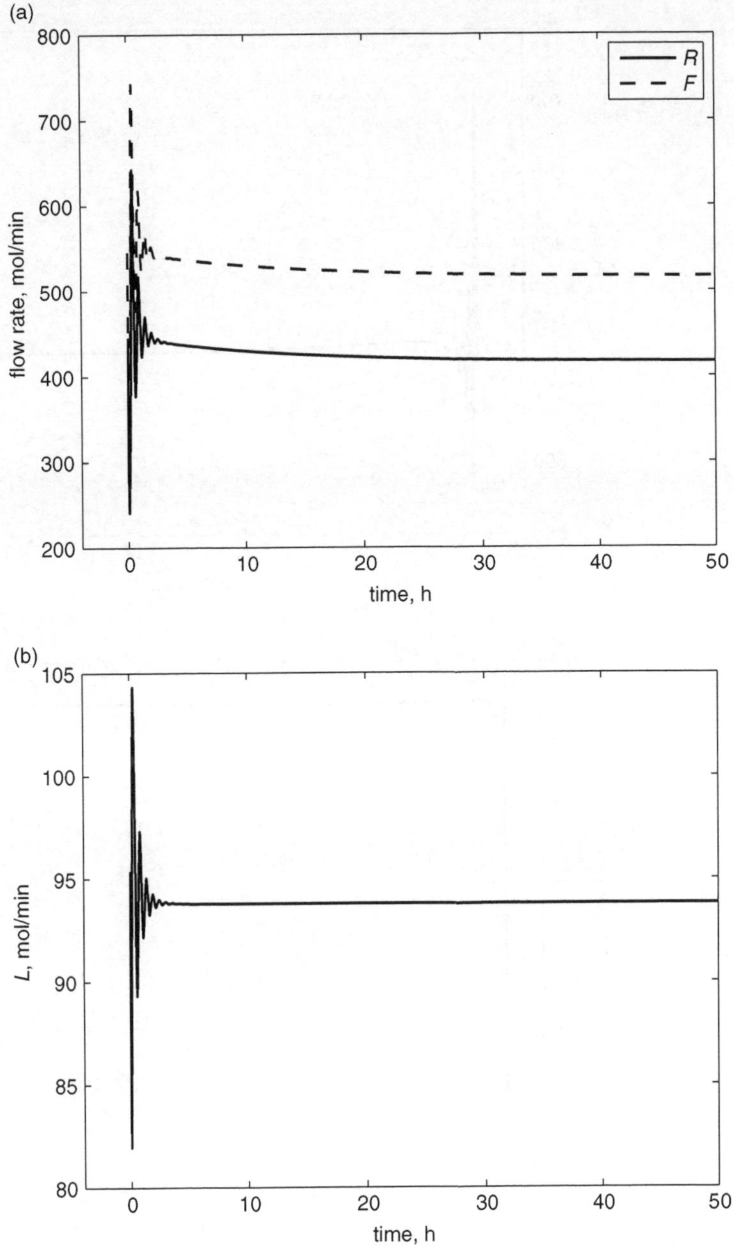

Figure 5.20 Evolution of the process stream flow rates for a 25% unmeasured increase in the inlet impurity levels y_{I0} occurring at $t = 0$, under plant–model parameter mismatch. The reaction rate k_1 and the mass-transfer coefficient $\mathcal{K}_B$ in the controller model are assumed to be overestimated by 10% compared with their values in the plant. (a) Effluent and recycle flow rates, and (b) product flow rate.

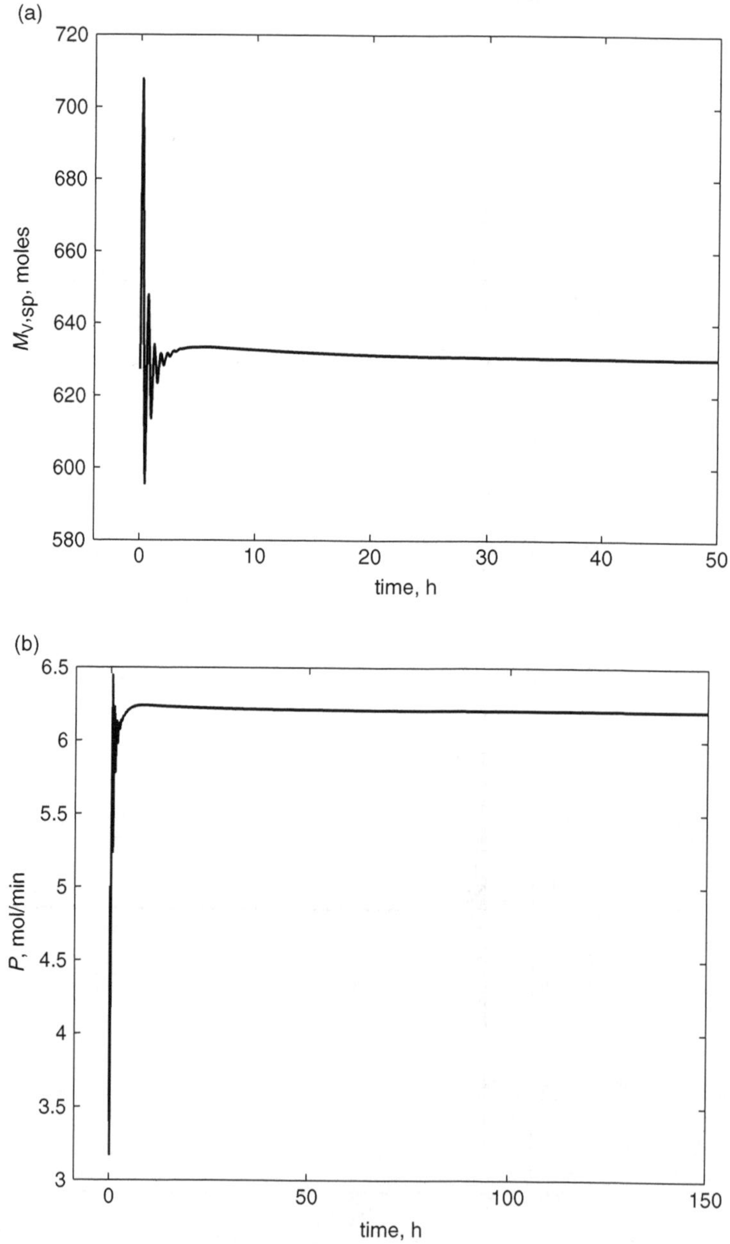

Figure 5.21 Evolution of (a) the condenser vapor holdup setpoint and (b) the purge flow rate for a 25% unmeasured increase in the inlet impurity levels y_{I0} occurring at $t = 0$, under plant–model parameter mismatch. The reaction rate k_1 and the mass-transfer coefficient $\mathcal{K}_B$ in the controller model are assumed to be overestimated by 10% compared with their values in the plant.

5.6 Synopsis

In this chapter, we developed a unified framework for analyzing the dynamic behavior of integrated processes. Using a completely generic description, we investigated the dynamic behavior that emerges as a result of process integration through significant material recycling *and* impurity purging.

Drawing on the results derived in the first two chapters in this part of the book, we showed that the dynamics of these processes evolve over three distinct time horizons:

- a *fast* time scale, which corresponds to the response of the individual process units in the recycle loop
- an *intermediate* time scale, corresponding to the dynamic response of the *overall* process
- a *slow* time scale that reflects the presence of process impurities and of the purge stream that is used to eliminate them.

We showed that this dynamic behavior originates, in effect, from the large discrepancies between the flow rates of the different material streams in the process (captured by the recycle and purge numbers), thereby establishing a clear and causal connection between the *steady-state design* and the *dynamic response* of an integrated process.

We proposed a method for deriving nonlinear low-dimensional models for the dynamics in each time scale. Subsequently, we proposed a hierarchical controller design framework that takes advantage of the time-scale multiplicity, and relies on a multi-tiered structure of coordinated decentralized and supervisory controllers in order to address distributed and process-level control objectives.

Part III

Process systems with energy integration

6 Process systems with energy recycling

6.1 Introduction

The previous chapters have concentrated on analyzing the material-balance dynamics of several classes of integrated process systems. We demonstrated that the dynamic behavior of the systems considered exhibits several time scales and described a method for the derivation of reduced-order models describing the dynamics in each time scale. Also, a hierarchical controller design framework was introduced, with distributed control of the fast dynamics and supervisory control of the dynamics at the systems level.

In what follows, we focus on the dynamic features associated with *energy integration*, a ubiquitous design feature in the process and energy industries. Energy-integrated designs are motivated by the high cost of energy and the corresponding need to minimize fuel and utility usage. Pairing energy generation and consumption within the same plant is an effective means to this end, and can be implemented in numerous ways, including the use of feed–effluent heat-exchange, plant-wide heat exchanger networks, heat pumping, and heat integration and thermal coupling of distillation columns.

Energy-integrated designs invariably introduce a dynamic coupling between the process units. Several authors have documented a positive-feedback effect due to recycling of energy, which can lead to an intricate dynamic behavior, featuring, e.g., an inverse response or open-loop instability (Morud and Skogestad 1994, 1996, Bildea and Dimian 1998, Jacobsen and Berezowski 1998, Mizsey *et al.* 1998, Morud and Skogestad 1998, Reyes and Luyben 2000a, Chen and Yu 2003).

The present chapter considers processes in which energy recycling and recovery is significant in comparison with any available energy sources and/or sinks, as well as with any energy input through the feed stream. Following the developments in (Jogwar *et al.* 2009), we construct a prototype process and use it to identify and characterize the underlying dynamic structure, demonstrating that – in analogy with the case of processes with significant material recycling, discussed in Chapter 3 – the simultaneous presence of energy flows of different magnitudes causes model stiffness and is at the origin of a time-scale separation in the energy-balance dynamics. A model-reduction framework is developed and reduced-order descriptions of the dynamics in each time scale are derived. We then propose guidelines for taking these dynamic characteristics into consideration in controller design, and present several illustrative examples and a simulation case study.

6.2 Dynamics of processes with significant energy recovery

Let us consider the process in Figure 6.1, consisting of N units in series. Let H_0 denote an external energy input to the first unit and H_i, $i = 1, \ldots, N$, the outlet energy flow from the ith unit. While H_i are (intuitively) assumed to be associated with material streams, we also account for the potential presence of other sources of energy in the units, Q_i, originating, e.g., from direct heating or cooling, or from the energy effect of chemical reactions. Finally, we assume that the means exist (e.g., via direct heat exchange, transfer using a heat-transfer medium, heat pumping or material recycle) to recover energy from the last unit (or output stream) at a rate Q_{out} and recycle it to the first unit (or process input) at a rate $Q_{\text{in}} \equiv Q_{\text{out}}$.

The prototype process in Figure 6.1 captures the structural and dynamic properties of integrated process designs with significant energy recycling, most notably processes that rely on a feed–effluent heat exchanger (FEHE) to recover energy from the products. Typically, FEHEs (Figure 6.2) are used for *preheating* feed streams using the heat of reaction carried by the products, but they are equally effective in the recovery of refrigeration for feed *cooling* (see Example 4.1).

Assuming that individual process units are modeled as lumped-parameter systems and that kinetic and potential energy contributions are negligible, the

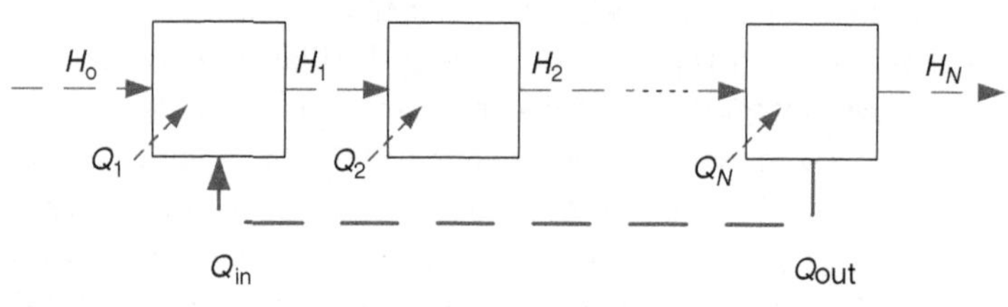

Figure 6.1 A prototype energy-integrated process.

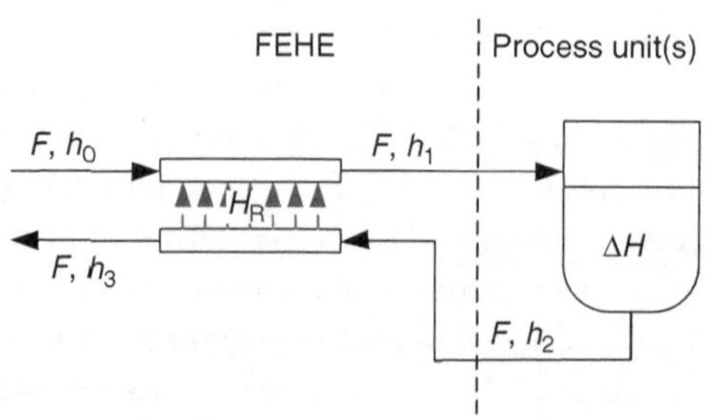

Figure 6.2 A process with a feed–effluent heat exchanger.

mathematical model that describes the energy balance of the process can be written in the form

$$\begin{aligned}\dot{\theta}_1 &= H_0 - H_1 + Q_1 + Q_{\text{in}} \\ &\vdots \\ \dot{\theta}_i &= H_{i-1} - H_i + Q_i \\ &\vdots \\ \dot{\theta}_N &= H_{N-1} - H_N + Q_N - Q_{\text{out}}\end{aligned} \tag{6.1}$$

with θ_i being the enthalpies of units $i = 1, \ldots, N$. Equation (6.1) can be rewritten in a compact form as

$$\dot{\boldsymbol{\theta}} = \sum_{i=0,N} \gamma_i H_i + \sum_{i=0,N} \phi_i(\mathbf{x}, \boldsymbol{\theta}) Q_i + \sum_{i=1}^{N-1} \gamma_i H_i + \gamma_q^{\text{in}} Q_{\text{in}} + \gamma_q^{\text{out}} Q_{\text{out}} \tag{6.2}$$

with $\boldsymbol{\theta} = [\theta_1 \ldots \theta_i \ldots \theta_N]^{\mathrm{T}}$, $\boldsymbol{\theta} \in \mathcal{Q} \subset \mathbb{R}^N$ being the vector of unit enthalpies and γ_i and γ_q^j being appropriately defined vector functions. For reasons that will become apparent later in the book, we adhere to using a separate notation for Q_{in} and Q_{out}, in spite of the fact that they represent, in effect, the same quantity.

We denote by $\omega_i = H_i/H_{i,\mathrm{s}}$ the dimensionless variables corresponding to the energy flow rates H_i, $i = 1, \ldots, N$ (the subscript s denotes steady-state values). Appending a generic representation of the overall and component material-balance equations, with $\mathbf{x} \in \mathcal{X} \subset \mathbb{R}^m$ being the material-balance variables, the overall mathematical model of the process in Figure 6.1 becomes

$$\begin{aligned}\dot{\mathbf{x}} &= \mathbf{f}(\mathbf{x}, \boldsymbol{\theta}) \\ \dot{\boldsymbol{\theta}} &= \sum_{i=0,N} \gamma_i(\mathbf{x}, \boldsymbol{\theta}) H_i + \sum_{i=0,N} \phi_i(\mathbf{x}, \boldsymbol{\theta}) Q_i \\ &\quad + \sum_{i=1}^{N-1} \gamma_i(\mathbf{x}, \boldsymbol{\theta}) H_i + \gamma_q^{\text{in}} Q_{\text{in}} + \gamma_q^{\text{out}} Q_{\text{out}}\end{aligned} \tag{6.3}$$

Let us now concentrate on the energy dynamics of the system (6.3), for which we make the following steady-state assumptions.

Assumption 6.1. *The internal energy flow rates $H_{i,\mathrm{s}}$, $i = 1, \ldots, N-1$, are of similar magnitudes (i.e., $l_i = H_{i,\mathrm{s}}/H_{1,\mathrm{s}} = \mathcal{O}(1)$) and much larger than the inlet and outlet energy flows $H_{0,\mathrm{s}}$ and $H_{N,\mathrm{s}}$, that is*

$$\frac{H_{0,\mathrm{s}}}{H_{1,\mathrm{s}}} \ll 1 \tag{6.4}$$

Assumption 6.2. *The energy inputs Q_i, $i = 1, \ldots, N$, to the individual units are of similar magnitude to the inlet energy flow. Equivalently, $Q_{i,\mathrm{s}}/H_{0,\mathrm{s}} = \mathcal{O}(1)$ and $Q_{i,\mathrm{s}}/H_{1,\mathrm{s}} \ll 1$.*

Assumption 6.3. *The magnitude of the energy flow rate $Q_{\mathrm{in}} \equiv Q_{\mathrm{out}}$ is similar to the magnitude of the energy flows H_i, $i = 1, \ldots, N-1$, and we have*

$$\frac{Q_{\mathrm{in,s}}}{H_{1,\mathrm{s}}} \equiv \frac{Q_{\mathrm{out,s}}}{H_{1,\mathrm{s}}} = m_{\mathrm{in}} = m_{\mathrm{out}} = \mathcal{O}(1) \tag{6.5}$$

We also define $m_{\mathrm{in}} = Q_{\mathrm{in,s}}/H_{1,\mathrm{s}}$ and $m_{\mathrm{out}} = Q_{\mathrm{out,s}}/H_{1,\mathrm{s}}$.

Assumptions 6.1–6.3 ensure that, in steady-state operation, the rate of energy recovery from the effluent stream(s) is larger than the amount of energy input to the system from the outside, thereby guaranteeing a tight energy integration. These assumptions are not expected to remain true during inherently transient events such as plant start-ups, when a significant amount of energy needs to be accumulated in the process. In such cases, plant operation cannot rely on significant energy recovery, and external energy sources are required.

The above assumptions also suggest that the degree of energy integration of a process can be quantified by means of a dimensionless number.

Definition 6.1. *The **energy recovery number** is the ratio of the rate at which energy is recovered from the process effluent stream(s) to the rate at which energy is input to the process through the material feed streams:*

$$\mathsf{Erc} = \frac{H_{1,\mathrm{s}}}{H_{0,\mathrm{s}}} \tag{6.6}$$

Remark 6.1. *The definition of the energy recovery number follows the same principle as that of the recycle number* **Rc** *(Definition 3.1). Both numbers characterize the intensity of recycling/recovery of a process **inventory** (see, e.g., Farschman* et al. *1998) – that is, energy and mass, respectively. From the perspective of inventory recycling, the two numbers* **Erc** *and* **Rc** *are, in effect, particular cases of the same dimensionless quantity.*

With the above definition, we can rewrite the model (6.3) as

$$\begin{aligned} \dot{\mathbf{x}} &= \mathbf{f}(\mathbf{x}, \boldsymbol{\theta}) \\ \dot{\boldsymbol{\theta}} &= \sum_{i=0,N} \gamma_i(\mathbf{x}, \boldsymbol{\theta}) H_i + \sum_{i=0,N} \phi_i(\mathbf{x}, \boldsymbol{\theta}) Q_i + \mathsf{Erc} H_{0,\mathrm{s}} \\ &\quad \times \left(\sum_{i=1}^{N-1} \gamma_i(\mathbf{x}, \boldsymbol{\theta}) l_i \omega_i + \gamma_q^{\mathrm{in}}(\mathbf{x}, \boldsymbol{\theta}) m_{\mathrm{in}} \omega_{\mathrm{in}} + \gamma_q^{\mathrm{out}} m_{\mathrm{out}}(\mathbf{x}, \boldsymbol{\theta}) \omega_{\mathrm{out}} \right) \end{aligned} \tag{6.7}$$

where $\omega_{\text{in}} = Q_{\text{in}}/Q_{\text{in,s}}$ and $\omega_{\text{out}} = Q_{\text{out}}/Q_{\text{out,s}}$. More generally, we can write (6.7) as

$$\begin{aligned} \dot{\mathbf{x}} &= \mathbf{f}(\mathbf{x}, \boldsymbol{\theta}) \\ \dot{\boldsymbol{\theta}} &= \boldsymbol{\Phi}(\mathbf{x}, \boldsymbol{\theta}) + \boldsymbol{\Gamma}^{\text{s}}(\mathbf{x}, \boldsymbol{\theta})\boldsymbol{\omega}^{\text{s}} + \mathsf{Erc}\boldsymbol{\Gamma}^{\text{l}}(\mathbf{x}, \boldsymbol{\theta})\boldsymbol{\omega}^{\text{l}} \end{aligned} \tag{6.8}$$

where $\boldsymbol{\omega}^{\text{s}} \in \mathcal{U}^{\text{s}} \subset \mathbb{R}^{m^{\text{s}}}$ is a vector of scaled variables that correspond to the small inlet and outlet energy flows, and $\boldsymbol{\omega}^{\text{l}} \in \mathcal{U}^{\text{l}} \subset \mathbb{R}^{m^{\text{l}}}$ is a vector of scaled variables corresponding to the large internal and recycle energy flows. $\mathbf{f}$ and $\boldsymbol{\Phi}$ are vector functions, and $\mathbf{G}^{\text{s}}$ and $\mathbf{G}^{\text{l}}$ are matrices of appropriate dimensions.

Our interest in this chapter concentrates on process systems with tight energy integration, or, according to Assumption 6.1, processes for which the energy recovery number is large. For mathematical convenience and in keeping with the analysis framework developed throughout the book, we will define the small parameter

$$\varepsilon = \frac{1}{\mathsf{Erc}} \ll 1 \tag{6.9}$$

which allows us to rewrite (6.8) as

$$\begin{aligned} \dot{\mathbf{x}} &= \mathbf{f}(\mathbf{x}, \boldsymbol{\theta}) \\ \dot{\boldsymbol{\theta}} &= \boldsymbol{\Phi}(\mathbf{x}, \boldsymbol{\theta}) + \boldsymbol{\Gamma}^{\text{s}}(\mathbf{x}, \boldsymbol{\theta})\boldsymbol{\omega}^{\text{s}} + \frac{1}{\varepsilon}\boldsymbol{\Gamma}^{\text{l}}(\mathbf{x}, \boldsymbol{\theta})\boldsymbol{\omega}^{\text{l}} \end{aligned} \tag{6.10}$$

Owing to the presence of the small parameter ε, the generic model of Equation (6.10) is stiff. According to the theory developed in Chapter 2, the dynamics of integrated processes with energy recycling have the potential to evolve over multiple time scales. As was shown in the previous chapters, the rational approach to addressing the control of such systems involves the properly coordinated synthesis of separate fast and slow controllers, so that overall stability, output tracking, and disturbance rejection can be achieved. The design of such controllers and the closed-loop analysis must be performed on the basis of separate reduced-order models that describe the dynamics in the fast and slow time scales. The derivation of reduced-order, non-stiff models for the energy dynamics of the class of process systems considered is addressed in the next section.

6.3 Model reduction

We proceed by defining the fast time scale $\tau = t/\varepsilon$ and rewriting the model (6.10) in this fast time scale to obtain

$$\begin{aligned} \frac{d\mathbf{x}}{d\tau} &= \mathbf{f}(\mathbf{x}, \boldsymbol{\theta}) \\ \frac{d\boldsymbol{\theta}}{d\tau} &= \varepsilon(\boldsymbol{\Phi}(\mathbf{x}, \boldsymbol{\theta}) + \boldsymbol{\Gamma}^{\text{s}}(\mathbf{x}, \boldsymbol{\theta})\boldsymbol{\omega}^{\text{s}}) + \boldsymbol{\Gamma}^{\text{l}}(\mathbf{x}, \boldsymbol{\theta})\boldsymbol{\omega}^{\text{l}} \end{aligned} \tag{6.11}$$

Let us now consider the limit $\varepsilon \to 0$, i.e., the ideal case of total heat integration via an infinitely high energy recycle. In this limit, we obtain

$$\begin{aligned} \frac{d\mathbf{x}}{d\tau} &= \mathbf{0} \\ \frac{d\boldsymbol{\theta}}{d\tau} &= \boldsymbol{\Gamma}^{\mathrm{l}}(\mathbf{x}, \boldsymbol{\theta})\boldsymbol{\omega}^{\mathrm{l}} \end{aligned} \tag{6.12}$$

which represents a description of the fast dynamics of the process, involving only the variables $\boldsymbol{\theta}$ that pertain to the energy balance.

It is evident that the fast dynamics described by Equation (6.12) are only influenced by the large energy flows $\boldsymbol{\omega}^{\mathrm{l}}$. However, it is easy to verify (e.g., by summing the equations in (6.1)) that the large internal flows do not affect the total enthalpy of the process,

$$\theta_{\mathrm{tot}} = \sum_{i=1}^{N} \theta_i \tag{6.13}$$

and, furthermore, that the transient behavior of θ_{tot} is governed exclusively by the small external energy flows and the (small) energy-generation terms in the energy balance. In turn, this indicates that the differential equations in (6.12) are not linearly independent. Thus, the solution of the algebraic equation system consisting of the (linearly dependent) quasi-steady-state conditions

$$0 = \boldsymbol{\Gamma}^{\mathrm{l}}(\mathbf{x}, \boldsymbol{\theta})\boldsymbol{\omega}^{\mathrm{l}} \tag{6.14}$$

corresponding to the fast component of the dynamics does not consist of a set of isolated equilibrium enthalpy values; rather, it consists of an equilibrium manifold, or subspace, in which a slower dynamic component evolves. From the physical arguments presented above, this manifold is at most 1D.

In pursuing the derivation of a description of the slow component of the system dynamics, we make the following (typically true, as we will see in the examples below) assumption.

Assumption 6.4. *There exist a full column rank matrix* $\mathbf{B}(\mathbf{x}, \boldsymbol{\theta}) \in \mathcal{B} \subset \mathbb{R}^{\mathcal{N}\times\mathcal{N}-1}$ *and a matrix* $\tilde{\boldsymbol{\Gamma}}^{\mathrm{l}}(\mathbf{x}, \boldsymbol{\theta}) \subset \mathbb{R}^{(\mathcal{N}-1)\times m^{\mathrm{l}}}$ *with linearly independent rows, such that* $\boldsymbol{\Gamma}^{\mathrm{l}}(\mathbf{x}, \boldsymbol{\theta})$ *can be rewritten as*

$$\boldsymbol{\Gamma}^{\mathrm{l}}(\mathbf{x}, \boldsymbol{\theta}) = \mathbf{B}(\mathbf{x}, \boldsymbol{\theta})\tilde{\boldsymbol{\Gamma}}^{\mathrm{l}}(\mathbf{x}, \boldsymbol{\theta}) \tag{6.15}$$

Assumption 6.4 allows us to isolate a set of linearly independent constraints

$$0 = \tilde{\boldsymbol{\Gamma}}^{\mathrm{l}}(\mathbf{x}, \boldsymbol{\theta})\boldsymbol{\omega}^{\mathrm{l}} \tag{6.16}$$

corresponding to the fast dynamics. Subsequently, we consider the same limit case of an infinite energy recycle (i.e., $\varepsilon \to 0$) in the original time scale t, under the constraints (6.16), giving

$$\begin{aligned}\dot{\mathbf{x}} &= \mathbf{f}(\mathbf{x}, \boldsymbol{\theta}) \\ \dot{\boldsymbol{\theta}} &= \boldsymbol{\Phi}(\mathbf{x}, \boldsymbol{\theta}) + \boldsymbol{\Gamma}^{\mathrm{s}}(\mathbf{x}, \boldsymbol{\theta})\boldsymbol{\omega}^{\mathrm{s}} + \mathbf{B}(\mathbf{x}, \boldsymbol{\theta}) \lim_{\varepsilon \to 0} \frac{1}{\varepsilon} \tilde{\boldsymbol{\Gamma}}^{\mathrm{l}}(\mathbf{x}, \boldsymbol{\theta})\boldsymbol{\omega}^{\mathrm{l}} \\ 0 &= \tilde{\boldsymbol{\Gamma}}^{\mathrm{l}}(\mathbf{x}, \boldsymbol{\theta})\boldsymbol{\omega}^{\mathrm{l}}\end{aligned} \tag{6.17}$$

The terms $\lim_{\varepsilon \to 0}(1/\varepsilon)\tilde{\boldsymbol{\Gamma}}^{\mathrm{l}}(\mathbf{x}, \boldsymbol{\theta})\boldsymbol{\omega}^{\mathrm{l}}$ (which, being based on Equation (6.7), represent differences between large internal energy flows), become indeterminate in the slow time scale. These terms do, however, remain finite, and constitute an additional set of algebraic (rather than differential) variables in the model of the slow dynamics. On defining $\mathbf{z} = \lim_{\varepsilon \to 0}(1/\varepsilon)\tilde{\boldsymbol{\Gamma}}^{\mathrm{l}}(\mathbf{x}, \boldsymbol{\theta})\boldsymbol{\omega}^{\mathrm{l}}$, the reduced-order representation of the slow dynamics becomes

$$\begin{aligned}\dot{\mathbf{x}} &= \mathbf{f}(\mathbf{x}, \boldsymbol{\theta}) \\ \dot{\boldsymbol{\theta}} &= \boldsymbol{\Phi}(\mathbf{x}, \boldsymbol{\theta}) + \boldsymbol{\Gamma}^{\mathrm{s}}(\mathbf{x}, \boldsymbol{\theta})\boldsymbol{\omega}^{\mathrm{s}} + \mathbf{B}(\mathbf{x}, \boldsymbol{\theta})\mathbf{z} \\ 0 &= \tilde{\boldsymbol{\Gamma}}^{\mathrm{l}}(\mathbf{x}, \boldsymbol{\theta})\boldsymbol{\omega}^{\mathrm{l}}\end{aligned} \tag{6.18}$$

Equation (6.18) constitutes a differential algebraic representation of the slow dynamics of the process. This DAE is of high index: the algebraic variables $\mathbf{z}$ cannot be computed directly by solving the algebraic equations of the model. Rather, in order to compute $\mathbf{z}$ and, implicitly, to obtain an ODE representation of the slow dynamics, the algebraic constraints of Equation (6.18) must be differentiated with respect to the state variables once or several times. Prior to proceeding with this operation, we make the following remark and assumption.

Remark 6.2. *As mentioned above, the large energy flow rates $\boldsymbol{\omega}^{\mathrm{l}}$ present in the algebraic equations of (6.18) are assumed to be associated with material flow rates, whereby the material streams act as energy carriers. In light of these facts, $\boldsymbol{\omega}^{\mathrm{l}}$ can be viewed as a product of two separate terms, i.e., $\omega_i^{\mathrm{l}} = h_i F_i$, where F_i represents the material flow rate and h_i the specific (e.g., molar) enthalpy of stream i. Significant energy flow can thus originate either from a large (compared with the process feed flow rate) material flow or from a material stream having a much higher enthalpy (or temperature) than the feed and output streams.*

Assumption 6.5. *In view of Remark 6.2, it is assumed that all the material flow rates associated with $\boldsymbol{\omega}^{\mathrm{l}}$ are determined by appropriate functions of the process state variables (e.g., via feedback control laws, constitutive relations or pressure–flow correlations).*

The above assumption allows us to differentiate the algebraic equations in (6.18) with respect to the states. One differentiation yields

$$0 = L_{\boldsymbol{\Phi}(\mathbf{x},\boldsymbol{\theta})}\left(\tilde{\boldsymbol{\Gamma}}^{\mathrm{l}}(\mathbf{x},\boldsymbol{\theta})\boldsymbol{\omega}^{\mathrm{l}}\right) + L_{\boldsymbol{\Gamma}^{\mathrm{s}}(\mathbf{x},\boldsymbol{\theta})}\left(\tilde{\boldsymbol{\Gamma}}^{\mathrm{l}}(\mathbf{x},\boldsymbol{\theta})\boldsymbol{\omega}^{\mathrm{l}}\right)\boldsymbol{\omega}^{\mathrm{s}} + L_{\mathbf{B}(\mathbf{x},\boldsymbol{\theta})}\left(\tilde{\boldsymbol{\Gamma}}^{\mathrm{l}}(\mathbf{x},\boldsymbol{\theta})\boldsymbol{\omega}^{l}\right)\mathbf{z} \tag{6.19}$$

Assuming that the matrix $L_{\mathbf{B}(\mathbf{x},\boldsymbol{\theta})}(\mathbf{x},\boldsymbol{\theta})\left(\tilde{\boldsymbol{\Gamma}}^{\mathrm{l}}(\mathbf{x},\boldsymbol{\theta})\boldsymbol{\omega}^{\mathrm{l}}\right)$ is invertible, $\mathbf{z}$ can be computed as

$$\mathbf{z} = -L_{\mathbf{B}(\mathbf{x},\boldsymbol{\theta})}\left(\tilde{\boldsymbol{\Gamma}}^{\mathrm{l}}(\mathbf{x},\boldsymbol{\theta})\boldsymbol{\omega}^{l}\right)^{-1}\left[L_{\boldsymbol{\Phi}(\mathbf{x},\boldsymbol{\theta})}\left(\tilde{\boldsymbol{\Gamma}}^{\mathrm{l}}(\mathbf{x},\boldsymbol{\theta})\boldsymbol{\omega}^{\mathrm{l}}\right) + L_{\boldsymbol{\Gamma}^{\mathrm{s}}(\mathbf{x},\boldsymbol{\theta})}\left(\tilde{\boldsymbol{\Gamma}}^{\mathrm{l}}(\mathbf{x},\boldsymbol{\theta})\boldsymbol{\omega}^{\mathrm{l}}\right)\boldsymbol{\omega}^{\mathrm{s}}\right] \tag{6.20}$$

On substituting (6.20) into (6.18), we obtain a (non-minimal-order) ODE representation of the slow dynamics. A minimal-order ODE system can be obtained via a coordinate transformation (Section 2.3) involving the energy-balance equations and constraints:

$$\begin{bmatrix}\boldsymbol{\zeta}\\ \boldsymbol{\eta}\end{bmatrix} = \boldsymbol{T}(\mathbf{x},\boldsymbol{\theta})\begin{bmatrix}\boldsymbol{\delta}(\mathbf{x},\boldsymbol{\theta})\\ \tilde{\boldsymbol{\Gamma}}^{\mathrm{l}}(\mathbf{x},\boldsymbol{\theta})\boldsymbol{\omega}^{\mathrm{l}}\end{bmatrix} \tag{6.21}$$

under which the reduced-order model of the slow dynamics becomes

$$\begin{aligned}
\dot{\mathbf{x}} &= \mathbf{f}(\mathbf{x},\boldsymbol{\theta})\,|_{\boldsymbol{\theta}=\boldsymbol{T}^{-1}(\mathbf{x},\boldsymbol{\zeta})}\\
\dot{\boldsymbol{\zeta}} &= \frac{\partial\boldsymbol{\delta}}{\partial\boldsymbol{\theta}}\boldsymbol{\Phi}(\mathbf{x},\boldsymbol{\theta})\,|_{\boldsymbol{\theta}=\boldsymbol{T}^{-1}(\mathbf{x},\boldsymbol{\zeta})} + \frac{\partial\boldsymbol{\delta}}{\partial\boldsymbol{\theta}}\boldsymbol{\Gamma}^{\mathrm{s}}(\mathbf{x},\boldsymbol{\theta})\,|_{\boldsymbol{\theta}=\boldsymbol{T}^{-1}(\mathbf{x},\boldsymbol{\zeta})}\,\boldsymbol{\omega}^{\mathrm{s}}\\
&\quad + \frac{\partial\boldsymbol{\delta}}{\partial\boldsymbol{\theta}}\mathbf{B}(\mathbf{x},\boldsymbol{\theta})\mathbf{z}\,|_{\boldsymbol{\theta}=\boldsymbol{T}^{-1}(\mathbf{x},\boldsymbol{\zeta})}\\
\boldsymbol{\eta} &= 0
\end{aligned} \tag{6.22}$$

From the preceding discussion regarding the equilibrium manifold of the fast component of the system dynamics having a maximum dimension of one, we can infer that, in effect, $\boldsymbol{\zeta} \in \mathcal{Q}^1 \subset \mathbb{R}^1$.

Remark 6.3. *The vector function $\boldsymbol{\delta}(\mathbf{x},\boldsymbol{\theta})$ can be arbitrarily chosen (as long as the invertibility of $\boldsymbol{T}(\mathbf{x},\boldsymbol{\theta})$ is preserved), which allows us to describe the slow component of the energy dynamics in terms of the enthalpy/temperature of any one of the units. Furthermore, $\boldsymbol{\delta}(\mathbf{x},\boldsymbol{\theta})$ may be chosen in such a way that $(\partial\boldsymbol{\delta}/\partial\boldsymbol{\theta})\mathbf{B}(\mathbf{x},\boldsymbol{\theta}) = 0$. In this case, the model (6.18) will be independent of $\mathbf{z}$ and the corresponding ζ represents a true "slow" variable in the system (whereas the original state variables evolve both in the fast and in the slow time scales). For example, on choosing $\boldsymbol{\delta}(\mathbf{x},\boldsymbol{\theta})$ as the sum of all the unit enthalpies (Equation (6.13)), it can be shown that indeed $(\partial\boldsymbol{\delta}/\partial\boldsymbol{\theta})\mathbf{B}(\mathbf{x},\boldsymbol{\theta}) = 0$. Thus, the total enthalpy of the process evolves only over a slow time scale.*

6.4 Control implications

The analysis presented in Section 6.3 demonstrates that the variables in the energy-balance equations of integrated process systems with significant energy recycling evolve over two time horizons. The transient behavior of the enthalpies of the individual units that are part of the energy recycle loop is fast, whereas the evolution of the total enthalpy of the process occurs over a slower time scale. While these results are analogous to those derived in Chapter 3 for processes with *material* recycling, the control insights that we derive from the reduced-order models (6.12) and (6.22) are somewhat different.

- The fast model (6.12) describes the evolution of the enthalpies/temperatures of the individual units. Thus, control objectives related to the individual units (e.g., reactor temperature control) should be addressed in this time scale. The significant energy flows $\boldsymbol{\omega}^1$ associated with the internal streams are available as manipulated inputs to this end. Note that it is often practical to vary $\boldsymbol{\omega}^1$ by modifying a material flow rate, rather than by varying a stream's enthalpy/temperature.
- According to our analysis, the slow component of the energy dynamics can be characterized in terms of the enthalpy (or temperature) of any of the individual units, which, in turn, can constitute a control objective in the slow time scale. However, Remark 6.3 allows us to regard the control problem in the slow time scale in a more general context. The fact that the total process enthalpy (which is itself difficult to measure and, thus, impractical to control) is a true "slow" variable of the process suggests that the control objectives to be addressed over a long time horizon should instead pertain to energy management. Applications include high-value activities such as monitoring the energy efficiency of a process (or even an entire manufacturing site) as well as process-wide energy optimization. Intuitively, Equation (6.22) indicates that the small inlet and outlet energy flows are available and should be used as manipulated inputs to this end.

6.5 Illustrative examples

As demonstrated in Chapters 3–5, the variables of the material balance of an integrated process system can themselves exhibit a dynamic behavior with up to three time scales. As such, it is intuitive that, when considering both the energy-balance and the material-balance equations, the dynamics of integrated process systems can span *several* (i.e., more than three) time scales. For clarity, this aspect was not directly accounted for in the theoretical analysis presented above. We utilize the following examples to illustrate the concepts developed so far, as well as to confirm this observation.

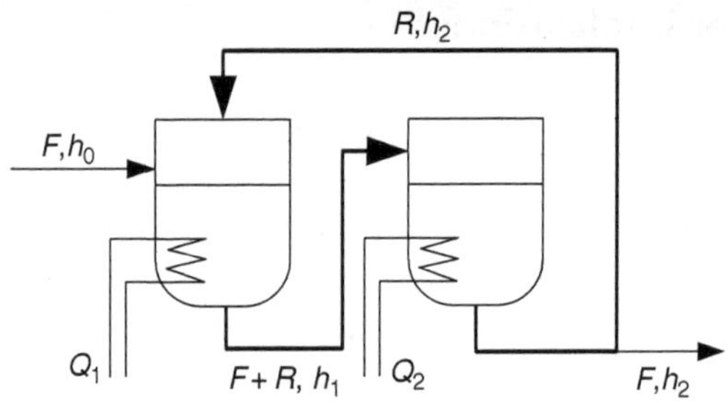

Figure 6.3 Two heated tanks in series with recycle.

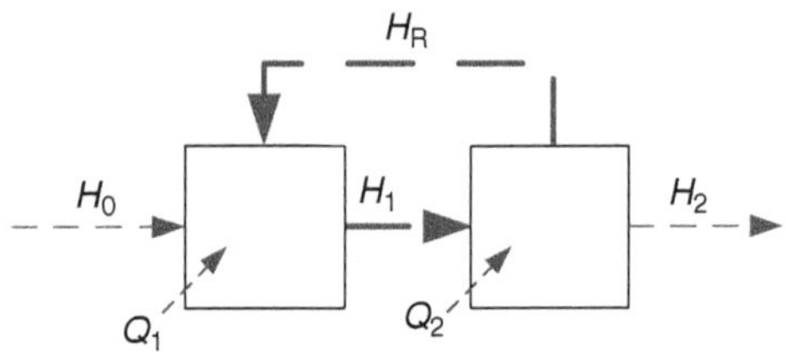

Figure 6.4 Energy flows for heated tanks in series.

6.5.1 Cascade of heated tanks

We begin with the analysis of a basic system consisting of two heated tanks interconnected via a material recycle stream, which acts as an energy carrier, as in Figure 6.3. Let F be the molar feed flow rate to the first tank (with molar enthalpy h_0), R the molar recycle flow rate, h_1 and h_2 the molar enthalpies, and Q_1 and Q_2 the rates at which heat is input to tanks 1 and 2.

Considering that the holdups in the two vessels remain constant and that physical properties, such as the heat capacity and density, are temperature-independent, and using the notation in Figure 6.4, we can write the energy balance of the system as

$$\begin{aligned}\dot{\theta}_1 &= H_0 + H_{\mathrm{R}} - H_1 + Q_1 \\ \dot{\theta}_2 &= H_1 - H_{\mathrm{R}} - H_2 + Q_2\end{aligned} \tag{6.23}$$

where θ_1 and θ_2 are the enthalpies of tanks 1 and 2.

In this case, Assumption 6.2 translates to $Q_1/H_0 = \mathcal{O}(1)$, $Q_2/H_0 = \mathcal{O}(1)$. Let us now assume that the rate at which material is recycled is very high, i.e., $F_{\mathrm{s}}/R_{\mathrm{s}} = \varepsilon \ll 1$. Observing that the energy flows H_i are, in effect, a product of a material flow rate and a specific enthalpy, i.e., $H_i = F_i h_i$, (6.23) becomes

$$\begin{aligned}\dot{\theta}_1 &= Fh_0 - Fh_1 - \frac{1}{\varepsilon}F_{\mathrm{s}}u_{\mathrm{r}}(h_1 - h_2) + Q_1 \\ \dot{\theta}_2 &= Fh_1 + \frac{1}{\varepsilon}F_{\mathrm{s}}u_{\mathrm{r}}(h_1 - h_2) - Fh_2 + Q_2\end{aligned} \tag{6.24}$$

where $u_{\mathrm{r}} = R/R_{\mathrm{s}}$.

Proceeding with the analysis as outlined in Section 6.3, we derive a fast component of the energy dynamics in the form

$$\begin{aligned}\frac{d\theta_1}{d\tau} &= F_s u_r (h_2 - h_1)\\ \frac{d\theta_2}{d\tau} &= F_s u_r (h_1 - h_2)\end{aligned} \tag{6.25}$$

with the associated quasi-steady-state constraint

$$0 = h_1 - h_2 \tag{6.26}$$

and a 1D slow component of the dynamics described by

$$\frac{d(\theta_1 + \theta_2)}{dt} = F h_0 - F h_2 + Q_1 + Q_2 \tag{6.27}$$

Clearly, this very simple process belongs to the class considered in Section 6.2. From the defining assumptions, we can expect that the enthalpy in the second tank will not differ significantly from the enthalpy of the feed stream. Thus, the energy recovery number of the process,

$$\textsf{Erc} = \frac{R_s h_2}{F_s h_s} \cong \frac{R_s}{F_s} \gg 1 \tag{6.28}$$

is very large. Moreover, we have shown that the enthalpies of the individual tanks (θ_1, and θ_2) evolve both in a fast and in a slow time scale, while the total enthalpy ($\theta_1 + \theta_2$) evolves only in the slow time scale.

A similar result was reported earlier (Georgakis 1986), when an eigenvalue analysis was used to prove that a time-scale separation is present in the transient evolution of the states θ_1 and θ_2. It is noteworthy, however, that, in contrast to the approach presented in this chapter, an eigenvalue analysis does not provide a means by which to derive physically meaningful reduced-order models for the dynamics in each time scale.

Remark 6.4. *In this case, significant energy recycling is achieved through the presence of a large material recycle stream acting as an energy carrier. This situation may naturally arise when (slow) chemical reactions occur in the vessels; the large material recycle stream would thus also introduce a time-scale separation in the material-balance dynamics, as pointed out in Chapter 3.*

6.5.2 Processes with feed–effluent heat exchange

Energy exchange between effluent and feed streams is a key feature in integrated process designs. Feed–effluent heat exchangers (FEHEs), such as the one in the generic process of Figure 6.2, are frequently used in the case of processes with mildly exothermic or endothermic effects, which operate at temperatures significantly different from that at which the feed streams are normally available. Typical examples include high-temperature reactions (in which heat is recovered

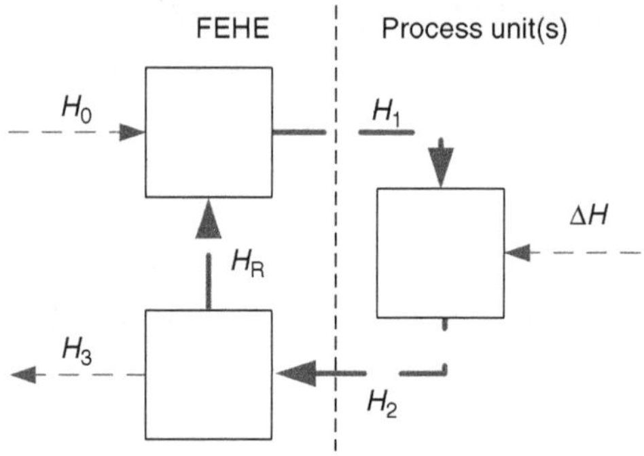

Figure 6.5 Energy flows in a process with a feed–effluent heat exchanger.

to warm the feed stream to the reaction temperature) and cryogenic processes, whereby refrigeration is recovered from the products to cool feed streams that are initially at ambient temperature.

The steady-state operation of process systems with FEHEs relies on recovering a large proportion of the energy of the effluent, effectively creating an energy recycle loop between the FEHE and the process units, as depicted in Figure 6.5. Evidently, this cannot be the case during significant transitions such as start-up procedures, when external energy input is required. Likewise, controllability considerations require that additional manipulated inputs (such as bypass flows or additional energy sources) be present. These aspects will be considered in a case study later in the book; in what follows, we analyze, without loss of generality, a process of the type depicted in Figure 6.5.

For simplicity, we assume that the holdups of all the units are constant and that the temperature dependence of physical properties such as density and heat capacity is negligible. Under these assumptions, the energy-balance equations for the process can be written as

$$
\begin{aligned}
\dot{\theta}_1 &= H_0 - H_1 + H_{\mathrm{R}} \\
\dot{\theta}_2 &= H_1 - H_2 + \Delta H \\
\dot{\theta}_3 &= H_2 - H_3 - H_{\mathrm{R}}
\end{aligned}
\tag{6.29}
$$

with θ_1, θ_2, and θ_3 being the enthalpies of the cold leg of the FEHE, overall process units, and the warm leg of the FEHE, respectively. Prior to proceeding with our analysis, we make the following set of assumptions regarding the steady-state operation of the process under consideration.[1]

- The process units operate at an elevated temperature (much higher than the temperature at which the feed is available). Hence, the specific enthalpy h_1 of

[1] Note that, since in this case the material flow rate is the same through all the units of the process, we can formulate most arguments in terms of specific stream enthalpies h_i, rather than resorting to energy flows.

the feed stream of the process units is much higher than the specific enthalpy h_0 of the FEHE feed stream, and the process has a high energy recovery number,

$$\mathsf{Erc} = \frac{Fh_{1,\mathrm{s}}}{Fh_{0,\mathrm{s}}} \gg 1 \tag{6.30}$$

or, equivalently, $\varepsilon = h_{0,\mathrm{s}}/h_{1,\mathrm{s}} \ll 1$.

- The thermal effect of any reactions is mild and does not induce a phase change or a significant change in the temperature of the process units. Hence, the specific enthalpies of the feed and effluent streams, h_1 and h_2, of the process units are of comparable magnitude: $k_2 = h_{2,\mathrm{s}}/h_{1,\mathrm{s}} = \mathcal{O}(1)$. (Note that this assumption is by no means restrictive; rather, it reflects current industrial practice. For example, the use of adiabatic units in highly exothermic processes is avoided for safety reasons, and external cooling systems are preferred. This issue is addressed in detail in Chapter 7.)
- The FEHE achieves significant energy recovery. Hence, the energy flow H_R is of comparable magnitude to H_1, $k_\mathrm{R} = H_{\mathrm{R,s}}/H_{1,\mathrm{s}} = \mathcal{O}(1)$, and the steady-state specific enthalpies h_0 and h_3 of the inlet and outlet streams are of comparable magnitude, $h_{0,\mathrm{s}}/h_{3,\mathrm{s}} = \mathcal{O}(1)$.

Under these assumptions (and accounting for the fact that the energy flows H_i associated with the material streams can be written as products of the material flow rate F and a specific enthalpy, $H_i = Fh_i$), the model of Equation (6.29) becomes

$$\begin{aligned}
\dot{\theta}_1 &= Fh_0 + \frac{1}{\varepsilon}Fh_{0\mathrm{s}}(-u_{H1} + k_\mathrm{R}u_{H\mathrm{R}}) \\
\dot{\theta}_2 &= \frac{1}{\varepsilon}Fh_{0\mathrm{s}}(u_{H1} - k_2u_{H2}) + \Delta H \\
\dot{\theta}_3 &= \frac{1}{\varepsilon}Fh_{0\mathrm{s}}(k_2u_{H2} - k_\mathrm{R}u_{H\mathrm{R}}) - Fh_3
\end{aligned} \tag{6.31}$$

with $u_{H1} = h_1/h_{1,\mathrm{s}}$, $u_{H\mathrm{R}} = H_\mathrm{R}/H_{\mathrm{R,s}}$ and $u_{H2} = h_2/h_{2,\mathrm{s}}$. We can now apply the method developed in Section 6.3 to obtain a description of the fast component of the energy dynamics:

$$\begin{aligned}
\frac{d\theta_1}{d\tau} &= Fh_{0\mathrm{s}}(-u_{H1} + k_\mathrm{R}u_{H\mathrm{R}}) \\
\frac{d\theta_2}{d\tau} &= Fh_{0\mathrm{s}}(u_{H1} - k_2u_{H2}) \\
\frac{d\theta_3}{d\tau} &= Fh_{0\mathrm{s}}(k_2u_{H2} - k_\mathrm{R}u_{H\mathrm{R}})
\end{aligned} \tag{6.32}$$

having an equilibrium manifold described by the (linearly dependent) equations

$$\begin{aligned}
0 &= -u_{H1} + k_\mathrm{R}u_{H\mathrm{R}} \\
0 &= u_{H1} - k_2u_{H2} \\
0 &= k_2u_{H2} - k_\mathrm{R}u_{H\mathrm{R}}
\end{aligned} \tag{6.33}$$

Finally, the total enthalpy $\theta_1 + \theta_2 + \theta_3$ can be used to derive a representation of the slow component of the process dynamics:

$$\frac{d(\theta_1 + \theta_2 + \theta_3)}{dt} = F(h_0 - h_3) + \Delta H \tag{6.34}$$

The results above indicate that, in the case of processes that rely on a FEHE to achieve a high degree of energy integration, the two-time-scale behavior of the variables in the energy balance is due to large discrepancies between the *temperatures* of the internal streams and those of the inlet/outlet streams. Thus (and in contrast to the example discussed in Section 6.5.1), the presence of a large energy recycle is, in itself, unlikely to be directly related to stiffness at the level of the *material-balance* equations.

Note that these results translate directly to cryogenic processes with FEHE, for which the temperature (enthalpy) of the internal material streams is significantly lower than the temperature (enthalpy) of the process inlet and outlet streams (see Example 4.1).

Remark 6.5. *With a slight modification of Definition 6.1, that is,*

$$\mathsf{Erc} = \frac{Fh_{1,\mathrm{s}}}{Fh_{0,\mathrm{s}} + \Delta H_\mathrm{s}} \tag{6.35}$$

the energy recovery number reflects the ratio of the rate at which energy is recovered from the effluent of the process units via the FEHE and the sum of the rates at which energy is supplied to the plant via the inlet stream and generated internally. Thus, **Erc** *emerges naturally as a figure of economic merit for energy integration: a process with a high* **Erc** *can be regarded as more effectively integrated and will likely have lower operating costs (owing, e.g., to a diminished requirement on the amount of external energy used to preheat the feed) than a similar process with a lower* **Erc**. *Conversely, improved energy recovery may entail a higher capital cost (e.g., by requiring a larger heat-transfer area), and it is to be expected that the economically optimal plant design for a given process will entail a capital and operating cost tradeoff.*

6.5.3 Energy-integrated distillation

Distillation is one of the most energy-intensive operations in the process industries, and the prospect of improving the energy efficiency of distillation columns by integrating the condensation of the vapor stream with boiling the column bottoms has been considered early on (King 1980). One such integrated scheme, vapor recompression distillation (VRD), relies on heat pumping (Figure 6.6) for transferring heat from the top (vapor) and bottoms streams of a column. To make heat transfer thermodynamically possible, the vapor stream exiting the top of the distillation column is compressed in the compressor CC, then condensed in

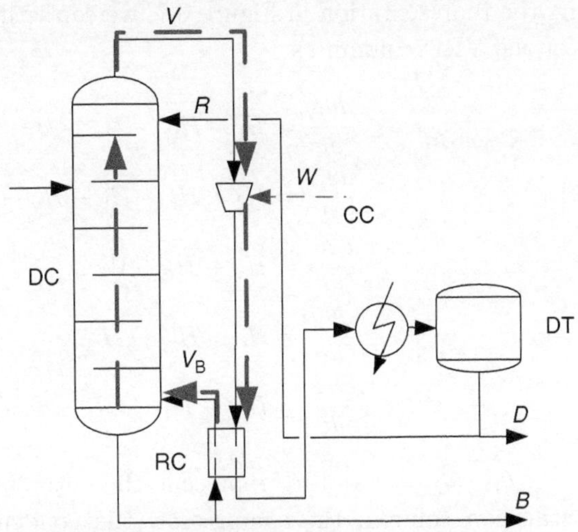

Figure 6.6 A vapor recompression distillation (VRD) column.

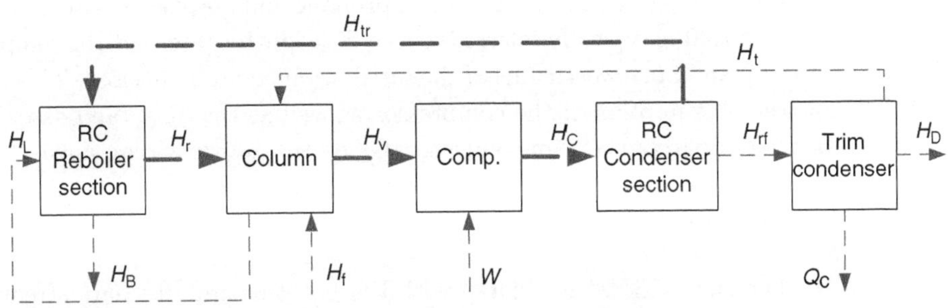

Figure 6.7 A block diagram of energy flows in the VRD column.

the reboiler–condenser RC against the boiling bottoms stream. The vaporized bottoms stream then enters the stripping section of the column. A trim condenser TC is used to adjust the temperature of the reflux to the required value as well as to condense any residual vapor. The condensed distillate is stored in the overhead accumulator DT.

A block diagram representation of the energy flows in the process is presented in Figure 6.7. For modeling purposes, the condenser section and the reboiler section of the reboiler–condenser are represented separately and connected via the corresponding exchanger duty H_{tr}. H_t represents energy flow associated with the column reflux (which can be regarded as a material recycle stream), H_l represents the energy flow associated with the liquid entering the reboiler, H_f, H_r, H_v, H_c, H_{rf}, H_D, and H_B represent, respectively, the energy flows associated with the feed, the vapor leaving the reboiler–condenser, the vapor leaving the column, the vapor leaving the compressor, the condensate leaving the reboiler–condenser, the bottoms stream, and the distillate stream. W is the compressor power input and Q_c is the trim condenser duty.

Using the representation in Figure 6.7, we can write the energy-balance equations for the VRD column as

$$
\begin{aligned}
\frac{d\theta_{\mathrm{R}}}{dt} &= H_{\mathrm{l}} + H_{\mathrm{tr}} - H_{\mathrm{r}} - H_{\mathrm{B}} \\
\frac{d\theta_{\mathrm{D}}}{dt} &= H_{\mathrm{f}} + H_{\mathrm{r}} + H_{\mathrm{t}} - H_{\mathrm{v}} - H_{\mathrm{l}} \\
\frac{d\theta_{\mathrm{C}}}{dt} &= H_{\mathrm{v}} - H_{\mathrm{c}} + W \\
\frac{d\theta_{\mathrm{F}}}{dt} &= H_{\mathrm{c}} - H_{\mathrm{tr}} - H_{\mathrm{rf}} \\
\frac{d\theta_{\mathrm{T}}}{dt} &= H_{\mathrm{rf}} - H_{\mathrm{D}} - H_{\mathrm{t}} - Q_{\mathrm{c}}
\end{aligned}
\tag{6.36}
$$

where θ_{R}, θ_{D}, θ_{C}, θ_{F}, and θ_{T} represent the enthalpies of the reboiler section, the distillation column, the compressor, the condenser section, and the trim condenser, respectively.

The economics of VRD favors separations involving components with similar boiling points (e.g., the separation of propane and propylene in an oil refinery, in a column that is typically referred to as a "C3 splitter") so that the temperatures of the top and bottom streams of the distillation column are close. This reduces the power consumption of the compressor as well as the duty (and associated heat-transfer area) of the trim condenser. Given the above, we can make the following observations and assumptions concerning the various steady-state energy flows in the process.

- The energy flows associated with the bottoms and distillate streams, respectively $H_{\mathrm{B,s}}$ and $H_{\mathrm{D,s}}$, are of the same order of magnitude as the energy flow associated with the feed, $H_{\mathrm{f,s}}$.
- The amounts of work done by the compressor, W, and the trim condenser duty Q_{c} are small and can be assumed to be of the same order of magnitude as that of the feed, i.e., $W_{\mathrm{s}}/H_{\mathrm{f,s}} = \mathcal{O}(1)$ and $Q_{\mathrm{c,s}}/H_{\mathrm{f,s}} = \mathcal{O}(1)$.
- The latent-heat contribution to the energy flow associated with a material stream is generally significantly larger than the sensible-heat contribution. The heat transfer across the reboiler–condenser is dominated by latent heat, and the reboiler–condenser duty H_{tr} is therefore much larger than the energy flow associated with the feed stream. We can thus compute the energy recovery number for VRD as

$$
\mathsf{Erc}_{\mathrm{VRD}} = \frac{H_{\mathrm{tr,s}}}{H_{\mathrm{in,s}}} \gg 1 \tag{6.37}
$$

  This observation also implies that the internal energy flows ($H_{\mathrm{f,s}}$, $H_{\mathrm{v,s}}$, and $H_{\mathrm{c,s}}$) are comparable in magnitude to $H_{\mathrm{tr,s}}$.
- The energy flow associated with the material recycle stream $H_{\mathrm{t,s}}$ can be of larger magnitude than the energy flow associated with the feed, depending on

the magnitude of the reflux ratio. However, it will always be of lower magnitude than $H_{\mathrm{tr,s}}$ due to the latent-heat contribution. In the present analysis we assume for simplicity that $H_{\mathrm{t,s}}$ is of comparable magnitude to the feed, and we make a similar argument for $H_{\mathrm{l,s}}$.

From the above, we can define the $\mathcal{O}(1)$ steady-state ratios $k_{\mathrm{r}} = H_{\mathrm{r,s}}/H_{\mathrm{tr,s}}$, $k_{\mathrm{c}} = H_{\mathrm{c,s}}/H_{\mathrm{tr,s}}$, and $k_{\mathrm{v}} = H_{\mathrm{v,s}}/H_{\mathrm{tr,s}}$, and the scaled energy flows $u_{\mathrm{tr}} = H_{\mathrm{tr}}/H_{\mathrm{tr,s}}$, $u_{\mathrm{r}} = H_{\mathrm{r}}/H_{\mathrm{r,s}}$, $u_{\mathrm{c}} = H_{\mathrm{c}}/H_{\mathrm{c,s}}$, and $u_{\mathrm{v}} = H_{\mathrm{v}}/H_{\mathrm{v,s}}$. We also define

$$\varepsilon = \frac{1}{\mathsf{Erc}_{\mathrm{VRD}}} \ll 1 \tag{6.38}$$

With this notation, (6.36) becomes

$$\begin{aligned}
\frac{d\theta_{\mathrm{R}}}{dt} &= H_{\mathrm{f,s}}\left[\frac{u_{\mathrm{tr}}}{\varepsilon} - \frac{k_{\mathrm{r}}u_{\mathrm{r}}}{\varepsilon}\right] + H_{\mathrm{l}} - H_{\mathrm{B}} \\
\frac{d\theta_{\mathrm{D}}}{dt} &= H_{\mathrm{f,s}}\left[\frac{k_{\mathrm{r}}u_{\mathrm{r}}}{\varepsilon} - \frac{k_{\mathrm{v}}u_{\mathrm{v}}}{\varepsilon}\right] + H_{\mathrm{f}} + H_{\mathrm{t}} - H_{\mathrm{l}} \\
\frac{d\theta_{\mathrm{C}}}{dt} &= H_{\mathrm{f,s}}\left[\frac{k_{\mathrm{v}}u_{\mathrm{v}}}{\varepsilon} - \frac{k_{\mathrm{c}}u_{\mathrm{c}}}{\varepsilon}\right] + W \\
\frac{d\theta_{\mathrm{F}}}{dt} &= H_{\mathrm{f,s}}\left[\frac{k_{\mathrm{c}}u_{\mathrm{c}}}{\varepsilon} - \frac{u_{\mathrm{tr}}}{\varepsilon}\right] - H_{\mathrm{rf}} \\
\frac{d\theta_{\mathrm{T}}}{dt} &= H_{\mathrm{rf}} - H_{\mathrm{D}} - H_{\mathrm{t}} - Q_{\mathrm{C}}
\end{aligned} \tag{6.39}$$

Equation (6.39) is in the form of the energy balance in (6.10), with

$$\begin{aligned}
\boldsymbol{\theta} &= \left[\theta_{\mathrm{R}}\ \theta_{\mathrm{D}}\ \theta_{\mathrm{C}}\ \theta_{\mathrm{F}}\ \theta_{\mathrm{T}}\right]^{\mathrm{T}} \\
\boldsymbol{\omega}^{\mathrm{s}} &= \left[H_{\mathrm{f}}\ H_{\mathrm{D}}\ H_{\mathrm{B}}\ W\ Q_{\mathrm{c}}\ H_{\mathrm{l}}\ H_{\mathrm{t}}\ H_{\mathrm{rf}}\right]^{\mathrm{T}} \\
\boldsymbol{\omega}^{\mathrm{l}} &= \left[u_{\mathrm{tr}}\ u_{\mathrm{r}}\ u_{\mathrm{v}}\ u_{\mathrm{c}}\right]^{\mathrm{T}}
\end{aligned} \tag{6.40}$$

6.6 Case study: control of a reactor–FEHE process

6.6.1 Process description

We consider a plant designed to convert a feed stream rich in compound A (of molar concentration c_{A0}) into compound B in a high-temperature, mildly exothermic, first-order reaction carried out in an adiabatic reactor (Figure 6.8). For improved operability, the plant features a heater that is used at full capacity in startup mode and as a trim heater during operation, as well as a bypass stream that is used to regulate heat recovery in the FEHE.

The operational objectives for this process are the stabilization of the reactor holdup and operating temperature, as well as controlling the product stream's temperature and maintaining its purity within the specifications of a downstream

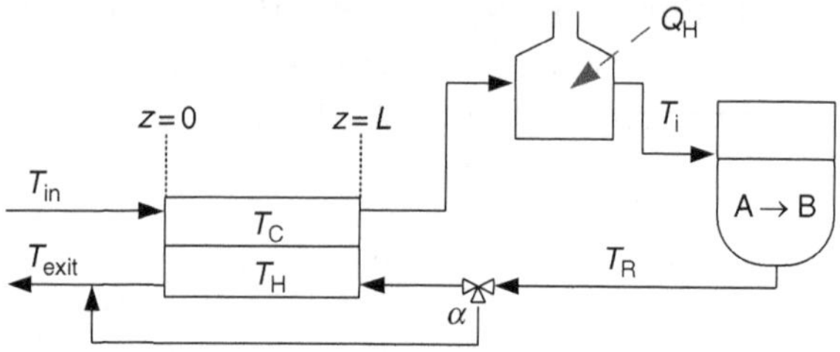

Figure 6.8 A reactor–FEHE process with a fired furnace and an exchanger bypass stream.

separation system. The reactor effluent flow rate, the bypass flow rate, and the heater duty are available as manipulated inputs.

Assuming that the temperature and composition dependence of the heat capacity and density of the process streams is not significant, that there is no phase change in the heat exchanger, and that the material holdup in both legs of the heat exchanger and in the heater is constant, the model of the process can be written as

$$
\begin{aligned}
\frac{dV}{dt} &= F_{\mathrm{in}} - F \\
\frac{dc_{\mathrm{A}}}{dt} &= \frac{F_{\mathrm{in}}}{V}(c_{\mathrm{A0}} - c_{\mathrm{A}}) - k_0 e^{-E/(RT_{\mathrm{R}})} c_{\mathrm{A}} \\
\frac{dT_{\mathrm{R}}}{dt} &= \frac{F_{\mathrm{in}}}{V}(T_{\mathrm{i}} - T_{\mathrm{R}}) - \frac{1}{\rho C_p} k_0 e^{-E/(RT_{\mathrm{R}})} c_{\mathrm{A}} \, \Delta H \\
\frac{\partial T_{\mathrm{H}}}{\partial t} &= -v_{\mathrm{H}} \frac{\partial T_{\mathrm{H}}}{\partial z} - \frac{UA}{\rho C_p} \frac{T_{\mathrm{H}} - T_{\mathrm{C}}}{V_{\mathrm{H}}} \\
\frac{\partial T_{\mathrm{C}}}{\partial t} &= -v_{\mathrm{C}} \frac{\partial T_{\mathrm{C}}}{\partial z} + \frac{UA}{\rho C_p} \frac{T_{\mathrm{H}} - T_{\mathrm{C}}}{V_{\mathrm{C}}} \\
\frac{dT_{\mathrm{i}}}{dt} &= \frac{F_{\mathrm{in}}}{V_{\mathrm{f}}}(T_{\mathrm{C},z=L} - T_{\mathrm{i}}) + \frac{Q_{\mathrm{H}}}{\rho C_p V_{\mathrm{f}}}
\end{aligned} \tag{6.41}
$$

with the boundary conditions

$$T_{\mathrm{H},z=L} = T_{\mathrm{R}} \tag{6.42}$$

$$T_{\mathrm{C},z=0} = T_{\mathrm{in}} \tag{6.43}$$

In the model above, F_{in} is the volumetric flow rate of the input stream to the system and F is the reactor outlet flow rate; V, V_{H}, V_{C}, and V_{f} are the holdups of the reactor, the FEHE hot and cold legs, and the furnace, respectively. T_{in} is the cold-leg inlet temperature to the FEHE. T_{H} and T_{C} are the hot- and cold-leg temperatures for the FEHE. T_{i} and T_{R} are the reactor-inlet and -outlet temperatures, and T_{exit} is the temperature of the product stream exiting the process. Q_{H} is the furnace heat duty, v_{H} and v_{C} are the velocities of the fluid in the hot and cold compartments of FEHE, L is the length of the exchanger,

Table 6.1. Nominal values of process parameters for the reactor–FEHE process

k_0	1.26×10^7 s^{-1}	c_A	55.19 mol/l
E	1.42×10^5 J/mol	T_R	922.00 K
ΔH	-54.82 kJ/mol	T_i	909.62 K
ρC_p	4.184×10^6 J m^{-3} K^{-1}	T_{exit}	363.53 K
UA	83.68×10^3 W/K	$T_{C,z=0}$	860.53 K
α	0.10	Q_H	118.42 kW
V	0.10 m^3	T_{in}	300.00 K
V_H	0.10 m^3	F_{in}	5.76×10^{-4} m^3/s
V_C	0.09 m^3	F	5.76×10^{-4} m^3/s
V_f	0.01 m^3	c_{A0}	1000.00 mol/m^3
L	5.00 m		

and z is the spatial coordinate. R is the universal gas constant, U is the heat-transfer coefficient and A is the heat-transfer area in the FEHE, and k_0 is the pre-exponential factor of the reaction rate constant; $k_0 e^{-E/(RT_R)}$ is the reaction rate constant. The nominal values of the process variables and parameters are presented in Table 6.1.

The FEHE bypass ratio, α, is defined as

$$\alpha = \frac{\text{flow rate of the bypass stream}}{\text{flow rate of the feed stream } (F)} \tag{6.44}$$

Using this definition and the assumption that the heat capacity and density of the streams do not change, we can calculate the temperature of the stream exiting the process as

$$T_{exit} = \alpha T_R + (1 - \alpha) T_{H,z=0} \tag{6.45}$$

6.6.2 System analysis

In order to carry out numerical simulations, the spatial derivatives in Equation (6.41) were discretized using a backwards finite-difference scheme on a grid of 1001 nodes (i.e., by dividing the exchanger length into 1000 discretization intervals). It was verified that increasing the number of nodes does not result in significant changes in the simulation results and that the energy balance at the process level is closed (a potential pitfall associated with the discretization of partial differential equations).

The temperature profiles of the heat exchanger (Figure 6.9) show a pinch at the cold end (the temperature of the hot stream leaving the exchanger $T_{H,z=0} =$ 301.48 K is very close to the inlet temperature). This indicates that the process is tightly integrated, with a high degree of energy recovery.

A dynamic simulation of the 2006-dimensional ODE system from initial conditions slightly perturbed from the values in Table 6.1 confirmed the theoretical

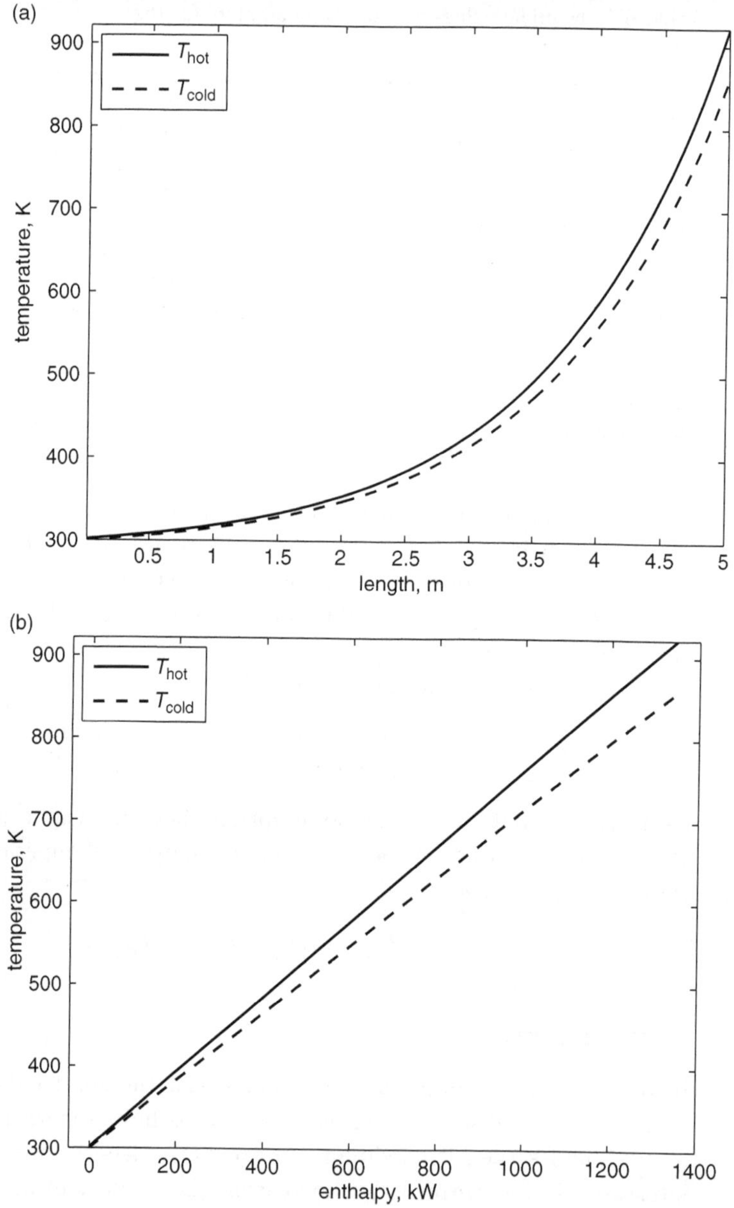

Figure 6.9 Temperature profiles in the FEHE: (a) longitudinal temperature profile and (b) temperature–enthalpy diagram.

analysis developed earlier in this chapter. As is typical of processes with tight energy integration, the temperatures of the individual units exhibit a two-time-scale behavior: the corresponding plots (Figures 6.10 and 6.11(a)) show a fast transient, followed by a slow approach to steady state. Conversely, as anticipated in Section 6.5.2, the variation of the total enthalpy (Figure 6.11(b)) is slow.

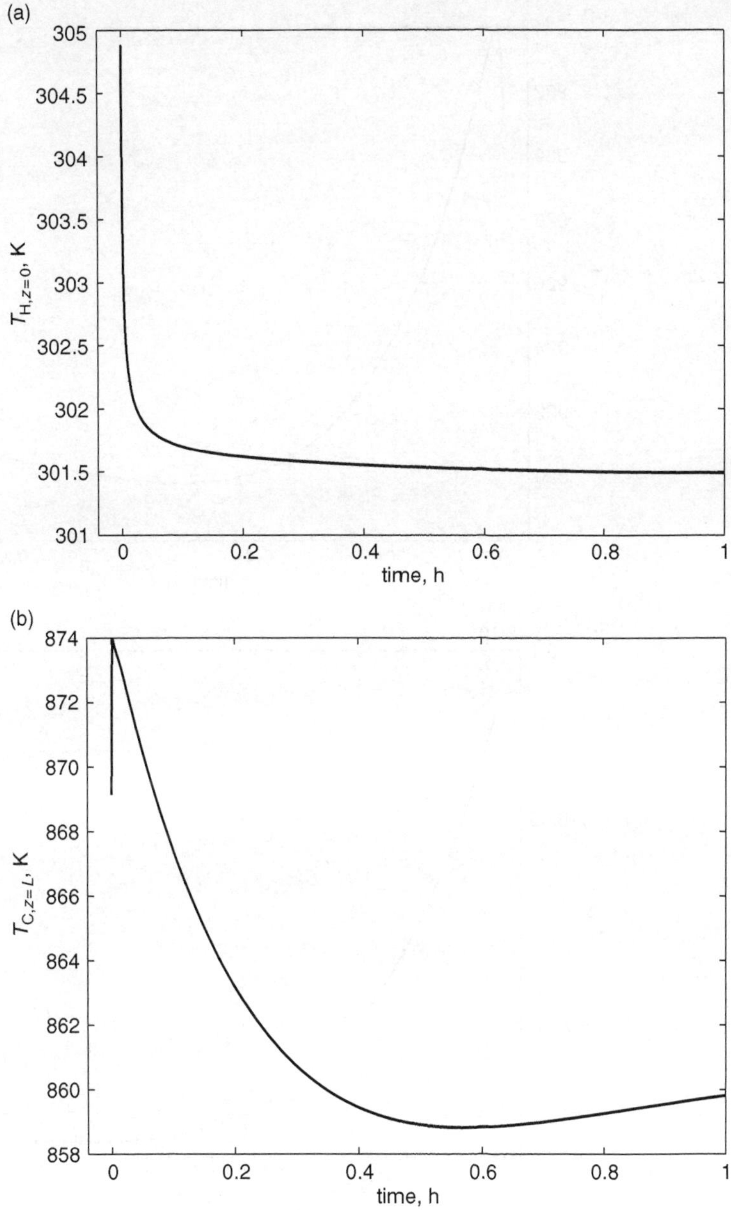

Figure 6.10 Evolution of (a) the hot-leg and (b) the cold-leg outlet temperatures for the perturbed system.

6.6.3 Reduced-order modeling

The results of the analysis above indicate that the process under consideration belongs to the general category of processes with high energy recycle. The rational development of a control framework would therefore entail a time-scale

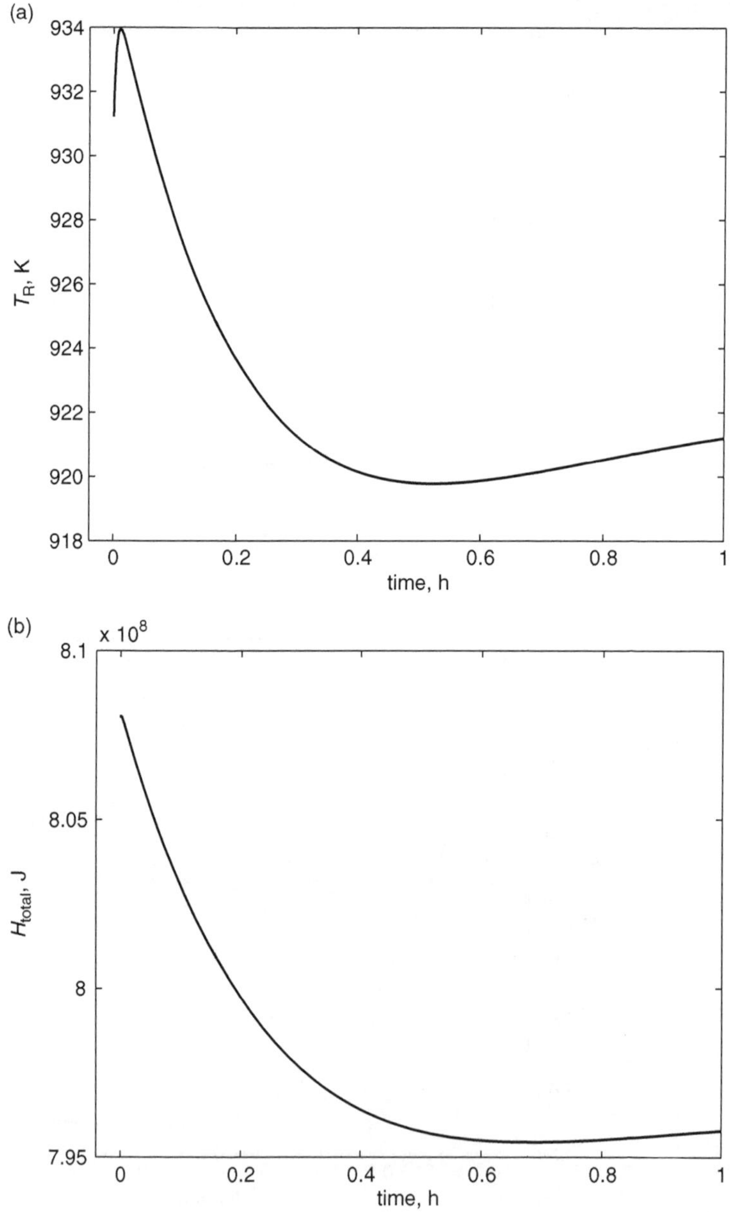

Figure 6.11 Evolution of (a) the reactor temperature and (b) the total enthalpy for the perturbed system.

decomposition and model reduction, as discussed earlier in this chapter. However, the presence of the partial-differential advection terms in Equation (6.41) prevents us from applying the methods developed in Section 6.3 directly. In order to circumvent this difficulty, let us first develop an alternative modeling construct that is based entirely on ODEs.

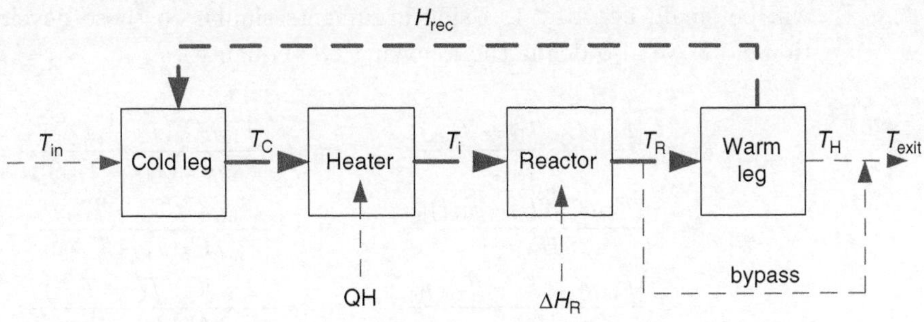

Figure 6.12 Block diagram of the reactor–FEHE process. H_{rec} represents the duty of the heat exchanger.

Figure 6.12 provides a block-diagram representation of the energy flows in the process. Postponing the derivation of an expression of the heat-exchanger duty H_{rec} (which is, evidently, central to this process and was originally captured in the partial differential equations) until later in this section, we can use this representation to develop a lumped-parameter model of the process:

$$\begin{aligned}
\frac{dV}{dt} &= F_{in} - F \\
\frac{dc_A}{dt} &= \frac{F_{in}}{V}(c_{A0} - c_A) - k_0 e^{-E/(RT_R)} c_A \\
\frac{dT_R}{dt} &= \frac{F_{in}}{V}(T_i - T_R) - \frac{1}{\rho C_p} k_0 e^{-E/(RT_R)} c_A \,\Delta H \\
\frac{dT_H}{dt} &= \frac{F_{in}(1-\alpha)}{V_H}(T_R - T_H) - \frac{H_{rec}}{\rho C_p V_H} \\
\frac{dT_C}{dt} &= \frac{F_{in}}{V_C}(T_{in} - T_C) + \frac{H_{rec}}{\rho C_p V_C} \\
\frac{dT_i}{dt} &= \frac{F_{in}}{V_f}(T_C - T_i) + \frac{Q_H}{\rho C_p V_f}
\end{aligned} \tag{6.46}$$

On the basis of this representation, we can define the energy recovery number:

$$\mathbf{Erc} = \frac{H_{rec,s}}{[F_{in}\rho C_p(T_{in} - T_{ref})]_s} \tag{6.47}$$

with T_{ref} being a reference temperature. In order to accomplish a high degree of thermal integration, **Erc** should be significant; conversely, we can expect that

$$\varepsilon = \frac{1}{\mathbf{Erc}} = \frac{[F_{in}\rho C_p(T_{in} - T_{ref})]_s}{H_{rec,s}} \tag{6.48}$$

will be small, i.e., $\varepsilon \ll 1$. Using arguments similar to those developed in Section 6.5.2, we also define the following $\mathcal{O}(1)$ quantities:

$$k_{\rm R} = \frac{[F_{\rm in}\rho C_p(T_{\rm R} - T_{\rm ref})]_{\rm s}}{H_{\rm rec,s}} \qquad u_{\rm R} = \frac{F_{\rm in}\rho C_p(T_{\rm R} - T_{\rm ref})}{[F_{\rm in}\rho C_p(T_{\rm R} - T_{\rm ref})]_{\rm s}} \tag{6.49}$$

$$k_{\rm i} = \frac{[F_{\rm in}\rho C_p(T_{\rm i} - T_{\rm ref})]_{\rm s}}{H_{\rm rec,s}} \qquad u_{\rm i} = \frac{F_{\rm in}\rho C_p(T_{\rm i} - T_{\rm ref})}{[F_{\rm in}\rho C_p(T_{\rm i} - T_{\rm ref})]_{\rm s}} \tag{6.50}$$

$$k_{\rm C} = \frac{[F_{\rm in}\rho C_p(T_{\rm c} - T_{\rm ref})]_{\rm s}}{H_{\rm rec,s}} \qquad u_{\rm C} = \frac{F_{\rm in}\rho C_p(T_{\rm C} - T_{\rm ref})}{[F_{\rm in}\rho C_p(T_{\rm C} - T_{\rm ref})]_{\rm s}} \tag{6.51}$$

Furthermore, we define

$$u_{\rm rec} = \frac{H_{\rm rec}}{H_{\rm rec,s}} \tag{6.52}$$

With this notation, we can rewrite the model in Equations (6.46) as

$$\begin{aligned}
\frac{dV}{dt} &= F_{\rm in} - F \\
\frac{dc_{\rm A}}{dt} &= \frac{F_{\rm in}}{V}(c_{\rm A0} - c_{\rm A}) - k_0 e^{-E/(RT_{\rm R})} c_{\rm A} \\
\frac{dT_{\rm R}}{dt} &= -\frac{1}{\rho C_p} k_0 e^{-E/(RT_{\rm R})} c_{\rm A}\,\Delta H + \frac{1}{\varepsilon}\frac{[F_{\rm in}(T_{\rm in} - T_{\rm ref})]_{\rm s}}{V_{\rm f}} V(k_{\rm i}u_{\rm i} - k_{\rm R}u_{\rm R}) \\
\frac{dT_{\rm H}}{dt} &= -\frac{F(1-\alpha)}{V_{\rm H}} T_{\rm H} - \frac{\alpha}{\varepsilon}\frac{[F_{\rm in}T_{\rm in}]_{\rm s}}{V_{\rm H}} k_{\rm R}u_{\rm R} \\
&\quad + \frac{1}{\varepsilon}\frac{[F_{\rm in}(T_{\rm in} - T_{\rm ref})]_{\rm s}}{V_{\rm H}}(k_{\rm R}u_{\rm R} - u_{\rm rec}) \\
\frac{dT_{\rm C}}{dt} &= \frac{F_{\rm in}}{V_{\rm C}} T_{\rm in} + \frac{1}{\varepsilon}\frac{[F_{\rm in}(T_{\rm in} - T_{\rm ref})]_{\rm s}}{V_{\rm C}}(-k_{\rm C}u_{\rm C} + u_{\rm rec}) \\
\frac{dT_{\rm i}}{dt} &= \frac{Q_{\rm H}}{\rho C_p V_{\rm f}} + \frac{1}{\varepsilon}\frac{[F_{\rm in}(T_{\rm in} - T_{\rm ref})]_{\rm s}}{V_{\rm f}}(k_{\rm C}u_{\rm C} - k_{\rm i}u_{\rm i})
\end{aligned} \tag{6.53}$$

An inspection of the process parameters in Table 6.1 reveals that α/ε is of $\mathcal{O}(1)$ (i.e., the energy flow associated with the bypass stream is much smaller than the amount of energy recycled to the process). Equations (6.53) can thus be rewritten as

$$\begin{aligned}
\frac{dV}{dt} &= F_{\rm in} - F \\
\frac{dc_{\rm A}}{dt} &= \frac{F_{\rm in}}{V}(c_{\rm A0} - c_{\rm A}) - k_0 e^{-E/(RT_{\rm R})} c_{\rm A} \\
\frac{dT_{\rm R}}{dt} &= -\frac{1}{\rho C_p} k_0 e^{-E/(RT_{\rm R})} c_{\rm A}\,\Delta H + \frac{1}{\varepsilon}\frac{[F_{\rm in}(T_{\rm in} - T_{\rm ref})]_{\rm s}}{V_f} V(k_{\rm i}u_{\rm i} - k_{\rm R}u_{\rm R})
\end{aligned}$$

$$
\begin{aligned}
\frac{dT_{\mathrm{H}}}{dt} &= -\frac{F(1-\alpha)}{V_H}T_{\mathrm{H}} - \frac{1}{V_{\mathrm{H}}}\alpha F T_{\mathrm{R}} + \frac{1}{\varepsilon}\frac{[F_{\mathrm{in}}(T_{\mathrm{in}}-T_{\mathrm{ref}})]_{\mathrm{s}}}{V_{\mathrm{H}}}(k_{\mathrm{R}}u_{\mathrm{R}} - u_{\mathrm{rec}}) \qquad (6.54)\\
\frac{dT_{\mathrm{C}}}{dt} &= \frac{F_{\mathrm{in}}}{V_{\mathrm{C}}}T_{\mathrm{in}} + \frac{1}{\varepsilon}\frac{[F_{\mathrm{in}}(T_{\mathrm{in}}-T_{\mathrm{ref}})]_{\mathrm{s}}}{V_{\mathrm{C}}}(-k_{\mathrm{C}}u_{\mathrm{C}} + u_{\mathrm{rec}})\\
\frac{dT_{\mathrm{i}}}{dt} &= \frac{Q_{\mathrm{H}}}{\rho C_p V_{\mathrm{f}}} + \frac{1}{\varepsilon}\frac{[F_{\mathrm{in}}(T_{\mathrm{in}}-T_{\mathrm{ref}})]_{\mathrm{s}}}{V_{\mathrm{f}}}(k_{\mathrm{C}}u_{\mathrm{C}} - k_{\mathrm{i}}u_{\mathrm{i}})
\end{aligned}
$$

which is in the form of Equation (6.10).

We can now proceed with the analysis as described in Section 6.3. Let us define the stretched, fast time variable $\tau = t/\varepsilon$. On rewriting the model in terms of τ and considering the limit $\varepsilon \to 0$, we obtain a description of the fast dynamics of the process:

$$
\begin{aligned}
\frac{dV}{d\tau} &= 0\\
\frac{dc_{\mathrm{A}}}{d\tau} &= 0\\
\frac{dT_{\mathrm{R}}}{d\tau} &= \frac{[F_{\mathrm{in}}(T_{\mathrm{in}}-T_{\mathrm{ref}})]_{\mathrm{s}}}{V}(k_{\mathrm{i}}u_{\mathrm{i}} - k_{\mathrm{R}}u_{\mathrm{R}})\\
\frac{dT_{\mathrm{H}}}{d\tau} &= \frac{[F_{\mathrm{in}}(T_{\mathrm{in}}-T_{\mathrm{ref}})]_{\mathrm{s}}}{V_{\mathrm{H}}}(k_{\mathrm{R}}u_{\mathrm{R}} - u_{\mathrm{rec}})\\
\frac{dT_{\mathrm{C}}}{d\tau} &= \frac{[F_{\mathrm{in}}(T_{\mathrm{in}}-T_{\mathrm{ref}})]_{\mathrm{s}}}{V_{\mathrm{C}}}(-k_{\mathrm{C}}u_{\mathrm{C}} + u_{\mathrm{rec}})\\
\frac{dT_{\mathrm{i}}}{d\tau} &= \frac{[F_{\mathrm{in}}(T_{\mathrm{in}}-T_{\mathrm{ref}})]_{\mathrm{s}}}{V_{\mathrm{f}}}(k_{\mathrm{C}}u_{\mathrm{C}} - k_{\mathrm{i}}u_{\mathrm{i}})
\end{aligned} \qquad (6.55)
$$

with the corresponding (linearly independent) quasi-steady-state constraints

$$
\begin{aligned}
0 &= k_{\mathrm{R}}u_{\mathrm{R}} - u_{\mathrm{rec}}\\
0 &= k_{\mathrm{i}}u_{\mathrm{i}} - k_{\mathrm{R}}u_{\mathrm{R}}\\
0 &= k_{\mathrm{C}}u_{\mathrm{C}} - k_{\mathrm{i}}u_{\mathrm{i}}
\end{aligned} \qquad (6.56)
$$

Subsequently, the slow component of the process dynamics takes the form of Equations (6.18):

$$
\begin{aligned}
\frac{dV}{dt} &= F_{\mathrm{in}} - F\\
\frac{dc_{\mathrm{A}}}{dt} &= \frac{F_{\mathrm{in}}}{V}(c_{\mathrm{A0}} - c_{\mathrm{A}}) - k_0 e^{-E/(RT_{\mathrm{R}})}c_{\mathrm{A}}\\
\frac{dT_{\mathrm{R}}}{dt} &= -\frac{1}{\rho C_p}k_0 e^{-E/(RT_{\mathrm{R}})}c_{\mathrm{A}}\,\Delta H + \frac{[F_{\mathrm{in}}(T_{\mathrm{in}}-T_{\mathrm{ref}})]_{\mathrm{s}}}{V_{\mathrm{f}}}V z_2\\
\frac{dT_{\mathrm{H}}}{dt} &= -\frac{F(1-\alpha)}{V_{\mathrm{H}}}T_{\mathrm{H}} - \frac{1}{V_{\mathrm{H}}}\alpha F T_{\mathrm{R}} + \frac{[F_{\mathrm{in}}(T_{\mathrm{in}}-T_{\mathrm{ref}})]_{\mathrm{s}}}{V_{\mathrm{H}}}z_1 \qquad (6.57)
\end{aligned}
$$

$$
\begin{aligned}
\frac{dT_\mathrm{C}}{dt} &= \frac{F_\mathrm{in}}{V_\mathrm{C}} T_\mathrm{in} - \frac{[F_\mathrm{in}(T_\mathrm{in} - T_\mathrm{ref})]_\mathrm{s}}{V_\mathrm{C}} (z_1 + z_2 + z_3) \\
\frac{dT_\mathrm{i}}{dt} &= \frac{Q_\mathrm{H}}{\rho C_p V_\mathrm{f}} + \frac{[F_\mathrm{in}(T_\mathrm{in} - T_\mathrm{ref})]_\mathrm{s}}{V_\mathrm{f}} z_3 \\
0 &= F_\mathrm{in} \rho C_p T_\mathrm{R} - H_\mathrm{rec} \\
0 &= T_\mathrm{i} - T_\mathrm{R} \\
0 &= T_\mathrm{C} - T_\mathrm{i}
\end{aligned}
$$

where

$$
\begin{aligned}
z_1 &= \lim_{\varepsilon \to 0} \frac{1}{\varepsilon} (k_\mathrm{R} u_\mathrm{R} - u_\mathrm{rec}) \\
z_2 &= \lim_{\varepsilon \to 0} \frac{1}{\varepsilon} (k_\mathrm{i} u_\mathrm{i} - k_\mathrm{R} u_\mathrm{R}) \\
z_3 &= \lim_{\varepsilon \to 0} \frac{1}{\varepsilon} (k_\mathrm{C} u_\mathrm{C} - k_\mathrm{i} u_\mathrm{i})
\end{aligned}
$$

The expressions of the algebraic constraints in Equations (6.57) have been obtained by substituting the definitions in Equations (6.49)–(6.52) into Equations (6.56). In order to determine the algebraic variables z_i, we must differentiate these algebraic constraints and, to this end, we must provide an expression for H_rec, the duty of the FEHE.

The process conditions in the FEHE (counter-current flow with no phase change and constant physical properties) lend themselves naturally to calculating the duty on the basis of the log mean temperature difference (LMTD):

$$
H_\mathrm{rec} = UA\,\mathrm{LMTD} = UA \frac{(T_\mathrm{R} - T_\mathrm{C}|_{z=L}) - (T_\mathrm{H}|_{z=0} - T_\mathrm{in})}{\ln \left(\dfrac{T_\mathrm{R} - T_\mathrm{C}|_{z=L}}{T_\mathrm{H}|_{z=0} - T_\mathrm{in}} \right)} \tag{6.58}
$$

or, in terms of the alternative construct developed in this section,

$$
H_\mathrm{rec} = UA \frac{(T_\mathrm{R} - T_\mathrm{C}) - (T_\mathrm{H} - T_\mathrm{in})}{\ln \left(\dfrac{T_\mathrm{R} - T_\mathrm{C}}{T_\mathrm{H} - T_\mathrm{in}} \right)} \tag{6.59}
$$

Solving dynamic models that employ the LMTD can, however, be problematic because the LMTD is not well defined when the temperature gradient along the heat exchanger is constant i.e., $T_\mathrm{R} = T_\mathrm{C}$ and $T_\mathrm{H} = T_\mathrm{in}$ (note that heat transfer in a physical exchanger would still take place under these circumstances). Moreover, the LMTD is not well defined in the case of a temperature cross-over (e.g., $T_\mathrm{R} > T_\mathrm{C}$ and $T_\mathrm{H} < T_\mathrm{in}$), a situation that can arise temporarily during transient operation. These issues were recognized relatively early (Paterson 1984), and several approximate formulations with improved numerical properties have been proposed (Chen 1987). In what follows, we will use the expression in Underwood (1970), which was verified to provide a satisfactory degree of accuracy for our process:

$$\mathrm{LMTD}_U = \left[\frac{(T_\mathrm{R} - T_\mathrm{C})^{1/3} + (T_\mathrm{H} - T_\mathrm{in})^{1/3}}{2}\right]^3 \tag{6.60}$$

By substituting this definition into Equations (6.57), after differentiating the constraints once to obtain the algebraic variables and applying a coordinate change of the type in Equation (6.21), namely

$$\delta = T_\mathrm{R} \tag{6.61}$$

and

$$\begin{aligned} \eta_1 &= F_\mathrm{in}\rho C_p T_\mathrm{R} - UA\left[\frac{(T_\mathrm{R} - T_\mathrm{C})^{1/3} + (T_\mathrm{H} - T_\mathrm{in})^{1/3}}{2}\right]^3 \\ \eta_2 &= T_\mathrm{i} - T_\mathrm{R} \\ \eta_3 &= T_\mathrm{C} - T_\mathrm{i} \end{aligned} \tag{6.62}$$

we obtained the following minimal-order state-space realization of the slow dynamics:

$$\begin{aligned} \frac{dV}{dt} &= F_\mathrm{in} - F \\ \frac{dc_\mathrm{A}}{dt} &= \frac{F_\mathrm{in}}{V}(c_\mathrm{A0} - c_\mathrm{A}) - k_0 e^{-E/(R\delta)} c_\mathrm{A} \\ \dot{\delta} &= \frac{1}{\mathrm{DEN}}(-kc_\mathrm{A}\,\Delta H\ VUA + \rho C_p T_\mathrm{in}\ UA(F_\mathrm{in} - F) \\ &\quad + 8\,(\rho C_p)^2 F F_\mathrm{in}\delta(\alpha - 1) + \rho C_p F\, T_\mathrm{in}\alpha UA \\ &\quad - \alpha\rho C_p \delta F\, UA + QHUA) \end{aligned} \tag{6.63}$$

with

$$\mathrm{DEN} = \rho C_p[UA(V_\mathrm{C} + V_\mathrm{f} + V) + 8V_\mathrm{h}F_\mathrm{in}\rho C_p(1 - \alpha)]$$

6.6.4 Controller design

According to the theoretical framework developed in this chapter, the design of a control system for this process should be carried out in a hierarchical fashion, starting with the stabilization of the unit temperatures. On examining the fast submodel in Equations (6.55), it is easy to observe that it does not contain any of the manipulated inputs of the process. At a first glance, this could constitute a matter of concern, since there are seemingly no means to control the fast dynamics. However, (6.55) contains no energy-source or energy-sink terms, describing, in effect, the flow of a fluid through the path comprised of the cold leg of the heat exchanger (where it is heated), the heater (where it undergoes no temperature change), the reactor (where, again, it undergoes no temperature change) and, finally, the warm leg of the heat exchanger, where it is cooled. From physical considerations (even under the ideal assumption that no energy is lost to the environment), the fast component of the process dynamics is clearly stable[2] and requires no feedback control action.

Turning to the slow dynamics, we are interested in stabilizing the reactor holdup, controlling the process outlet temperature, and controlling the temperature in the reactor to manage the conversion. The first two objectives can be addressed using simple linear controllers. In order to stabilize the reactor holdup, we implemented a PI controller using the reactor outlet flow rate, F, as a manipulated input:

$$F = F_{\mathrm{s}} + K_{\mathrm{v}} \left[(V_{\mathrm{set}} - V) + \frac{1}{\tau_{\mathrm{v}}} \int_0^t (V_{\mathrm{set}} - V) d\tau \right] \tag{6.64}$$

with $F_{\mathrm{s}} = 5.7667 \times 10^{-4}$ m^3/s, $K_{\mathrm{v}} = 5$ s^{-1}, and $\tau_{\mathrm{v}} = 18$ min.

The control of T_{exit} was addressed using α as a manipulated input and the PI control law

$$\alpha = \alpha_0 + K_{\mathrm{c}} \left[(T_{\mathrm{exit,set}} - T_{\mathrm{exit}}) + \frac{1}{\tau_I} \int_0^t (T_{\mathrm{exit,set}} - T_{\mathrm{exit}}) d\tau \right] \tag{6.65}$$

with $\alpha_0 = 0.1$, $K_{\mathrm{c}} = 0.0018$ K^{-1}, and $\tau_{\mathrm{I}} = 10$ s. The low reset time used for this controller is justified by the fast nature of the mixing process.[3]

Implementing the reactor temperature controller merits some discussion. While T_{R} is not a true slow variable (it has a two-time-scale behavior, as illustrated in Figure 6.11(a)), as we argued above, the fast transient of the process (and, inherently, of T_{R}) is stable. We are thus interested in controlling the *slow* component of the reactor temperature, which in effect governs the behavior of the entire process. To this end, we conveniently chose the coordinate transformation (6.61)–(6.62) so that the energy balance in Equations (6.63) is written in terms of the reactor temperature T_{R}, rather than in terms of the total enthalpy of the process.

The non-stiff, low-dimensional model (6.63) is ideally suited for control purposes. We used it to design an input–output linearizing controller with integral action that manipulates QH and enforces a first-order response in the T_{R} dynamics, namely

$$\beta d\frac{T_{\mathrm{R}}}{dt} + T_{\mathrm{R}} = T_{\mathrm{R,set}} \tag{6.66}$$

with $\beta = 15$ min.

[2] This argument was also verified through simulations, whereby the eigenvalues of the Jacobian matrix of the system in Equation (6.55) were shown to be negative.

[3] Equation (6.45) assumes instant and perfect mixing at the exit of the warm leg of the heat exchanger. In order to facilitate numerical simulations, this output was filtered with a first-order filter with a time constant $\tau_{\mathrm{filter}} = 2$ s prior to being used in the feedback controller.

The aforementioned controllers were implemented on the full-order 2006-dimensional discretization of the original distributed-parameter model, and their performance was tested through simulations. The relevant Matlab codes are presented in Appendix C.

6.6.5 Simulation results and discussion

We considered two scenarios that are typical for the operation of reactors with feed–effluent heat exchange. The first set of simulations traced the response of the closed-loop system to a 10% increase in the production rate, imposed at $t = 1\,\text{h}$ by increasing the feed flow rate. Subsequently, we analyzed the response of the same situation, but with the added complexity of an unmeasured 10 K increase in the feed temperature occurring at $t = 1\,\text{h}$. In both cases, the setpoint of the reactor temperature controller $T_{\text{R,set}}$ was increased by 2 K at $t = 1\,\text{h}$ in order to maintain reactor conversion at the higher production rate.

The simulation results in Figures 6.13–6.22 show that the proposed controller has excellent performance. This is all the more remarkable given the inherent mismatch between the (high-dimensional) plant model and the (low-dimensional) model used for controller design. These results further confirm the wide applicability of the theoretical framework proposed in this chapter for developing an understanding of the energy dynamics of integrated processes, and for designing effective temperature control and energy management systems.

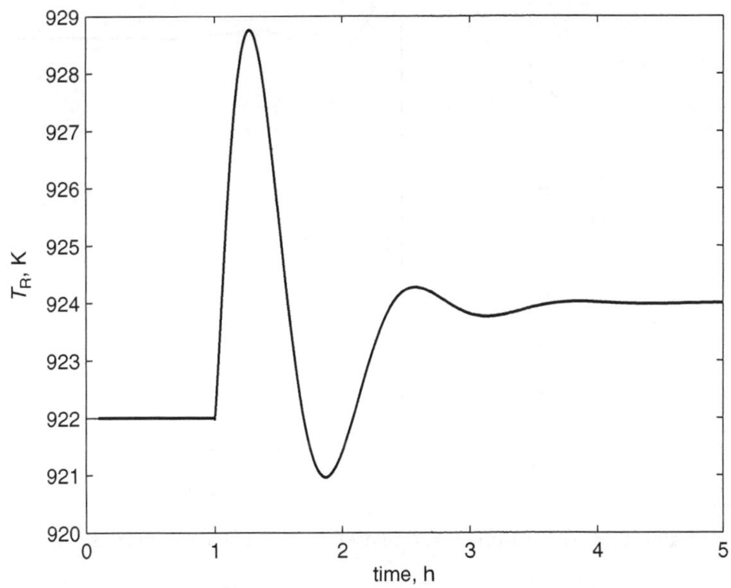

Figure 6.13 Evolution of the reactor temperature for a 10% increase in the production rate.

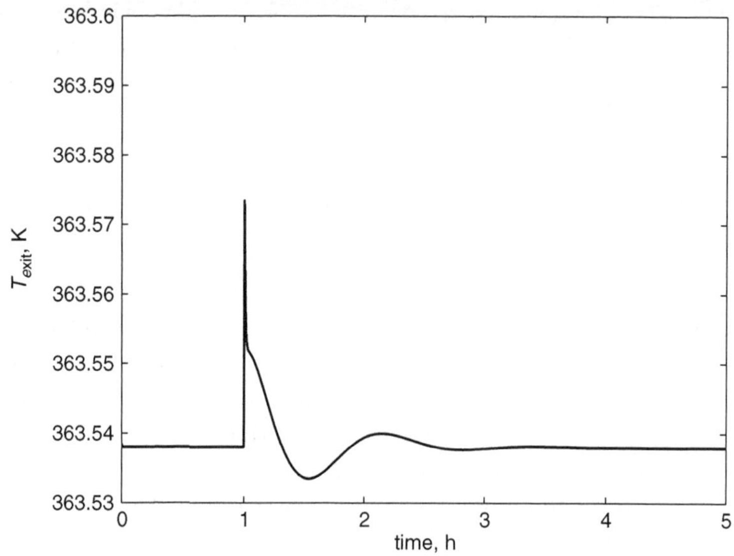

Figure 6.14 Evolution of the product temperature for a 10% increase in the production rate.

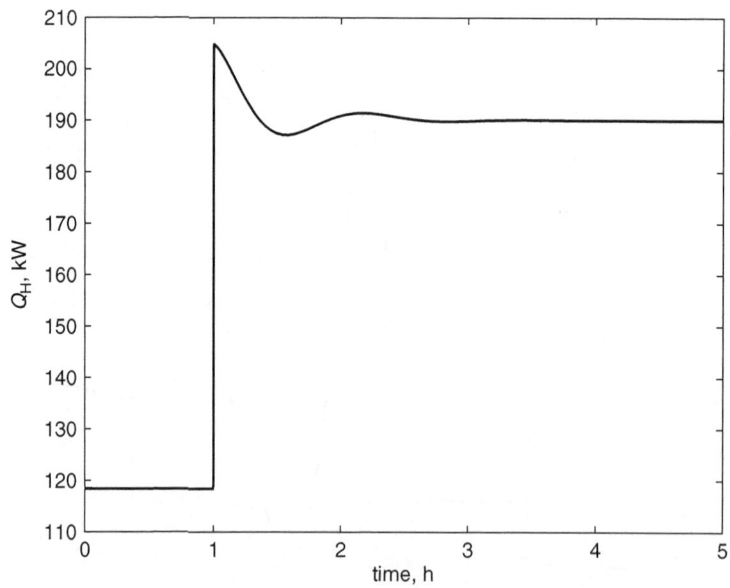

Figure 6.15 Evolution of the heater duty for a 10% increase in the production rate.

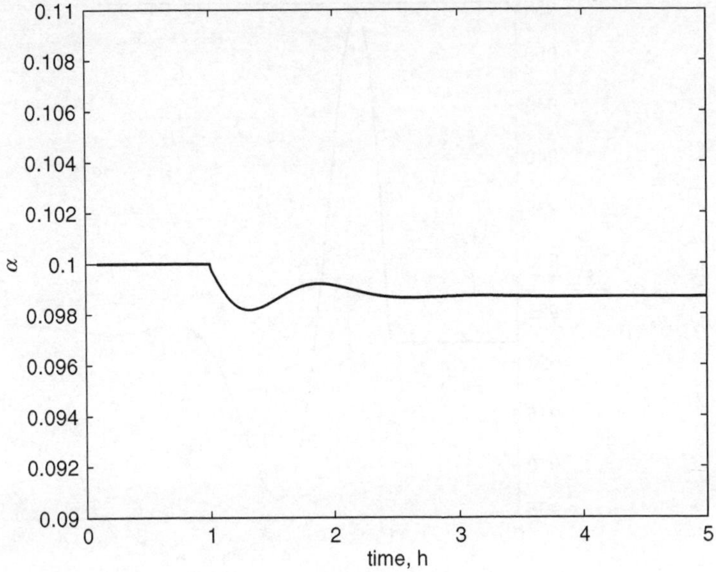

Figure 6.16 Evolution of the bypass ratio for a 10% increase in the production rate.

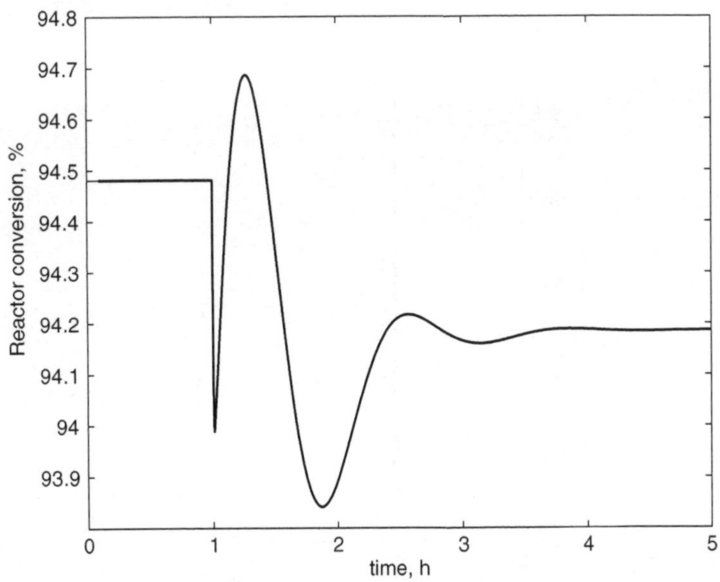

Figure 6.17 Evolution of the reactor conversion for a 10% increase in the production rate.

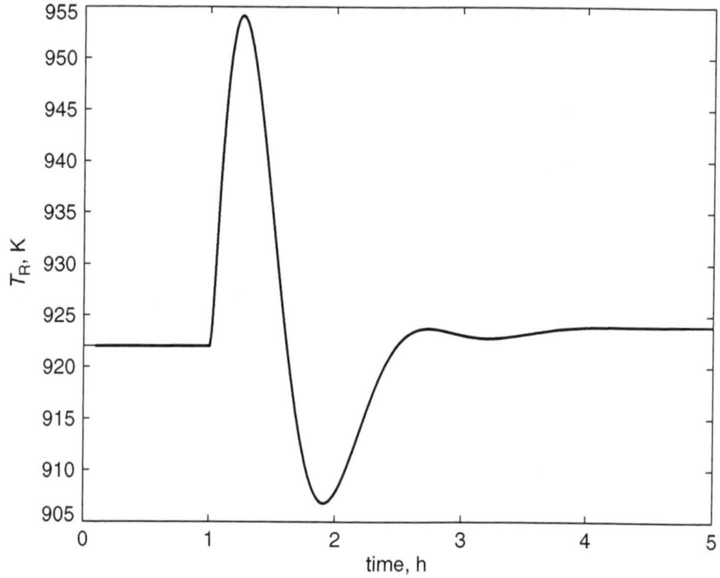

Figure 6.18 Evolution of the reactor temperature for a 10% increase in the production rate, in the presence of unmeasured disturbances in the feed temperature.

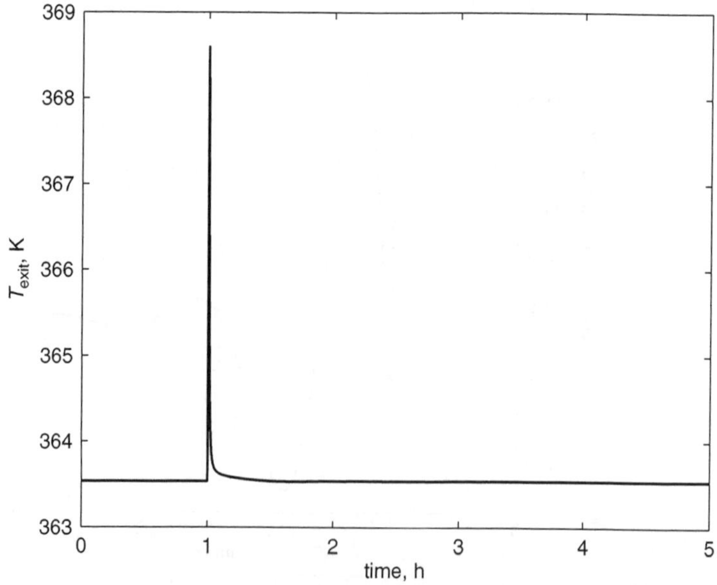

Figure 6.19 Evolution of the product temperature for a 10% increase in the production rate, in the presence of unmeasured disturbances in the feed temperature.

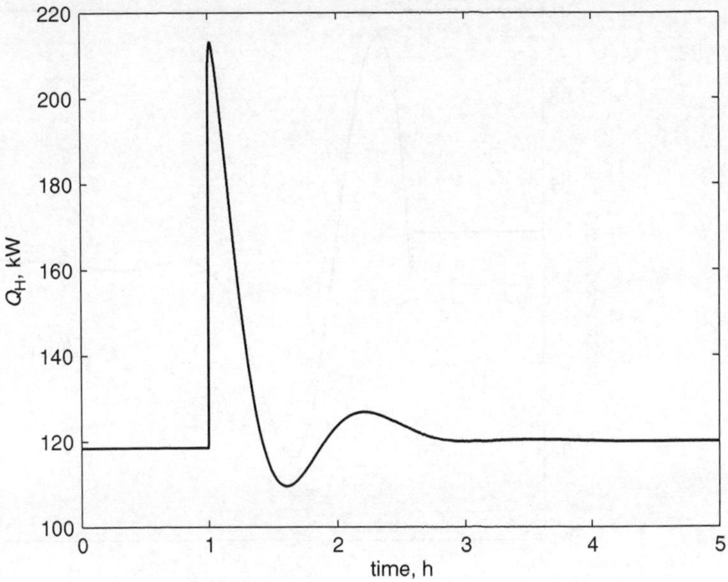

Figure 6.20 Evolution of the heater duty for a 10% increase in the production rate, in the presence of unmeasured disturbances in the feed temperature.

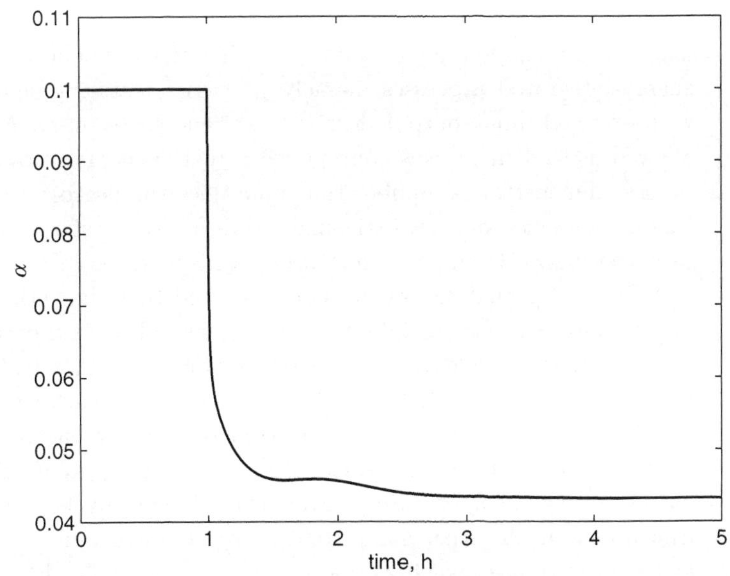

Figure 6.21 Evolution of the bypass ratio for a 10% increase in the production rate, in the presence of unmeasured disturbances in the feed temperature.

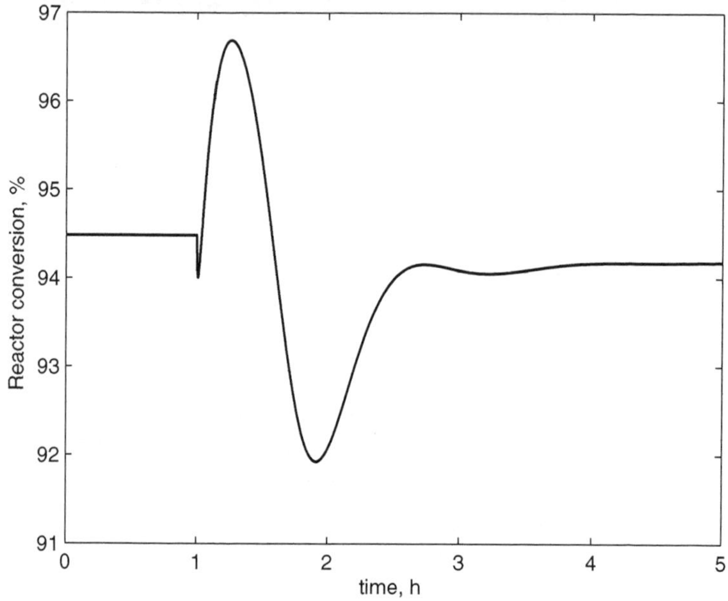

Figure 6.22 Evolution of the reactor conversion for a 10% increase in the production rate, in the presence of unmeasured disturbances in the feed temperature.

6.7 Synopsis

The present chapter focused on analyzing the dynamic aspects of a class of energy-integrated processes, namely processes featuring significant (compared with external input–output energy flows) energy recovery. A prototype system allowed us to define a new concept, namely, the energy recovery number, **Erc**, a process dimensionless number that quantifies the degree of energy integration. Using arguments concerning the magnitude of **Erc**, we showed that, in the general case, the energy dynamics of such processes is modeled by a set of stiff differential equations. Our analysis further demonstrated that the enthalpies of individual process units are fast variables, while the overall process enthalpy evolves only over a slow time scale. Singular perturbation tools aided us in deriving generic reduced-order models that describe the dynamics in each time scale.

Subsequently, we considered the control implications of our findings, and showed that control objectives related to the energy dynamics of the individual units (e.g., temperature control) should typically be addressed in the fast time scale. On the other hand, control objectives related to the energy dynamic at the process level (such as managing energy use) should be addressed in the slow time scale. These concepts were illustrated through several examples and a simulation case study.

7 Process systems with high energy throughput

7.1 Introduction

In this chapter we will expand the analysis presented in Chapter 6 by considering processes in which significant *external* energy sources and/or sinks are present. Examples include multiple-effect evaporators (Seider *et al.* 1999), stand-alone distillation columns and distillation column trains (Jogwar and Daoutidis 2010), and exothermic reactors with external heat exchanger (Baldea and Daoutidis 2006). Understanding the energy dynamics of such processes is beneficial both from a control point of view and from an economic point of view, given today's rising energy and utility costs. We will follow the results in (Baldea and Daoutidis 2006) and (Baldea and Daoutidis 2008), and show that the variables that pertain to the energy balance of such processes evolve in a faster time scale than the variables in the material-balance equations. Subsequently, we investigate the impact of this dynamic feature on controller design and demonstrate that the control of process systems with high energy throughput requires several tiers of control action. Specifically, the variables in the energy balance should be controlled in the fast time scale (over a short time horizon) and control objectives related to the material balance should be fulfilled in the slow time scale(s).

7.2 Modeling of process systems with high energy throughput

We consider the process system in Figure 7.1, featuring N process units in series. The first unit exchanges energy with the environment at a rate $H_0 + Q_{\text{in}}$, with H_0 being associated with the inlet material stream of the process and Q_{in} corresponding to an energy source. Unit N of the process exchanges heat with the

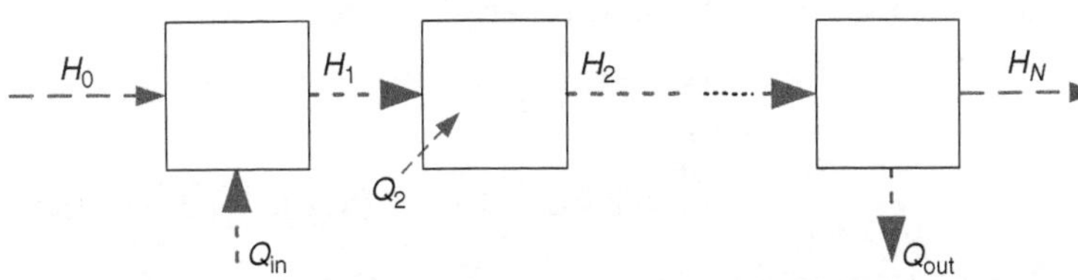

Figure 7.1 A process system with high energy throughput.

environment at a rate $H_N + Q_{\text{out}}$ such that, as for unit 1, H_N is associated with the process outlet material stream and Q_{out} corresponds to an energy sink.

We focus on the case in which the steady-state energy flows associated with the source Q_{in} and sink Q_{out} are of similar magnitude and significantly larger than the energy flows associated with the inlet and outlet material streams, i.e.,

$$\varepsilon = \frac{H_{0\text{s}}}{Q_{\text{in}}} \ll 1$$

as well as larger than the energy input–output flows Q_i to any of the units $2, \dots, N$. Under such circumstances, H_i, $i = 1, \dots, N-1$, act as an energy conduit between the source and the sink, having similar magnitudes to Q_{in} and Q_{out} at steady state. We can thus refer to such processes as having a *high energy throughput*. As we will demonstrate later in the chapter, ubiquitous units and processes such as distillation columns, multiple-effect evaporators, and reactors with external heat exchangers belong to this category.

To proceed with the analysis, we note that the structure of the process in Figure 7.1 is very similar to that of the processes with significant energy recycle considered in the previous chapter (Figure 6.1), with the obvious and necessary distinction that in the present case $Q_{\text{in}} \not\equiv Q_{\text{out}}$. Thus, using Assumptions 6.1–6.3, the dynamic behavior of the process in Figure 7.1 can be modeled by a system of equations of the type (6.10), that is

$$\begin{aligned} \dot{\mathbf{x}} &= \mathbf{f}(\mathbf{x}, \boldsymbol{\theta}) \\ \dot{\boldsymbol{\theta}} &= \boldsymbol{\Phi}(\mathbf{x}, \boldsymbol{\theta}) + \boldsymbol{\Gamma}^{\text{s}}(\mathbf{x}, \boldsymbol{\theta})\boldsymbol{\omega}^{\text{s}} + \frac{1}{\varepsilon}\boldsymbol{\Gamma}^{\text{l}}(\mathbf{x}, \boldsymbol{\theta})\boldsymbol{\omega}^{\text{l}} \end{aligned} \tag{7.1}$$

where $\boldsymbol{\omega}^{\text{s}} \in \mathcal{U}^{\text{s}} \subset \mathbb{R}^{m^{\text{s}}}$ is a vector of scaled variables that correspond to the small inlet and outlet energy flows, and $\boldsymbol{\omega}^{\text{l}} \in \mathcal{U}^{\text{l}} \subset \mathbb{R}^{m^{\text{l}}}$ is a vector of scaled variables corresponding to the large, *external* energy source Q_{in} and sink Q_{out}, and *internal* energy flows H_i, $i = 1, \dots, N-1$. $\mathbf{f}$ and $\boldsymbol{\Phi}$ are vector functions and $\mathbf{G}^{\text{s}}$ and $\mathbf{G}^{\text{l}}$ are matrices of appropriate dimensions.

7.3 Nonlinear model reduction

Owing to the presence of the small parameter ε, the model in Equation (7.1) is stiff and can potentially exhibit a dynamic behavior with multiple time scales. Proceeding in a manner similar to the one adopted in Chapter 6, we define the fast time scale $\tau = t/\varepsilon$ and rewrite (7.1) as

$$\begin{aligned} \frac{d\mathbf{x}}{d\tau} &= \varepsilon \mathbf{f}(\mathbf{x}, \boldsymbol{\theta}) \\ \frac{d\boldsymbol{\theta}}{d\tau} &= \varepsilon\boldsymbol{\Phi}(\mathbf{x}, \boldsymbol{\theta}) + \varepsilon\boldsymbol{\Gamma}^{\text{s}}(\mathbf{x}, \boldsymbol{\theta})\boldsymbol{\omega}^{\text{s}} + \boldsymbol{\Gamma}^{\text{l}}(\mathbf{x}, \boldsymbol{\theta})\boldsymbol{\omega}^{\text{l}} \end{aligned} \tag{7.2}$$

We continue to refer to the analysis in Chapter 6 and consider the limit $\varepsilon \to 0$, making the important distinction that in this case the limit corresponds to an infinitely high energy *throughput* rather than to an infinitely high energy *recovery*. In this limit, we obtain

$$\begin{aligned}\frac{d\mathbf{x}}{d\tau} &= 0 \\ \frac{d\boldsymbol{\theta}}{d\tau} &= \mathbf{\Gamma}^{\mathrm{l}}(\mathbf{x}, \boldsymbol{\theta})\boldsymbol{\omega}^{\mathrm{l}}\end{aligned} \tag{7.3}$$

which represents a description of the fast dynamics of the process, involving only the variables $\boldsymbol{\theta}$ that pertain to the energy balance.

In Chapter 6, we argued that large, internal energy recycle flows *do not affect* the total enthalpy of the process. Conversely, it is intuitive that, in the case of processes with high energy throughput, the total enthalpy *will be affected* by the large flows $\boldsymbol{\omega}^{\mathrm{l}}$. Therefore, it can be verified that, in antithesis to the developments in the previous chapter, the steady-state conditions that correspond to Equations (7.3) *are linearly independent.* Also, it is evident that the fast energy dynamics described by Equations (7.3) are influenced only by the large energy flows $\boldsymbol{\omega}^{\mathrm{l}}$.

The observations above indicate that, upon defining $\boldsymbol{\omega}^{\mathrm{l}}$ as a function of the state variables $\boldsymbol{\theta}$ (e.g., via feedback control laws), the Jacobian matrix

$$\frac{\partial}{\partial \boldsymbol{\theta}}\mathbf{\Gamma}^{\mathrm{l}}(\mathbf{x}, \boldsymbol{\theta})\boldsymbol{\omega}^{\mathrm{l}}$$

is nonsingular, and the equations $\mathbf{0} = \mathbf{\Gamma}^{\mathrm{l}}(\mathbf{x}, \boldsymbol{\theta})\boldsymbol{\omega}^{\mathrm{l}}$ can be solved for the quasi-steady-state values $\boldsymbol{\theta}^{\star} = \mathbf{k}(\mathbf{x}, \boldsymbol{\omega}^{\mathrm{l}})$ of the enthalpies (or temperatures) of each unit.

On substituting the solution $\boldsymbol{\theta}^{\star}$ into the model, we obtain a description of the dynamics of the process after the fast temperature "boundary layer":

$$\begin{aligned}\dot{\mathbf{x}} &= \mathbf{f}(\mathbf{x}, \mathbf{k}(\mathbf{x}, \boldsymbol{\omega}^{\mathrm{l}})) \\ \dot{\boldsymbol{\theta}} &= 0\end{aligned} \tag{7.4}$$

which constitutes a representation of the slow dynamics of the process.

Remark 7.1. *The energy flow rates $\{H_1 \ldots H_{N-1}\} \subset \boldsymbol{\omega}^{\mathrm{l}}$ are typically functions of the flow rates of (internal) material streams and cannot be set independently. In order to preserve the simplicity of the presentation, this dependence has not been explicitly accounted for; the results presented thus far are, however, independent of this consideration. It is also intuitive that the following statements apply.*

- *The flow rates of the material streams are likely to belong to different groups of manipulated inputs (as demonstrated in Chapter 5). Further, only some of these groups will have any impact on the fast energy dynamics.*
- *Only a subgroup of the material flow rates which are available as manipulated inputs in the fast time scale τ can actually be used to address control objectives related to the variables θ.*

The observations above are expanded upon in the subsequent section and illustrated in the examples introduced later in the chapter.

7.4 Control implications

As was the case in the previous chapters of the book, the potential presence of two distinct scales in the dynamic behavior of process systems with high energy throughput requires that the objectives pertaining to their operation and control be addressed using separate, coordinated fast and slow controllers.

Energy-related control objectives are to be addressed in the fast time scale τ, where $\boldsymbol{\omega}^1$ are available as manipulated inputs (Equations (7.3)). From a practical point of view, however, only a limited number of material flow rates may be available to address objectives related to temperature control; thus, the set of manipulated inputs in the fast time scale could often consist solely of the large energy flows Q_{in} and Q_{out}. Simple, distributed controllers for the stabilization (and fast disturbance rejection) of unit temperatures are a typical choice at this level.

The control objectives related to the material balance (e.g., stabilization of material holdups in the individual units, the evolution of product purity) should subsequently be addressed according to the developments in Chapter 5.

7.5 Case study 1: dynamics of high-purity distillation columns

7.5.1 System description

High-purity distillation columns are multi-staged separation systems that rely on a large internal material recycle for increasing the purity of the distillate/bottoms streams. The internal boilup (vapor) stream undergoes phase changes (evaporation/condensation), accumulating and, subsequently, releasing energy. This material stream thus acts as an energy carrier, contributing to the high energy throughput of the system. The differential equation system describing the *material* balance of high-purity distillation columns was shown to exhibit a two-time-scale behavior, and an explicit nonlinear low-order model of the slow input–output dynamics that is suitable for analysis and controller design was provided in Kumar and Daoutidis (2003). The present example aims to analyze the *energy* dynamics of high-purity distillation columns.

We consider a distillation column with N stages (numbered from top to bottom), into which a saturated liquid containing a mixture of two components (denoted by 1 and 2) with mole fractions $x_{1\text{f}}$ and $x_{2\text{f}}$, respectively, is fed at (molar) flow rate F_0 and temperature T_0 on stage N_{f} (we analyze the case of a binary mixture for simplicity; the results developed below can easily be generalized to mixtures of M components). The heavy component 2 is removed at the bottom from the reboiler at a flow rate B, while the light component 1 is removed at the top from the condenser at a flow rate D. In this column, a large (compared with the feed, distillate, and bottom product flow rates) vapor boilup V and liquid recycle R are required in order to attain a high purity of the products.

We denote by Q_{r} and Q_{C} the heat duties in the reboiler and condenser. For simplicity, we assume that the relative volatilities of the components are constant, and hence that the phase equilibrium on stage i is given by

$$y_{1,i} = \frac{\alpha_1 x_{1,i}}{1 + (\alpha_1 - 1)x_{1,i}} \tag{7.5}$$

We also assume that the heat capacities $C_{p,\mathrm{l}}$ and $C_{p,\mathrm{v}}$ of the liquid and vapor phases are constant. Under the above assumptions, the dynamic model of the column can be written as

$$\begin{aligned}
\dot{M}_{\mathrm{C}} &= V - R - D \\
\dot{x}_{1,\mathrm{D}} &= (V/M_{\mathrm{C}})(y_{1,1} - x_{1,\mathrm{D}}) \\
\dot{\theta}_{\mathrm{C}} &= V\left(C_{p,\mathrm{v}}T_1 + \sum_{j=1}^{2} y_{j,1}\lambda_j\right) - (R+D)C_{p,\mathrm{l}}T_{\mathrm{C}} - Q_{\mathrm{C}} \\
&\vdots \\
\dot{x}_{1,i} &= \frac{1}{M_i}[V(y_{1,i+1} - y_{1,i}) + R(x_{1,i-1} - x_{1,i})] \\
\dot{\theta}_i &= V\left(C_{p,\mathrm{v}}T_{i+1} + \sum_{j=1}^{2} y_{j,i+1}\lambda_j\right) - V\left(C_{p,\mathrm{v}}T_i + \sum_{j=1}^{2} y_{j,i}\lambda_j\right) \\
&\quad + RC_{p,\mathrm{l}}(T_{i-1} - T_i) \\
&\vdots \\
\dot{x}_{1,N_{\mathrm{f}}} &= \frac{1}{M_{N_{\mathrm{f}}}}[V(y_{1,N_{\mathrm{f}}+1} - y_{1,N_{\mathrm{f}}}) + R(x_{1,N_{\mathrm{f}}-1} - x_{1,N_{\mathrm{f}}}) \\
&\quad + F(x_{1,N_{\mathrm{f}}-1} - x_{1,N_{\mathrm{f}}})] \\
\dot{\theta}_{\mathrm{f}} &= V\left(C_{p,\mathrm{v}}T_{N_{\mathrm{f}}+1} + \sum_{j=1}^{2} y_{j,N_{\mathrm{f}}+1}\lambda_j\right) - V\left(C_{p,\mathrm{v}}T_{N_{\mathrm{f}}} + \sum_{j=1}^{2} y_{j,N_{\mathrm{f}}}\lambda_j\right) \\
&\quad + RC_{p,\mathrm{l}}(T_{N_{\mathrm{f}}-1} - T_{N_{\mathrm{f}}}) + FC_{p,\mathrm{l}}(T_0 - T_{N_{\mathrm{f}}}) \\
&\vdots \\
\dot{x}_{1,i} &= \frac{1}{M_i}[V(y_{1,i+1} - y_{1,i}) + R(x_{1,i-1} - x_{1,i}) + F(x_{1,i-1} - x_{1,i})]
\end{aligned} \tag{7.6}$$

$$\dot{\theta}_i = V\left(C_{p,\mathrm{v}}T_{i+1} + \sum_{j=1}^{2} y_{j,i+1}\lambda_j\right) - V\left(C_{p,\mathrm{v}}T_i + \sum_{j=1}^{2} y_{j,i}\lambda_j\right)$$
$$+ RC_{p,\mathrm{l}}(T_{i-1} - T_i) + FC_{p,\mathrm{l}}(T_{i-1} - T_i)$$

$$\vdots$$

$$\dot{M}_\mathrm{B} = R - V + F - B$$
$$\dot{x}_{1,\mathrm{B}} = \frac{1}{M_\mathrm{B}}[R(x_{1,N} - x_{1,\mathrm{B}}) - V(y_{1,\mathrm{B}} - x_{1,\mathrm{B}}) + F(x_{1,N} - x_{1,\mathrm{B}})]$$
$$\dot{\theta}_\mathrm{B} = (R + F)C_{p,\mathrm{l}}T_N - BC_{p,\mathrm{l}}T_\mathrm{B} - V\left(C_{p,\mathrm{v}}T_\mathrm{B} + \sum_{j=1}^{2} y_{j,\mathrm{B}}\lambda_j\right) + Q_\mathrm{B}$$

In Equations (7.6), M_C, $x_{1,\mathrm{D}}$, $y_{1,\mathrm{D}}$, and T_C are the molar liquid holdup, the liquid mole fraction and vapor mole fraction of component 1, and the temperature in the condenser, M_i, $x_{1,i}$, $y_{1,i}$, and T_i are the molar liquid holdup, the liquid mole fractions and vapor mole fractions of component 1, and the temperature on stage i, and M_B, $x_{1,\mathrm{B}}$, $y_{1,\mathrm{B}}$, and T_B are the corresponding holdup, liquid mole fractions, vapor mole fractions, and temperature in the reboiler. θ_i are the stage enthalpies and λ_j is the latent heat of vaporization of component j, $j = 1, 2$. By expressing the stage enthalpies as a function of the stage temperatures, $\theta_i = M_i C_{p,\mathrm{l}} T_i$, Equations (7.6) can be rewritten as

$$\dot{M}_\mathrm{C} = V - R - D$$
$$\dot{x}_{1,\mathrm{D}} = (V/M_\mathrm{C})(y_{1,1} - x_{1,\mathrm{D}})$$
$$\dot{T}_\mathrm{C} = \frac{1}{M_\mathrm{C}C_{p,\mathrm{l}}}\left[V\left(C_{p,\mathrm{v}}T_1 + \sum_{j=1}^{2} y_{j,1}\lambda_j - C_{p,\mathrm{l}}T_\mathrm{C}\right) - Q_\mathrm{C}\right]$$

$$\vdots$$

$$\dot{x}_{1,i} = \frac{1}{M_i}[V(y_{1,i+1} - y_{1,i}) + R(x_{1,i-1} - x_{1,i})]$$
$$\dot{T}_i = \frac{1}{M_i C_{p,\mathrm{l}}}\left[V\left(C_{p,\mathrm{v}}T_{i+1} + \sum_{j=1}^{2} y_{j,i+1}\lambda_j\right)\right.$$
$$\left. - V\left(C_{p,\mathrm{v}}T_i + \sum_{j=1}^{2} y_{j,i}\lambda_j\right) + RC_{p,\mathrm{l}}(T_{i-1} - T_i)\right]$$

$$\vdots$$

$$\dot{x}_{1,N_\mathrm{f}} = \frac{1}{M_{N_\mathrm{f}}}[V(y_{1,N_\mathrm{f}+1} - y_{1,N_\mathrm{f}}) + R(x_{1,N_\mathrm{f}-1} - x_{1,N_\mathrm{f}})$$
$$+ F(x_{1,N_\mathrm{f}-1} - x_{1,N_\mathrm{f}})]$$

$$\dot{T}_{N_\mathrm{f}} = \frac{1}{M_{N_\mathrm{f}} C_{p,\mathrm{l}}} \left[V \left(C_{p,\mathrm{v}} T_{N_\mathrm{f}+1} + \sum_{j=1}^{2} y_{j,N_\mathrm{f}+1} \lambda_j \right) - V \left(C_{p,\mathrm{v}} T_{N_\mathrm{f}} + \sum_{j=1}^{2} y_{j,N_\mathrm{f}} \lambda_j \right) + R C_{p,\mathrm{l}} (T_{N_\mathrm{f}-1} - T_{N_\mathrm{f}}) + F C_{p,\mathrm{l}} (T_0 - T_{N_\mathrm{f}}) \right] \tag{7.7}$$

$$\vdots$$

$$\dot{x}_{1,i} = \frac{1}{M_i} [V(y_{1,i+1} - y_{1,i}) + R(x_{1,i-1} - x_{1,i}) + F(x_{1,i-1} - x_{1,i})]$$

$$\dot{T}_i = \frac{1}{M_i C_{p,\mathrm{l}}} \left[V \left(C_{p,\mathrm{v}} T_{i+1} + \sum_{j=1}^{2} y_{j,i+1} \lambda_j \right) - V \left(C_{p,\mathrm{v}} T_i + \sum_{j=1}^{2} y_{j,i} \lambda_j \right) + R C_{p,\mathrm{l}} (T_{i-1} - T_i) + F C_{p,\mathrm{l}} (T_{i-1} - T_i) \right]$$

$$\vdots$$

$$\dot{M}_\mathrm{B} = R - V + F - B$$

$$\dot{x}_{1,\mathrm{B}} = \frac{1}{M_\mathrm{B}} [R(x_{1,N} - x_{1,\mathrm{B}}) - V(y_{1,\mathrm{B}} - x_{1,\mathrm{B}}) + F(x_{1,N} - x_{1,\mathrm{B}})]$$

$$\dot{T}_\mathrm{B} = \frac{1}{M_\mathrm{B} C_{p,\mathrm{l}}} \left[R C_{p,\mathrm{l}} (T_N - T_\mathrm{B}) + F C_{p,\mathrm{l}} (T_N - T_\mathrm{B}) + V C_{p,\mathrm{l}} T_\mathrm{B} - V \left(C_{p,\mathrm{v}} T_\mathrm{B} + \sum_{j=1}^{2} y_{j,\mathrm{B}} \lambda_j \right) + Q_\mathrm{B} \right]$$

7.5.2 Reduced-order modeling

At steady state, the presence of a large molar liquid recycle R implies an equally large molar vapor boilup V. On the other hand, the feed flow rate F, the distillate flow rate D, and the bottom-product flow rate B are of the same order of magnitude and much smaller than the flow rates of the internal streams. Therefore, we can define $\varepsilon_1 = (F_\mathrm{s}/R_\mathrm{s}) \ll 1$ and $\kappa = V_\mathrm{s}/R_\mathrm{s} = \mathcal{O}(1)$, where the subscript s refers to the nominal steady state. Let us also define the scaled vapor and reflux flow rates $\bar{V} = V/V_\mathrm{s}$ and $\bar{R} = R/R_\mathrm{s}$. With this notation, the model of the distillation column becomes

$$
\begin{aligned}
\dot{M}_\mathrm{C} &= \frac{1}{\varepsilon_1} F_\mathrm{s}(\kappa\bar{V} - \bar{R}) - D \\
\dot{x}_{1,\mathrm{D}} &= \frac{1}{\varepsilon_1}\frac{F}{M_\mathrm{C}}\kappa\bar{V}(y_{1,1} - x_{1,\mathrm{D}}) \\
\dot{T}_\mathrm{C} &= \frac{1}{M_\mathrm{C} C_{p,\mathrm{l}}}\left[\frac{1}{\varepsilon_1} F_\mathrm{s}\kappa\bar{V}\left(C_{p,\mathrm{v}} + \sum_{j=1}^{2} y_{j,1}\frac{\lambda_j}{T_1}\right) T_1 \right. \\
&\qquad \left. - \frac{1}{\varepsilon_1} F_\mathrm{s}\kappa\bar{V} C_{p,\mathrm{l}} T_\mathrm{C} - Q_\mathrm{C}\right] \\
&\vdots \\
\dot{x}_{1,i} &= \frac{1}{\varepsilon_1}\frac{F_\mathrm{s}}{M_i}\left[\kappa\bar{V}(y_{1,i+1} - y_{1,i}) + \bar{R}(x_{1,i-1} - x_{1,i})\right] \\
\dot{T}_i &= \frac{1}{\varepsilon_1}\frac{F_\mathrm{s}}{M_i C_{p,\mathrm{l}}}\left[\kappa\bar{V}\left(C_{p,\mathrm{v}} + \sum_{j=1}^{2} y_{j,i+1}\frac{\lambda_j}{T_{i+1}}\right) T_{i+1} \right. \\
&\qquad \left. - \kappa\bar{V}\left(C_{p,\mathrm{v}} + \sum_{j=1}^{2} y_{j,i}\frac{\lambda_j}{T_i}\right) T_i + \bar{R} C_{p,l}(T_{i-1} - T_i)\right] \\
&\vdots \\
\dot{x}_{1,N_\mathrm{f}} &= \frac{1}{M_{N_\mathrm{f}}}\left[\frac{1}{\varepsilon_1} F_\mathrm{s}\kappa\bar{V}(y_{1,N_\mathrm{f}+1} - y_{1,N_\mathrm{f}}) + \frac{1}{\varepsilon_1} F_\mathrm{s}\bar{R}(x_{1,N_\mathrm{f}-1} - x_{1,N_\mathrm{f}}) \right. \\
&\qquad \left. + F(x_{1,N_\mathrm{f}-1} - x_{1,N_\mathrm{f}})\right] \\
\dot{T}_{N_\mathrm{f}} &= \frac{1}{M_{N_\mathrm{f}} C_{p,\mathrm{l}}}\left[\frac{1}{\varepsilon_1} F_\mathrm{s}\kappa\bar{V}\left(C_{p,\mathrm{v}} + \sum_{j=1}^{2} y_{j,N_\mathrm{f}+1}\frac{\lambda_j}{T_{N_\mathrm{f}+1}}\right) T_{N_\mathrm{f}+1} \right. \\
&\qquad - \frac{1}{\varepsilon_1} F_\mathrm{s}\kappa\bar{V}\left(C_{p,\mathrm{v}} + \sum_{j=1}^{2} y_{j,N_\mathrm{f}}\frac{\lambda_j}{T_{N_\mathrm{f}}}\right) T_{N_\mathrm{f}} \\
&\qquad \left. + \frac{1}{\varepsilon_1} F_\mathrm{s}\bar{R} C_{p,\mathrm{l}}(T_{N_\mathrm{f}-1} - T_{N_\mathrm{f}}) + F C_{p,\mathrm{l}}(T_0 - T_{N_\mathrm{f}})\right] \qquad (7.8) \\
&\vdots \\
\dot{x}_{1,i} &= \frac{1}{M_i}\left[\frac{1}{\varepsilon_1} F_\mathrm{s}\kappa\bar{V}(y_{1,i+1} - y_{1,i}) + \frac{1}{\varepsilon_1} F_\mathrm{s}\bar{R}(x_{1,i-1} - x_{1,i}) \right. \\
&\qquad \left. + F(x_{1,i-1} - x_{1,i})\right]
\end{aligned}
$$

$$\dot{T}_i = \frac{1}{M_i C_{p,\mathrm{l}}} \left[\frac{1}{\varepsilon_1} F_\mathrm{s} \kappa \bar{V} \left(C_{p,\mathrm{v}} + \sum_{j=1}^{2} y_{j,i+1} \frac{\lambda_j}{T_{i+1}} \right) T_{i+1} \right.$$
$$- \frac{1}{\varepsilon_1} F_\mathrm{s} \kappa \bar{V} \left(C_{p,\mathrm{v}} + \sum_{j=1}^{2} y_{j,i} \frac{\lambda_j}{T_i} \right) T_i$$
$$\left. + \frac{1}{\varepsilon_1} F_\mathrm{s} \bar{R} C_{p,\mathrm{l}} (T_{i-1} - T_i) + F C_{p,\mathrm{l}} (T_{i-1} - T_i) \right]$$

$$\vdots$$

$$\dot{M}_\mathrm{B} = \frac{1}{\varepsilon_1} F_\mathrm{s} (\bar{R} - \kappa \bar{V}) + F - B$$
$$\dot{x}_{1,\mathrm{B}} = \frac{1}{M_\mathrm{B}} \left[\frac{1}{\varepsilon_1} F_\mathrm{s} \bar{R} (x_{1,N} - x_{1,\mathrm{B}}) - \frac{1}{\varepsilon_1} F_\mathrm{s} \kappa \bar{V} (y_{1,\mathrm{B}} - x_{1,\mathrm{B}}) \right.$$
$$\left. + F (x_{1,N} - x_{1,\mathrm{B}}) \right]$$
$$\dot{T}_\mathrm{B} = \frac{1}{M_\mathrm{B} C_{p,\mathrm{l}}} \left[\frac{1}{\varepsilon_1} F_\mathrm{s} \bar{R} C_{p,\mathrm{l}} (T_N - T_\mathrm{B}) + F C_{p,\mathrm{l}} (T_N - T_\mathrm{B}) \right.$$
$$+ \frac{1}{\varepsilon_1} F_\mathrm{s} \kappa \bar{V} C_{p,\mathrm{l}} T_\mathrm{B} - \frac{1}{\varepsilon_1} F_\mathrm{s} \kappa \bar{V}$$
$$\left. \times \left(C_{p,\mathrm{v}} + \sum_{j=1}^{2} y_{j,\mathrm{B}} \frac{\lambda_j}{T_\mathrm{B}} \right) T_\mathrm{B} + Q_\mathrm{B} \right]$$

As argued earlier in this chapter, the latent heat of the vapor phase is typically very large, and the term

$$C_{p,\mathrm{v}} + \sum_{j=1}^{2} y_j \frac{\lambda_j}{T} \tag{7.9}$$

in the above equation (with λ being the latent heat of vaporization) is much larger than the heat capacity of the liquid phase, $C_{p,\mathrm{l}}$. For example, for stage 1, we can write

$$\frac{C_{p,\mathrm{l}}}{\left(C_{p,\mathrm{v}} + \sum_{j=1}^{2} y_{j,1} \lambda_j / T_1 \right)_\mathrm{s}} = \varepsilon \ll 1 \tag{7.10}$$

with the index s again denoting steady-state values. A similar relation holds for the other stages, i.e.,

$$\frac{C_{p,\mathrm{l}}}{\left(C_{p,\mathrm{v}} + \sum_{j=1}^{2} y_{j,i} \lambda_j / T_i \right)_\mathrm{s}} = \frac{1}{\nu_i} \varepsilon \tag{7.11}$$

with $\nu_i = \mathcal{O}(1)$ and $i = 2, \ldots, N, B$.

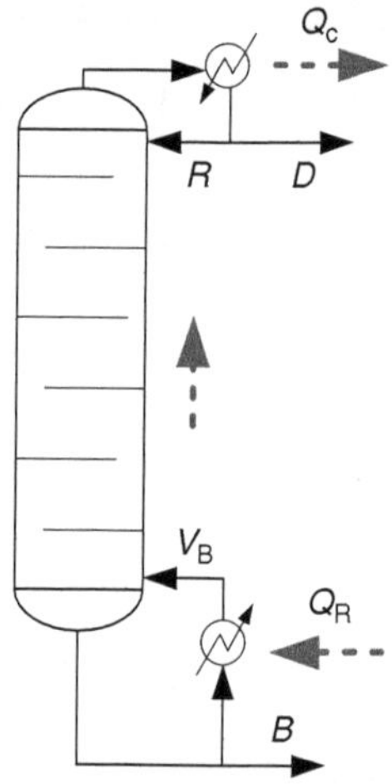

Figure 7.2 Large energy flows in the high-purity distillation column.

Using a simple energy balance, we can see that, in order to supply enough energy to vaporize the liquid, the energy input of the reboiler, Q_{B}, must be of the same order of magnitude as the energy flow associated with the vapor leaving the reboiler, that is

$$\frac{Q_{\mathrm{B,s}}}{\left((1/\varepsilon_1)F_{\mathrm{s}}\kappa(C_{p,\mathrm{v}} + \sum_{j=1}^{2} y_{j,\mathrm{B}}\lambda_j/T_{\mathrm{B}})\right)_{\mathrm{s}}} = \omega_{\mathrm{B}} = \mathcal{O}(1) \tag{7.12}$$

Likewise, in order to condense the column vapor stream in the condenser, its entire latent heat must be removed, and the condenser heat duty and the energy flow associated with the vapor phase leaving the first stage must be of similar magnitude:

$$\frac{Q_{\mathrm{C,s}}}{\left((1/\varepsilon_1)F_{\mathrm{s}}\kappa\left(C_{p,\mathrm{v}} + \sum_{j=1}^{2} y_{j,1}\lambda_j/T_1\right)\right)_{\mathrm{s}}} = \omega_{\mathrm{C}} = \mathcal{O}(1) \tag{7.13}$$

The evident implication of these observations is that the amount of energy carried by the vapor stream from the bottom of the column to the top is much larger than the amount of energy carried by the liquid reflux and, implicitly, than that carried by the streams F, D and B. Thus, the column has a high energy throughput from the reboiler to the condenser. Figure 7.2 presents a diagram of the energy flows in the distillation column. An illustration of the energy flows on an individual column tray is presented in Figure 7.3.

Let us now denote $\bar{Q}_{\mathrm{C}} = Q_{\mathrm{C}}/Q_{\mathrm{C,s}}$ and $\bar{Q}_{\mathrm{B}} = Q_{\mathrm{B}}/Q_{\mathrm{B,s}}$, and

$$\sigma_i = \frac{C_{p,\mathrm{v}} + \sum_{j=1}^{2} y_{j,i}\lambda_i/T_i}{\left(C_{p,\mathrm{v}} + \sum_{j=1}^{2} y_{j,i}\lambda_i/T_i\right)_{\mathrm{s}}} \tag{7.14}$$

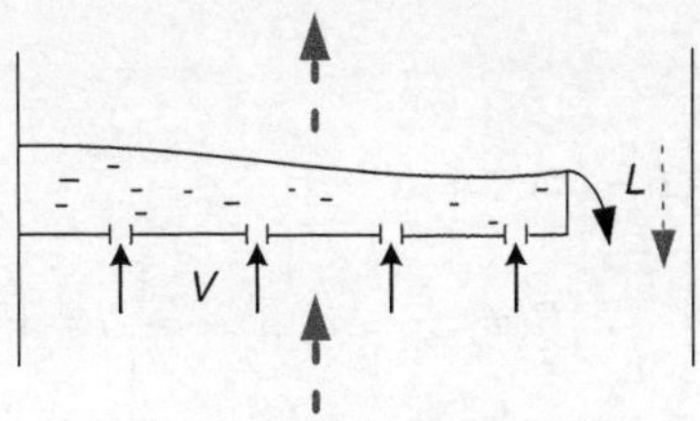

Figure 7.3 Material (solid) and energy (dashed) flows on a distillation-column tray.

With the above notation, the model of the distillation column becomes

$$
\begin{aligned}
\dot{M}_{\mathrm{C}} &= \frac{1}{\varepsilon_1} F_{\mathrm{s}}(\kappa\bar{V} - \bar{R}) - D \\
\dot{x}_{1,\mathrm{D}} &= \frac{1}{\varepsilon_1}\frac{F}{M_{\mathrm{C}}}\kappa\bar{V}(y_{1,1} - x_{1,\mathrm{D}}) \\
\dot{T}_{\mathrm{C}} &= \frac{1}{M_{\mathrm{C}}C_{p,\mathrm{l}}}\left(\frac{1}{\varepsilon_1}F_{\mathrm{s}}\kappa\bar{V}\frac{1}{\varepsilon}\sigma_1 C_{p,\mathrm{l}}T_1 - \frac{1}{\varepsilon_1}F_{\mathrm{s}}\kappa\bar{V}C_{p,\mathrm{l}}T_{\mathrm{C}} - \frac{1}{\varepsilon_1}\omega_{\mathrm{C}}F_{\mathrm{s}}\kappa\frac{1}{\varepsilon}C_{p,\mathrm{l}}\bar{Q}_{\mathrm{C}}\right) \\
&\;\vdots \\
\dot{x}_{1,i} &= \frac{1}{\varepsilon_1}\frac{F_{\mathrm{s}}}{M_i}[\kappa\bar{V}(y_{1,i+1} - y_{1,i}) + \bar{R}(x_{1,i-1} - x_{1,i})] \\
\dot{T}_i &= \frac{1}{\varepsilon_1}\frac{F_{\mathrm{s}}}{M_i C_{p,\mathrm{l}}}\left[\kappa\bar{V}\frac{1}{\varepsilon}\sigma_{i+1}\nu_{i+1}C_{p,\mathrm{l}}T_{i+1} - \kappa\bar{V}\frac{1}{\varepsilon}\sigma_i\nu_i C_{p,\mathrm{l}}T_i + \bar{R}C_{p,\mathrm{l}}(T_{i-1} - T_i)\right] \\
&\;\vdots \\
\dot{x}_{1,N_{\mathrm{f}}} &= \frac{1}{M_{N_{\mathrm{f}}}}\left[\frac{1}{\varepsilon_1}F_{\mathrm{s}}\kappa\bar{V}(y_{1,N_{\mathrm{f}}+1} - y_{1,N_{\mathrm{f}}}) + \frac{1}{\varepsilon_1}F_{\mathrm{s}}\bar{R}(x_{1,N_{\mathrm{f}}-1} - x_{1,N_{\mathrm{f}}})\right. \\
&\qquad \left. + F(x_{1,N_{\mathrm{f}}-1} - x_{1,N_{\mathrm{f}}})\right] \\
\dot{T}_{N_{\mathrm{f}}} &= \frac{1}{M_{N_{\mathrm{f}}}C_{p,\mathrm{l}}}\left[\frac{1}{\varepsilon_1}F_{\mathrm{s}}\kappa\bar{V}\frac{1}{\varepsilon}\sigma_{N_{\mathrm{f}}+1}\nu_{N_{\mathrm{f}}+1}C_{p,\mathrm{l}}T_{N_{\mathrm{f}}+1} - \frac{1}{\varepsilon_1}F_{\mathrm{s}}\kappa\bar{V}\frac{1}{\varepsilon}\sigma_{N_{\mathrm{f}}}\nu_{N_{\mathrm{f}}}C_{p,\mathrm{l}}T_{N_{\mathrm{f}}}\right. \\
&\qquad \left. + \frac{1}{\varepsilon_1}F_{\mathrm{s}}\bar{R}C_{p,\mathrm{l}}(T_{N_{\mathrm{f}}-1} - T_{N_{\mathrm{f}}}) + FC_{p,\mathrm{l}}(T_0 - T_{N_{\mathrm{f}}})\right] \\
&\;\vdots \\
\dot{x}_{1,i} &= \frac{1}{M_i}\left[\frac{1}{\varepsilon_1}F_{\mathrm{s}}\kappa\bar{V}(y_{1,i+1} - y_{1,i}) + \frac{1}{\varepsilon_1}F_{\mathrm{s}}\bar{R}(x_{1,i-1} - x_{1,i}) + F(x_{1,i-1} - x_{1,i})\right] \\
\dot{T}_i &= \frac{1}{M_i C_{p,\mathrm{l}}}\left[\frac{1}{\varepsilon_1}F_{\mathrm{s}}\kappa\bar{V}\frac{1}{\varepsilon}\sigma_{i+1}\nu_{i+1}C_{p,\mathrm{l}}T_{i+1} - \frac{1}{\varepsilon_1}F_{\mathrm{s}}\kappa\bar{V}\frac{1}{\varepsilon}\sigma_i\nu_i C_{p,\mathrm{l}}T_i\right. \\
&\qquad \left. + \frac{1}{\varepsilon_1}F_{\mathrm{s}}\bar{R}C_{p,\mathrm{l}}(T_{i-1} - T_i) + FC_{p,\mathrm{l}}(T_{i-1} - T_i)\right]
\end{aligned}
$$

$$
\begin{aligned}
&\vdots\\
\dot{M}_{\mathrm{R}} &= \frac{1}{\varepsilon_1} F_{\mathrm{s}}(\bar{R} - \kappa\bar{V}) + F - B\\
\dot{x}_{1,\mathrm{B}} &= \frac{1}{M_{\mathrm{B}}}\left[\frac{1}{\varepsilon_1} F_{\mathrm{s}}\bar{R}(x_{1,N} - x_{1,\mathrm{B}}) - \frac{1}{\varepsilon_1} F_{\mathrm{s}}\kappa\bar{V}(y_{1,\mathrm{B}} - x_{1,\mathrm{B}}) + F(x_{1,N} - x_{1,\mathrm{B}})\right]\\
\dot{T}_{\mathrm{B}} &= \frac{1}{M_{\mathrm{B}}C_{p,\mathrm{l}}}\left[\frac{1}{\varepsilon_1} F_{\mathrm{s}}\bar{R}C_{p,\mathrm{l}}(T_N - T_{\mathrm{B}}) + FC_{p,\mathrm{l}}(T_N - T_{\mathrm{B}}) + \frac{1}{\varepsilon_1} F_{\mathrm{s}}\kappa\bar{V}C_{p,\mathrm{l}}T_{\mathrm{B}}\right.\\
&\qquad\left. - \frac{1}{\varepsilon_1} F_{\mathrm{s}}\kappa\bar{V}\frac{1}{\varepsilon}\sigma_{\mathrm{R}}\nu_{\mathrm{R}}C_{p,\mathrm{l}}T_{\mathrm{B}} + \frac{1}{\varepsilon_1}\omega_{\mathrm{B}}F_{\mathrm{s}}\kappa\nu_{\mathrm{B}}\frac{1}{\varepsilon}C_{p,\mathrm{l}}\bar{Q}_{\mathrm{B}}\right]
\end{aligned}
$$

Finally, we define $\varepsilon_2 = \varepsilon_1 \cdot \varepsilon$ and rewrite the model as

$$
\begin{aligned}
\dot{M}_{\mathrm{C}} &= \frac{1}{\varepsilon_1} F_{\mathrm{s}}(\kappa\bar{V} - \bar{R}) - D\\
\dot{x}_{1,\mathrm{D}} &= \frac{1}{\varepsilon_1}\frac{1}{M_{\mathrm{C}}} F_{\mathrm{s}}\kappa\bar{V}(y_{1,1} - x_{1,\mathrm{D}})\\
\dot{T}_{\mathrm{C}} &= \frac{1}{M_{\mathrm{C}}C_{p,\mathrm{l}}}\left[\frac{1}{\varepsilon_2} F_{\mathrm{s}}\kappa C_{p,\mathrm{l}}(\bar{V}\sigma_1 T_1 - \omega_{\mathrm{C}}\bar{Q}_{\mathrm{C}}) - \frac{1}{\varepsilon_1} F_{\mathrm{s}}\kappa\bar{R}C_{p,\mathrm{l}}T_{\mathrm{C}}\right]\\
&\vdots\\
\dot{x}_{1,i} &= \frac{1}{\varepsilon_1}\frac{F_{\mathrm{s}}}{M_i}\left[\kappa\bar{V}(y_{1,i+1} - y_{1,i}) + \bar{R}(x_{1,i-1} - x_{1,i})\right]\\
\dot{T}_i &= \frac{1}{M_i C_{p,\mathrm{l}}}\left[\frac{1}{\varepsilon_2} F_{\mathrm{s}}\kappa\bar{V}(\sigma_{i+1}\nu_{i+1}C_{p,\mathrm{l}}T_{i+1} - \sigma_i\nu_i C_{p,\mathrm{l}}T_i)\right.\\
&\qquad\left. + \frac{1}{\varepsilon_1} F_{\mathrm{s}}\bar{R}C_{p,\mathrm{l}}(T_{i-1} - T_i)\right]\\
&\vdots\\
\dot{x}_{1,N_{\mathrm{f}}} &= \frac{1}{M_{N_{\mathrm{f}}}}\left[\frac{1}{\varepsilon_1} F_{\mathrm{s}}\kappa\bar{V}(y_{1,N_{\mathrm{f}}+1} - y_{1,N_{\mathrm{f}}}) + \frac{1}{\varepsilon_1} F_{\mathrm{s}}\bar{R}(x_{1,N_{\mathrm{f}}-1} - x_{1,N_{\mathrm{f}}})\right.\\
&\qquad\left. + F(x_{1,N_{\mathrm{f}}-1} - x_{1,N_{\mathrm{f}}})\right]\\
\dot{T}_{N_{\mathrm{f}}} &= \frac{1}{M_{N_{\mathrm{f}}}C_{p,\mathrm{l}}}\left[\frac{1}{\varepsilon_2} F_{\mathrm{s}}\kappa\bar{V}(\sigma_{N_{\mathrm{f}+1}}\nu_{N_{\mathrm{f}+1}}C_{p,\mathrm{l}}T_{N_{\mathrm{f}+1}} - \sigma_{N_{\mathrm{f}}}\nu_{N_{\mathrm{f}}}C_{p,\mathrm{l}}T_{N_{\mathrm{f}}})\right.\\
&\qquad\left. + \frac{1}{\varepsilon_1} F_{\mathrm{s}}\bar{R}C_{p,\mathrm{l}}(T_{N_{\mathrm{f}}-1} - T_{N_{\mathrm{f}}}) + FC_{p,\mathrm{l}}(T_0 - T_{N_{\mathrm{f}}})\right] \qquad (7.15)\\
&\vdots\\
\dot{x}_{1,i} &= \frac{1}{M_i}\left[\frac{1}{\varepsilon_1} F_{\mathrm{s}}\kappa\bar{V}(y_{1,i+1} - y_{1,i}) + \frac{1}{\varepsilon_1} F_{\mathrm{s}}\bar{R}(x_{1,i-1} - x_{1,i})\right.\\
&\qquad\left. + F(x_{1,i-1} - x_{1,i})\right]
\end{aligned}
$$

$$\dot{T}_i = \frac{1}{M_i C_{p,\mathrm{l}}} \left[\frac{1}{\varepsilon_2} F_\mathrm{s} \kappa \bar{V} (\sigma_{i+1} \nu_{i+1} C_{p,\mathrm{l}} T_{i+1} - \sigma_i \nu_i C_{p,\mathrm{l}} T_i) \right.$$
$$\left. + \frac{1}{\varepsilon_1} F_\mathrm{s} \bar{R} C_{p,\mathrm{l}} (T_{i-1} - T_i) + F C_{p,\mathrm{l}} (T_{i-1} - T_i) \right]$$

$$\vdots$$

$$\dot{M}_\mathrm{B} = \frac{1}{\varepsilon_1} F_\mathrm{s} (\bar{R} - \kappa \bar{V}) + F - B$$
$$\dot{x}_{1,\mathrm{B}} = \frac{1}{M_\mathrm{B}} \left[\frac{1}{\varepsilon_1} F_\mathrm{s} \bar{R} (x_{1,N} - x_{1,\mathrm{B}}) - \frac{1}{\varepsilon_1} F_\mathrm{s} \kappa \bar{V} (y_{1,\mathrm{B}} - x_{1,\mathrm{B}}) \right.$$
$$\left. + F (x_{1,N} - x_{1,\mathrm{B}}) \right]$$
$$\dot{T}_\mathrm{B} = \frac{1}{M_\mathrm{B} C_{p,\mathrm{l}}} \left[\frac{1}{\varepsilon_1} F_\mathrm{s} \bar{R} C_{p,\mathrm{l}} (T_N - T_\mathrm{B}) + F C_{p,\mathrm{l}} (T_N - T_\mathrm{B}) \right.$$
$$\left. + \frac{1}{\varepsilon_1} F_\mathrm{s} \kappa \bar{V} C_{p,\mathrm{l}} T_\mathrm{B} - \frac{1}{\varepsilon_2} F_\mathrm{s} \kappa C_{p,\mathrm{l}} \nu_\mathrm{B} (\bar{V} \sigma_\mathrm{R} T_\mathrm{B} - \omega_\mathrm{B} \bar{Q}_\mathrm{B}) \right]$$

Equations (7.15) are in the general form of Equations (7.1), with ε_2 playing the role of the small parameter in the sense used in (7.1). On the other hand, ε_1 captures the presence of material streams of vastly different flow rates, which, as we saw in Section 3.5, leads to a time-scale separation in the dynamics of the material-balance variables.

By definition, however, we have $\varepsilon_2 \ll \varepsilon_1$. We can thus proceed with the time-scale decomposition and model reduction of the high-purity distillation column model as outlined in Section 7.3, by defining the stretched, fast time scale $\tau_2 = t/\varepsilon_2$, in which the model becomes

$$\frac{dM_\mathrm{C}}{d\tau_2} = \frac{\varepsilon_2}{\varepsilon_1} F_\mathrm{s} (\kappa \bar{V} - \bar{R}) - \varepsilon_2 D$$
$$\frac{x_{1,\mathrm{D}}}{d\tau_2} = \frac{\varepsilon_2}{\varepsilon_1} \frac{1}{M_\mathrm{C}} F_\mathrm{s} \kappa \bar{V} (y_{1,1} - x_{1,\mathrm{D}})$$
$$\frac{dT_\mathrm{C}}{d\tau_2} = \frac{1}{M_\mathrm{C} C_{p,\mathrm{l}}} \left[F_\mathrm{s} \kappa C_{p,\mathrm{l}} (\bar{V} \sigma_1 T_1 - \omega_\mathrm{C} \bar{Q}_\mathrm{C}) - \frac{\varepsilon_2}{\varepsilon_1} F_\mathrm{s} \kappa \bar{V} C_{p,\mathrm{l}} T_\mathrm{C} \right]$$

$$\vdots$$

$$\frac{dx_{1,i}}{d\tau_2} = \frac{\varepsilon_2}{\varepsilon_1} \frac{F_\mathrm{s}}{M_i} \left[\kappa \bar{V} (y_{1,i+1} - y_{1,i}) + \bar{R} (x_{1,i-1} - x_{1,i}) \right]$$
$$\frac{dT_i}{d\tau_2} = \frac{1}{M_i C_{p,\mathrm{l}}} \left[F_\mathrm{s} \kappa \bar{V} (\sigma_{i+1} \nu_{i+1} C_{p,\mathrm{l}} T_{i+1} - \sigma_i \nu_i C_{p,\mathrm{l}} T_i) \right.$$
$$\left. + \frac{\varepsilon_2}{\varepsilon_1} F_\mathrm{s} \bar{R} C_{p,\mathrm{l}} (T_{i-1} - T_i) \right]$$

$$\vdots$$

$$
\begin{aligned}
\frac{dx_{1,N_\mathrm{f}}}{d\tau_2} &= \frac{1}{M_{N_\mathrm{f}}}\left[\frac{\varepsilon_2}{\varepsilon_1}F_\mathrm{s}\kappa\bar{V}(y_{1,N_\mathrm{f}+1}-y_{1,N_\mathrm{f}})\right.\\
&\qquad\left.+\frac{\varepsilon_2}{\varepsilon_1}F_\mathrm{s}\bar{R}(x_{1,N_\mathrm{f}-1}-x_{1,N_\mathrm{f}})+F(x_{1,N_\mathrm{f}-1}-x_{1,N_\mathrm{f}})\right]\\
\frac{dT_{N_\mathrm{f}}}{d\tau_2} &= \frac{1}{M_{N_\mathrm{f}}C_{p,\mathrm{l}}}\left[F_\mathrm{s}\kappa\bar{V}(\sigma_{N_{\mathrm{f}+1}}\nu_{N_{\mathrm{f}+1}}C_{p,\mathrm{l}}T_{N_{\mathrm{f}+1}}-\sigma_{N_\mathrm{f}}\nu_{N_\mathrm{f}}C_{p,\mathrm{l}}T_{N_\mathrm{f}})\right.\\
&\qquad\left.+\frac{\varepsilon_2}{\varepsilon_1}F_\mathrm{s}\bar{R}C_{p,\mathrm{l}}(T_{N_\mathrm{f}-1}-T_{N_\mathrm{f}})+\varepsilon_2FC_{p,\mathrm{l}}(T_0-T_{N_\mathrm{f}})\right] \qquad (7.16)\\
&\vdots\\
\frac{dx_{1,i}}{d\tau_2} &= \frac{1}{M_i}\left[\frac{\varepsilon_2}{\varepsilon_1}F_\mathrm{s}\kappa\bar{V}(y_{1,i+1}-y_{1,i})+\frac{\varepsilon_2}{\varepsilon_1}F_\mathrm{s}\bar{R}(x_{1,i-1}-x_{1,i})\right.\\
&\qquad\left.+F(x_{1,i-1}-x_{1,i})\right]\\
\frac{dT_i}{d\tau_2} &= \frac{1}{M_iC_{p,\mathrm{l}}}\left[F_\mathrm{s}\kappa\bar{V}(\sigma_{i+1}\nu_{i+1}C_{p,\mathrm{l}}T_{i+1}-\sigma_i\nu_iC_{p,\mathrm{l}}T_i)\right.\\
&\qquad\left.+\frac{\varepsilon_2}{\varepsilon_1}F_\mathrm{s}\bar{R}C_{p,\mathrm{l}}(T_{i-1}-T_i)+\varepsilon_2FC_{p,\mathrm{l}}(T_{i-1}-T_i)\right]\\
&\vdots\\
\frac{dM_\mathrm{B}}{d\tau_2} &= \frac{\varepsilon_2}{\varepsilon_1}F_\mathrm{s}(\bar{R}-\kappa\bar{V})+\varepsilon_2(F-B)\\
\frac{x_{1,\mathrm{B}}}{d\tau_2} &= \frac{1}{M_\mathrm{B}}\left[\frac{\varepsilon_2}{\varepsilon_1}F_\mathrm{s}\bar{R}(x_{1,N}-x_{1,\mathrm{B}})-\frac{1}{\varepsilon_1}F_\mathrm{s}\kappa\bar{V}(y_{1,\mathrm{B}}-x_{1,\mathrm{B}})\right.\\
&\qquad\left.+F(x_{1,N}-x_{1,\mathrm{B}})\right]\\
\frac{dT_\mathrm{B}}{d\tau_2} &= \frac{1}{M_\mathrm{B}C_{p,\mathrm{l}}}\left[\frac{\varepsilon_2}{\varepsilon_1}F_\mathrm{s}\bar{R}C_{p,\mathrm{l}}(T_N-T_\mathrm{B})+\varepsilon_2FC_{p,\mathrm{l}}(T_N-T_\mathrm{B})\right.\\
&\qquad\left.+\frac{\varepsilon_2}{\varepsilon_1}F_\mathrm{s}\kappa\bar{V}C_{p,\mathrm{l}}T_\mathrm{B}-F_\mathrm{s}\kappa C_{p,\mathrm{l}}\nu_\mathrm{B}(\bar{V}\sigma_\mathrm{R}T_\mathrm{B}-\omega_\mathrm{B}\bar{Q}_\mathrm{B})\right]
\end{aligned}
$$

Let us now consider the limiting case of an infinitely high energy throughput and set $\varepsilon_2 = 0$, for which the column model becomes

$$
\begin{aligned}
\frac{dM_\mathrm{C}}{d\tau_2} &= 0\\
\frac{x_{1,\mathrm{D}}}{d\tau_2} &= 0\\
\frac{dT_\mathrm{C}}{d\tau_2} &= \frac{1}{M_\mathrm{C}C_{p,\mathrm{l}}}\left[F_\mathrm{s}\kappa C_{p,\mathrm{l}}(\bar{V}\sigma_1T_1-\omega_\mathrm{C}\bar{Q}_\mathrm{C})\right]\\
&\vdots
\end{aligned}
$$

$$\begin{aligned}
\frac{dx_{1,i}}{d\tau_2} &= 0 \\
\frac{dT_i}{d\tau_2} &= \frac{1}{M_i C_{p,\mathrm{l}}} \left[F_\mathrm{s} \kappa \bar{V} (\sigma_{i+1} \nu_{i+1} C_{p,\mathrm{l}} T_{i+1} - \sigma_i \nu_i C_{p,\mathrm{l}} T_i) \right] \\
&\vdots \\
\frac{dx_{1,N_\mathrm{f}}}{d\tau_2} &= 0 \\
\frac{dT_{N_\mathrm{f}}}{d\tau_2} &= \frac{1}{M_{N_\mathrm{f}} C_{p,\mathrm{l}}} \left[F_\mathrm{s} \kappa \bar{V} (\sigma_{N_{\mathrm{f}+1}} \nu_{N_{\mathrm{f}+1}} C_{p,\mathrm{l}} T_{N_{\mathrm{f}+1}} - \sigma_{N_\mathrm{f}} \nu_{N_\mathrm{f}} C_{p,\mathrm{l}} T_{N_\mathrm{f}}) \right] \\
&\vdots \\
\frac{dx_{1,i}}{d\tau_2} &= 0 \\
\frac{dT_i}{d\tau_2} &= \frac{1}{M_i C_{p,\mathrm{l}}} \left[F_\mathrm{s} \kappa \bar{V} (\sigma_{i+1} \nu_{i+1} C_{p,\mathrm{l}} T_{i+1} - \sigma_i \nu_i C_{p,\mathrm{l}} T_i) \right] \\
&\vdots \\
\frac{dM_\mathrm{B}}{d\tau_2} &= 0 \\
\frac{x_{1,\mathrm{B}}}{d\tau_2} &= 0 \\
\frac{dT_\mathrm{B}}{d\tau_2} &= \frac{1}{M_\mathrm{B} C_{p,\mathrm{l}}} \left[F_\mathrm{s} \kappa C_{p,\mathrm{l}} \nu_\mathrm{B} (-\bar{V} \sigma_\mathrm{R} T_\mathrm{B} + \omega_\mathrm{B} \bar{Q}_\mathrm{B}) \right]
\end{aligned} \tag{7.17}$$

Equation (7.17) is a description of the fast dynamics of the high-purity distillation column. It involves only the stage temperatures and it can be easily verified that the system of ODEs describing the fast dynamics (as well as the quasi-steady-state conditions that result from setting the left-hand side of (7.17) to zero) are linearly independent. The constraints arising from the fast dynamics can therefore be solved (typically numerically) for the quasi-steady-state values of the stage temperatures, $\mathbf{T}^\star = [T_\mathrm{C}^\star \; T_1^\star \; \ldots T_N^\star \; T_\mathrm{B}^\star]$, which can then be substituted into the ODE system (7.8) in order to obtain a description of the dynamics after the fast temperature transient:

$$\begin{aligned}
\dot{M}_\mathrm{C} &= \frac{1}{\varepsilon_1} F_\mathrm{s} (\kappa \bar{V} - \bar{R}) - D \\
\dot{x}_{1,\mathrm{D}} &= \frac{1}{\varepsilon_1} \frac{F}{M_\mathrm{C}} \kappa \bar{V} (y_{1,1} - x_{1,\mathrm{D}}) \\
&\vdots \\
\dot{x}_{1,i} &= \frac{1}{\varepsilon_1} \frac{F_\mathrm{s}}{M_i} \left[\kappa \bar{V} (y_{1,i+1} - y_{1,i}) + \bar{R} (x_{1,i-1} - x_{1,i}) \right] \\
&\vdots
\end{aligned}$$

$$
\begin{aligned}
\dot{x}_{1,N_\mathrm{f}} &= \frac{1}{M_{N_\mathrm{f}}}\left[\frac{1}{\varepsilon_1}F_\mathrm{s}\kappa\bar{V}(y_{1,N_\mathrm{f}+1} - y_{1,N_\mathrm{f}}) + \frac{1}{\varepsilon_1}F_\mathrm{s}\bar{R}(x_{1,N_\mathrm{f}-1} - x_{1,N_\mathrm{f}})\right.\\
&\qquad \left. + F(x_{1,N_\mathrm{f}-1} - x_{1,N_\mathrm{f}})\right] \qquad (7.18)\\
&\vdots\\
\dot{x}_{1,i} &= \frac{1}{M_i}\left[\frac{1}{\varepsilon_1}F_\mathrm{s}\kappa\bar{V}(y_{1,i+1} - y_{1,i}) + \frac{1}{\varepsilon_1}F_\mathrm{s}\bar{R}(x_{1,i-1} - x_{1,i})\right.\\
&\qquad \left. + F(x_{1,i-1} - x_{1,i})\right]\\
&\vdots\\
\dot{M}_\mathrm{R} &= \frac{1}{\varepsilon_1}F_\mathrm{s}(\bar{R} - \kappa\bar{V}) + F - B\\
\dot{x}_{1,\mathrm{B}} &= \frac{1}{M_\mathrm{B}}\left[\frac{1}{\varepsilon_1}F_\mathrm{s}\bar{R}(x_{1,N} - x_{1,\mathrm{B}}) - \frac{1}{\varepsilon_1}F_\mathrm{s}\kappa\bar{V}(y_{1,\mathrm{B}} - x_{1,\mathrm{B}})\right.\\
&\qquad \left. + F(x_{1,N} - x_{1,\mathrm{B}})\right]
\end{aligned}
$$

Remark 7.2. *The vapor flow rate $\bar{V}$ depends through a constitutive relation on the reboiler temperature $\bar{V} = \bar{V}\,(T_\mathrm{B}^\star)$, thus $\bar{V}$ cannot be set independently for control purposes, but only through manipulating the reboiler heat duty $\bar{Q}_\mathrm{B}$.*

As we anticipated at the beginning of this section, owing to the presence of the small parameter ε_1, the model in Equation (7.18) is still stiff. We will follow the developments in Section 3.5 to investigate its dynamics. To this end, let us define the intermediate stretched time scale τ_1, and consider the limit of an infinite recycle flow rate, or, equivalently, $\varepsilon_1 \rightarrow 0$:

$$
\begin{aligned}
\frac{dM_\mathrm{C}}{d\tau_1} &= F_\mathrm{s}(\kappa\bar{V} - \bar{R})\\
\frac{\dot{x}_{1,\mathrm{D}}}{d\tau_1} &= \frac{F_\mathrm{s}}{M_\mathrm{C}}\kappa\bar{V}(y_{1,1} - x_{1,\mathrm{D}})\\
&\vdots\\
\frac{\dot{x}_{1,i}}{d\tau_1} &= \frac{F_\mathrm{s}}{M_i}\left[\kappa\bar{V}(y_{1,i+1} - y_{1,i}) + \bar{R}(x_{1,i-1} - x_{1,i})\right]\\
&\vdots\\
\frac{\dot{x}_{1,N_\mathrm{f}}}{d\tau_1} &= \frac{1}{M_{N_\mathrm{f}}}\left[F_\mathrm{s}\kappa\bar{V}(y_{1,N_\mathrm{f}+1} - y_{1,N_\mathrm{f}}) + F_\mathrm{s}\bar{R}(x_{1,N_\mathrm{f}-1} - x_{1,N_\mathrm{f}})\right] \qquad (7.19)\\
&\vdots
\end{aligned}
$$

$$
\begin{aligned}
\frac{\dot{x}_{1,i}}{d\tau_1} &= \frac{1}{M_i}\left[F_s\kappa\bar{V}(y_{1,i+1}-y_{1,i}) + F_s\bar{R}(x_{1,i-1}-x_{1,i})\right] \\
&\vdots \\
\frac{\dot{M}_R}{d\tau_1} &= F_s(\bar{R}-\kappa\bar{V}) \\
\frac{\dot{x}_{1,B}}{d\tau_1} &= \frac{1}{M_B}\left[F_s\bar{R}(x_{1,N}-x_{1,B}) - F_s\kappa\bar{V}(y_{1,B}-x_{1,B})\right]
\end{aligned}
$$

We turn to the slow time scale t to obtain a description of the slow dynamics. In particular, on multiplying Equations (7.18) by ε_1 and considering the limit $\varepsilon_1 \to 0$, we obtain a set of algebraic constraints that need to be satisfied in the slow time scale:

$$
\begin{aligned}
\zeta_0 &\equiv \kappa\bar{V} - \bar{R} = 0 \\
\zeta_1 &\equiv \bar{V}(y_{1,1} - x_{1,D}) = 0 \\
&\vdots \\
\zeta_i &\equiv \kappa\bar{V}(y_{1,i+1}-y_{1,i}) + \bar{R}(x_{1,i-1}-x_{1,i}) = 0 \\
&\vdots \\
\zeta_{N_f} &\equiv \kappa\bar{V}(y_{1,N_f+1}-y_{1,N_f}) + \bar{R}(x_{1,N_f-1}-x_{1,N_f}) = 0 \\
&\vdots \\
\zeta_i &\equiv \kappa\bar{V}(y_{1,i+1}-y_{1,i}) + \bar{R}(x_{1,i-1}-x_{1,i}) = 0 \\
&\vdots \\
\zeta_N &\equiv \bar{R} - \kappa\bar{V} = 0 \\
\zeta_{N+1} &\equiv \bar{R}(x_{1,N}-x_{1,B}) - \kappa\bar{V}(y_{1,B}-x_{1,B}) = 0
\end{aligned}
\tag{7.20}
$$

The description of the dynamics in the intermediate time scale and the corresponding quasi-steady-state constraints involve only the large internal material flows of the column, i.e., $\bar{V}$ and $\bar{R}$. It is easy to verify that these flows do not influence the total material holdup of the column, or the holdup of any of the components in the column and, consequently, the constraints in Equations (7.20) are not linearly independent (more specifically, the last two constraints can be expressed as a linear combination of the remaining constraints).

Finally, let us consider the limit $\varepsilon_1 \to 0$ in the slow time scale t, where the ratios $\lim_{\varepsilon_1\to 0}\zeta_i/\varepsilon_1$ become indeterminate. By denoting these unknown, yet finite terms by z_i, $i = 0,\ldots,N+1$, we obtain a description of the slow dynamics of the column that captures its slow input–output behavior. This description is in the form of a DAE system of nontrivial index, since the algebraic constraints $\zeta_i = 0$ are singular with respect to the algebraic variables z_i; a state-space realization of the model can be obtained by applying the techniques presented in Chapter 3:

$$
\begin{aligned}
\dot{M}_{\mathrm{C}} &= F_{\mathrm{s}} z_0 - D \\
\dot{x}_{1,\mathrm{D}} &= \frac{1}{\varepsilon_1} \frac{F}{M_{\mathrm{C}}} \kappa z_1 \\
&\vdots \\
\dot{x}_{1,i} &= \frac{F_{\mathrm{s}}}{M_i} z_i \\
&\vdots \\
\dot{x}_{1,N_{\mathrm{f}}} &= \frac{1}{M_{N_{\mathrm{f}}}} \left[F_{\mathrm{s}} z_{N_{\mathrm{f}}} + F(x_{1,N_{\mathrm{f}}-1} - x_{1,N_{\mathrm{f}}})\right] \\
&\vdots \\
\dot{x}_{1,i} &= \frac{1}{M_i} \left[F_{\mathrm{s}} z_i + F(x_{1,i-1} - x_{1,i})\right] \\
&\vdots \\
\dot{M}_{\mathrm{B}} &= F_{\mathrm{s}} z_N + F - B \\
\dot{x}_{1,\mathrm{B}} &= \frac{1}{M_{\mathrm{B}}} \left[F_{\mathrm{s}} z_{N+1} + F(x_{1,N} - x_{1,\mathrm{B}})\right] \\
0 &= \kappa \bar{V} - \bar{R} \\
0 &= \bar{V}(y_{1,1} - x_{1,\mathrm{D}}) \\
&\vdots \\
0 &= \kappa \bar{V}(y_{1,i+1} - y_{1,i}) + \bar{R}(x_{1,i-1} - x_{1,i}) \\
&\vdots \\
0 &= \kappa \bar{V}(y_{1,N_{\mathrm{f}}+1} - y_{1,N_{\mathrm{f}}}) + \bar{R}(x_{1,N_{\mathrm{f}}-1} - x_{1,N_{\mathrm{f}}}) \\
&\vdots \\
0 &= \kappa \bar{V}(y_{1,i+1} - y_{1,i}) + \bar{R}(x_{1,i-1} - x_{1,i}) \\
&\vdots \\
0 &= \bar{R} - \kappa \bar{V} \\
0 &= \bar{R}(x_{1,N} - x_{1,\mathrm{B}}) - \kappa \bar{V}(y_{1,\mathrm{B}} - x_{1,\mathrm{B}})
\end{aligned}
\tag{7.21}
$$

Remark 7.3. *The analysis above indicates that high-purity distillation columns are systems with high energy throughput, their dynamic behavior featuring a fast component that involves the stage temperatures and is influenced only by the large energy inputs of the column, i.e., the reboiler and condenser heat duties. It is, however, easy to infer that the same behavior characterizes the dynamics of* *all distillation columns, since the derivation of Equations (7.17) and (7.18) is independent of the assumption concerning the presence of a large internal recycle (i.e., a high reflux ratio).*

7.5.3 Control implications

The developments in this subsection have revealed that high-purity distillation columns exhibit a dynamic behavior with three time scales. Thus, according to the results in Sections 7.4 and 3.4, the design of a control system involves the synthesis of a tiered structure featuring three levels of control action.

- Equations (7.17) single out the condenser and reboiler heat duties as the only available inputs in the **fast** time scale (also see Remark 7.2). Thus, the coolant and steam flows (i.e., the physical inputs that typically correspond to the condenser and reboiler duties) should be used for controlling the temperatures in the reboiler and the condenser. Simple linear controllers are generally sufficient at this level.
- Equations (7.19) indicate that the large reflux rate $\bar{R}$ is available as a manipulated input in the **intermediate** time scale and that, by employing a cascaded control configuration, the setpoint of the reboiler temperature can be used to determine the vapor boilup rate $\bar{V}$. Thus, the two large internal material flow rates $\bar{R}$ and $\bar{V}$ are at hand for addressing control objectives in the intermediate time scale, a time scale that captures the evolution of the compositions and holdups of the individual stages and of the reboiler and condenser. Thus, the large flow rates should be relied upon for controlling the material holdups in the reboiler and condenser, a task that is also commonly achieved with simple distributed controllers. Proper care should be taken in tuning the control loops (specifically, the temperature control loop in the fast time scale should be faster than the level loop in the intermediate time scale) in order to account for the potential use of a cascaded configuration.
- Once the large internal flow rates have been set via appropriate control laws, the index of the DAE system (7.21) is well defined, and a state-space realization (ODE representation) of the **slow** subsystem can be derived. This representation of the slow dynamics of the column can be used for the derivation of a model-based nonlinear controller to govern the input–output behavior of the column, namely to address the control of the product purity and of the overall material balance. To this end, the small distillate and bottoms flow rates as well as the setpoints of the level controllers are available as manipulated inputs.

7.5.4 Simulation results and discussion

We consider a distillation column for the separation of a mixture containing 80% (molar) normal-butane (n-butane) and 20% iso-butane, fed at a flow rate of 360

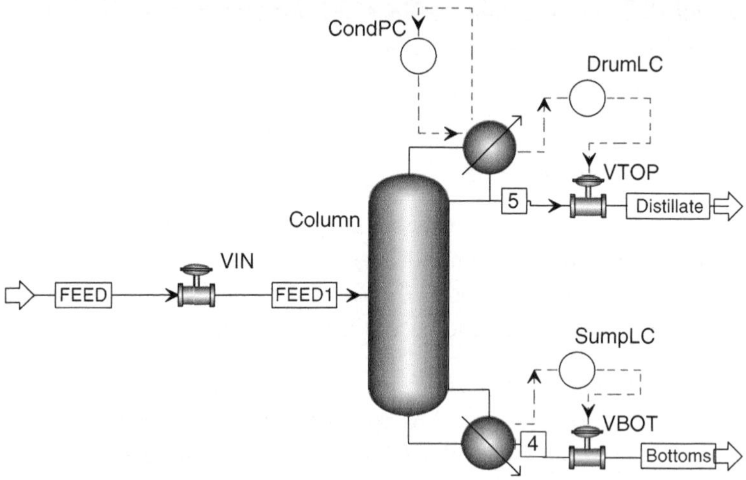

Figure 7.4 Distillation column for separation of n-butane and iso-butane.

kmol/h. It is desired that both distillation products be obtained at high purity (99.73% iso-butane in the distillate and 99.17% n-butane in the bottoms). The column has 39 stages (with the feed entering above stage 13) and is operated at a nominal top pressure of 1.5 atm. Stages are spaced at 0.7 m, the column is 1.5 m in diameter, and the weir height is 0.05 m. The nominal distillate flow rate is 69.8 kmol/h.

The column was modeled with AspenPlus®,[1] using the rigorous `radfrac` column model, in conjunction with the Redlich–Kwong–Soave equation of state for property estimation. Steady-state calculations indicated a reflux ratio of 87.67. This is a consequence of the difficult separation problem posed by the two close-boiling components.

Subsequently, we used Aspen Dynamics® for time-domain simulations. A basic control system was implemented with the sole purpose of stabilizing the (open-loop unstable) column dynamics. Specifically, the liquid levels in the reboiler and condenser are controlled using, respectively, the bottoms product flow rate and the distillate flow rate and two proportional controllers, while the total pressure in the column is controlled with the condenser heat duty and a PI controller (Figure 7.4). A controller for product purity was not implemented.

Dynamic simulations were aimed at capturing the multiple-time-scale behavior revealed by the theoretical developments presented above. Figures 7.5 and 7.6 show the evolution of the mole fraction of n-butane and of the temperature on selected column stages for a small step change in the reboiler duty. Visual inspection of the plots indicates that the temperatures exhibit a fast transient,

[1] AspenPlus and Aspen Dynamics are registered trademarks of Aspen Technology Inc.

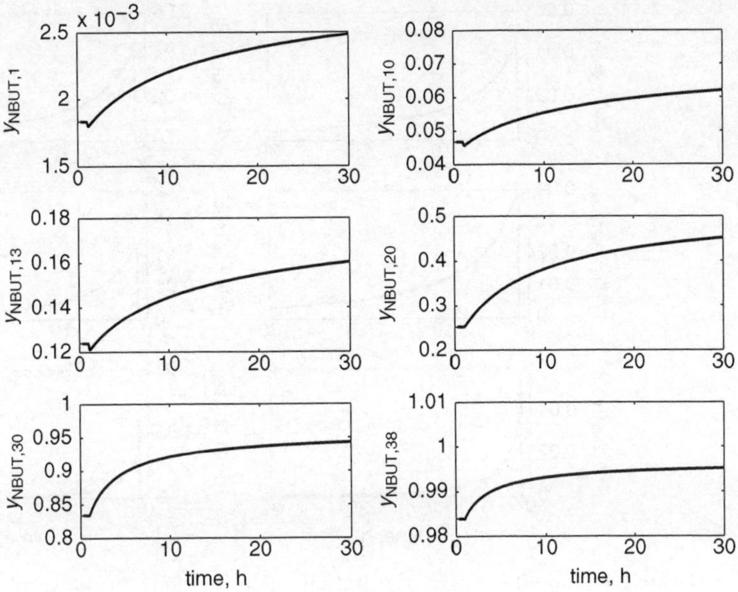

Figure 7.5 Mole fractions of n-butane on representative column stages for a 0.1% increase in column reboiler duty at time $t = 1\,\text{h}$.

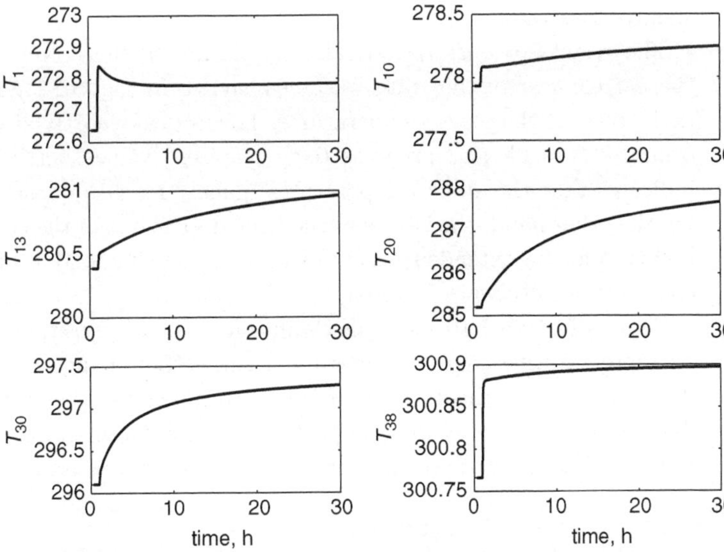

Figure 7.6 Temperatures of representative column stages for a 0.1% increase in column reboiler duty at time $t = 1\,\text{h}$.

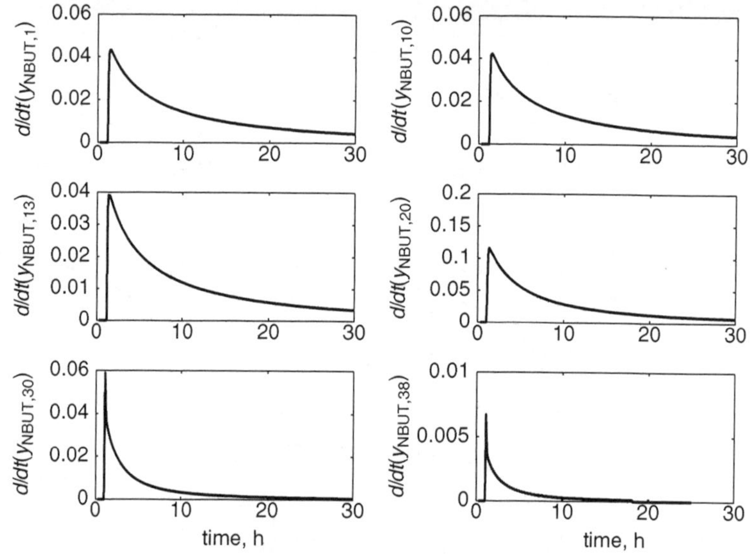

Figure 7.7 Time derivative of n-butane mole fractions on representative column stages for a 0.1% increase in column reboiler duty at time $t = 1\,\text{h}$.

while the mole fractions (slow variables) display only a slow evolution towards the new steady state. Evidently, compositions and temperatures cannot be fully decoupled, and the stage temperatures thus have a slow approach to a new steady-state value.

The results of our theoretical analysis are further confirmed by inspecting the *derivatives*[2] of the state variables, shown in Figures 7.7 and 7.8. The time derivatives of the stage temperatures show a peak shortly after the step change occurs at $t = 1\,\text{h}$ and progress to a near-zero value shortly thereafter, which indicates that the stage temperatures quickly reach a quasi-steady-state value. On the other hand, the values of the time derivatives of the compositions remain non-zero for an extended period of time, a clear indication of the fact that the compositions evolve over a long time horizon.

Figures 7.9 and 7.10 show the evolution of the compositions and temperatures on representative column stages for a small step change in the feed flow rate. According to our theory, this disturbance influences the slow material-balance dynamics and has very little impact on the fast energy dynamics of the column. Indeed, while there are significant (albeit slow) changes in the stage compositions,

[2] In order to calculate the time derivatives of the state variables, we used a finite-difference approach, sampling the simulation data with a sample time $\tau = 0.05\,\text{h}$:

$$\frac{d}{dt}y(i) = \frac{y(i) - y(i-1)}{\tau}\frac{1}{y(0)} \tag{7.22}$$

where $y(0)$ is the initial (steady-state) value of the variable and is used for scaling purposes.

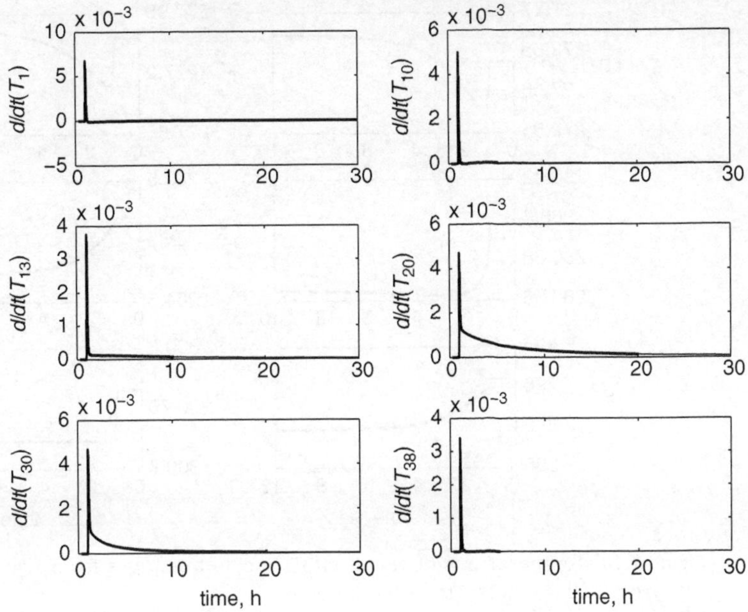

Figure 7.8 Time derivative of the temperatures of representative column stages for a 0.1% increase in column reboiler duty at time $t = 1\,\text{h}$.

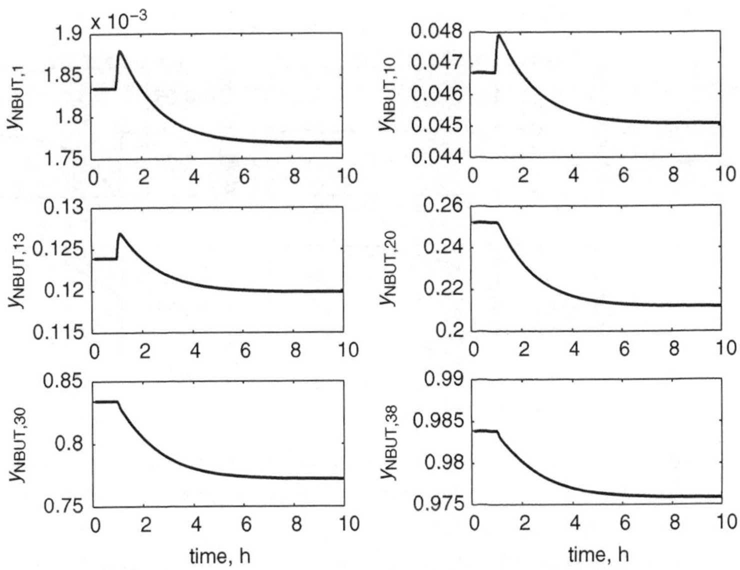

Figure 7.9 Mole fractions of n-butane on representative column stages for a rise of the feed flow rate from 360.0 kmol/h to 366.0 kmol/h.

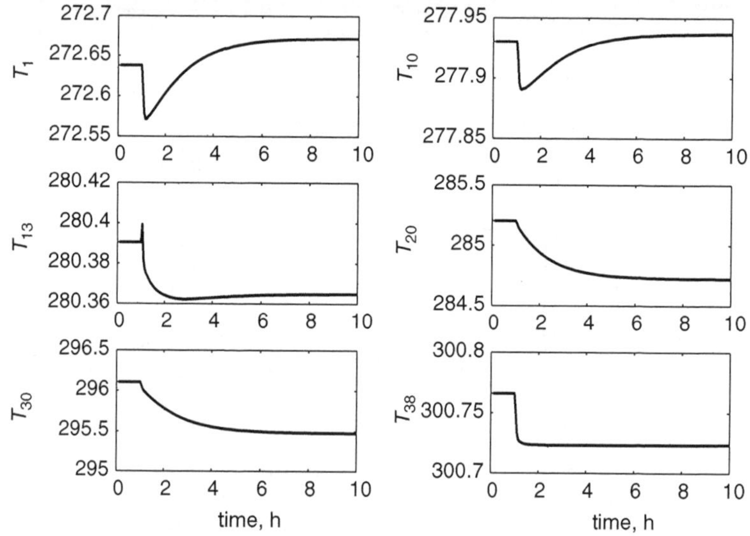

Figure 7.10 Temperatures of representative column stages for a rise of the feed flow rate from 360.0 kmol/h to 366.0 kmol/h.

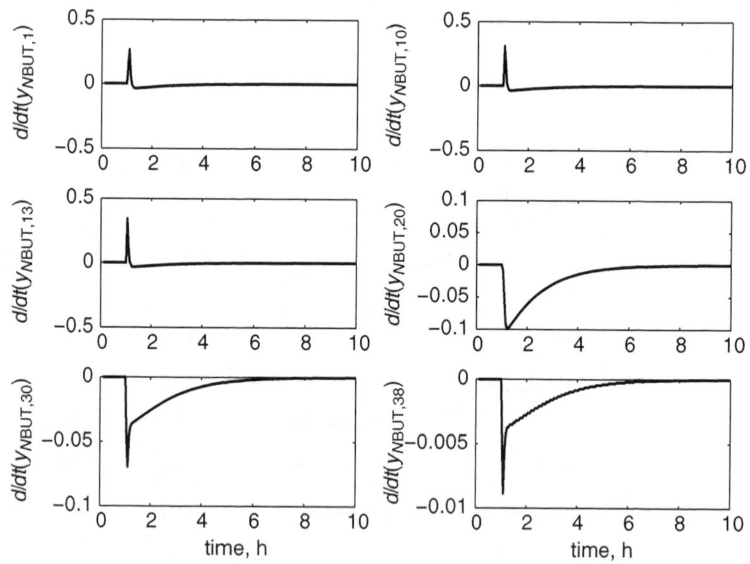

Figure 7.11 Time derivative of n-butane mole fractions on representative column stages for a rise of the feed flow rate from 360.0 kmol/h to 366.0 kmol/h.

the changes in the stage temperatures are very small. An examination of the time derivatives of the stage compositions and temperatures (Figures 7.11 and 7.12) further enforces these conclusions – note, in particular, that the values of the time derivative of the temperatures at the top and bottom of the column only briefly depart from zero and that their departure is very small. While the mid-point temperatures in the stripping section of the column (i.e., the stages

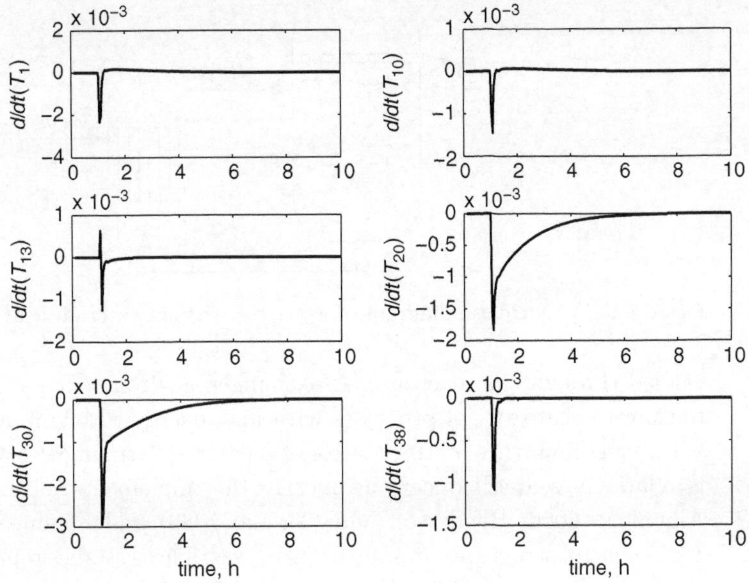

Figure 7.12 Time derivative of the temperatures of representative column stages for a rise of the feed flow rate from 360.0 kmol/h to 366.0 kmol/h.

below the feed point) do exhibit a slow transition to steady state, the temperature changes remain small. This behavior can be attributed to the dynamic influence of the liquid feed stream, which is stronger in the stripping section than in the rectifying section of the column.

7.6 Case study 2: control of a reactor with an external heat exchanger

7.6.1 Process description

In processes in which reactions with significant thermal effects are present, adiabatic reactor operation is not possible and direct heating/cooling for isothermal operation is often impractical or infeasible. In such cases, the reactor contents are recycled through an external heat exchanger (Seider *et al.* 1999). This configuration allows more efficient heat exchange and affords the process designer a choice of heat transfer area that is independent of the geometry of the reactor itself. The efficiency of the external heat exchanger can be increased further by increasing the heat capacity of the recycle stream, either by using excess quantities of a reactant or by introducing an inert diluent into the recycle loop, together with a separation unit. Such configurations can be used both in batch and in continuous processes, and are quite common in processes featuring fast, highly exothermic reactions (e.g., polymerization). The significant amount of energy

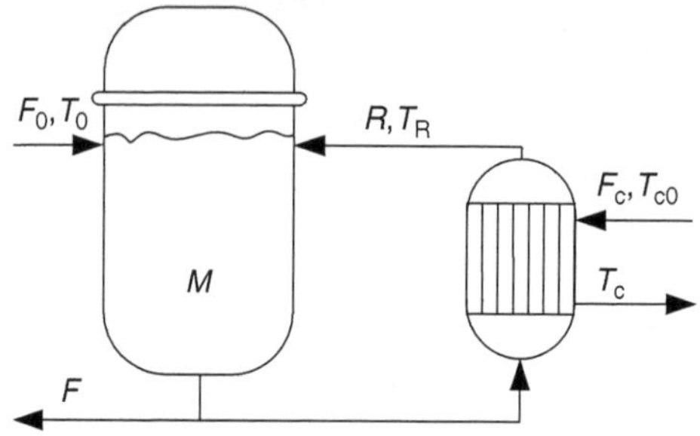

Figure 7.13 A schematic diagram of a reactor with an external heat exchanger.

carried through the reactor–heat exchanger–environment path places such systems in the category of processes with high energy throughput.

Existing literature on the control of reactor–external heat-exchanger processes is relatively scarce, concerning mostly the implementation of linear (Ali and Alhumaizi 2000, Henderson and Cornejo 1989) and nonlinear (Dadebo *et al.* 1997) control structures on specific processes. These studies report several control challenges, including difficult tuning of PID and model-based controllers due to the ill-conditioning of the process model.

In our analysis, we will consider a process consisting of a reactor with an external heat exchanger, as shown in Figure 7.13. The feed stream of flow rate F_0 contains the reactant A and its composition C_{A0} is assumed to be constant. Two consecutive first-order reactions take place in the reactor:

$$\mathrm{A} \xrightarrow{k_1} \mathrm{B} \xrightarrow{k_2} \mathrm{C} \tag{7.23}$$

The process produces either high-purity B (containing as little C as possible – operating point I (OP I)), or a mixture of B and C with $C_B/C_C = 1$ at operating point III (OP III) or $C_B/C_C = 1.5$ at operating point II (OP II), at a given production rate.

Since the thermal effect of the reactions is very high (see the process parameters in Tables 7.1 and 7.2), the adiabatic operation of the reactor is not possible. In order to control the reactor temperature, the reaction mass is recycled at a high rate through the heat exchanger.

The objectives for this process are thus the control of the reactor temperature T, of the reactor holdup M, and of the product purity C_B, at one of the operating points of interest.

7.6.2 System modeling and model reduction

In order to develop a model for this process, let M denote the reactor holdup, M_R the holdup in the tube side of the heat exchanger, and M_c the holdup in the shell side. We will denote by F_0 the feed flow rate to the reactor, by F the effluent flow rate, by F_c the coolant flow rate, and by R the recycle flow rate. Let T_0 be the temperature of the feed stream, T the reactor temperature, T_R the temperature

Table 7.1. Nominal parameter values for the reactor–heat-exchanger process (adapted from (Marroquin and Luyben 1973))

F	20.00 l/min	k_{10}	5.36×10^{10} min^{-1}
M	1200.00 l	k_{20}	4.61×10^{18} min^{-1}
M_c	68.80 l	C_p	4138.20 J l^{-1} K^{-1}
M_R	22.93 l	C_{pc}	4138.20 J l^{-1} K^{-1}
C_{A0}	2.00 mol/l	T_0	311.10 K
E_{a_1}	75.00 kJ/mol	T_{c0}	294.00 K
E_{a_2}	150.00 kJ/mol	U	1987.50 W m^{-2} K^{-1}
ΔH_1	−791.30 kJ/mol	A	11.14 m^2
ΔH_2	−527.50 kJ/mol		

Table 7.2. Temperatures, compositions, and flows for the reactor–heat-exchanger process at the operating points considered

	OP I	OP II	OP III
T, K	353.00	380.00	383.50
T_R, K	333.80	349.85	352.28
T_c, K	313.23	324.23	325.19
C_A, mol/l $\times 10^2$	7.43	1.45	1.031
C_B, mol/l	1.89	1.20	0.995
$F_c = R$, l/min	343.36	272.18	278.80

of the reaction mass in the tube side of the heat exchanger, and T_{c0} and T_c the inlet and outlet temperatures of the cooling medium, respectively. To preserve generality (we will revert to the case at hand later in this section), we will assume that $\mathcal{C}$ components participate in $\mathcal{R}$ stoichiometrically independent reactions, with reaction rates r_i, $i = 1, \ldots, \mathcal{R}$ and stoichiometric matrix $\mathbf{S} \in \mathbb{R}^{\mathcal{C}\times\mathcal{R}}$. We denote the heat-of-reaction vector by $\mathbf{\Delta H} = [\Delta H_1, \ldots, \Delta H_{\mathcal{R}}]^{\mathrm{T}}$. For simplicity, we consider the density and heat capacity of the reactants and products (ρ and C_p) and of the cooling medium used in the heat exchanger (ρ_c and C_{pc}) to be constant and temperature-independent, and C_p and C_{pc} to be of comparable magnitude, i.e., $C_p/C_{pc} = k_{cp} = \mathcal{O}(1)$. Assuming that all units are modeled as lumped-parameter systems and that the reactions take place only in the reactor, the process model becomes

$$\dot{M} = F_0 - F$$
$$\dot{\mathbf{C}} = \mathbf{Sr} + \frac{F_0}{M}(\mathbf{C}_0 - \mathbf{C})$$

$$\begin{aligned}\dot{T} &= -\frac{1}{C_p}\,\mathbf{\Delta H}^{\mathrm{T}}\,\mathbf{r} + \frac{F_0}{M}(T_0 - T) + \frac{\mathrm{R}}{M}(T_{\mathrm{R}} - T) \\ \dot{T}_{\mathrm{R}} &= \frac{R}{M_{\mathrm{R}}}(T - T_{\mathrm{R}}) - \frac{UA}{C_p M_{\mathrm{R}}}(T_{\mathrm{R}} - T_{\mathrm{c}}) \\ \dot{T}_{\mathrm{c}} &= \frac{F_{\mathrm{c}}}{M_{\mathrm{c}}}(T_{\mathrm{c}0} - T_{\mathrm{c}}) + \frac{UA}{C_{p\mathrm{c}} M_{\mathrm{c}}}(T_{\mathrm{R}} - T_{\mathrm{c}})\end{aligned} \tag{7.24}$$

where U denotes the overall heat transfer coefficient in the heat exchanger and A the heat transfer area.

Let us now define

$$\varepsilon = \frac{F_{0\mathrm{s}}}{R_{\mathrm{s}}} \tag{7.25}$$

where the subscript s denotes steady-state values. Since the recycle flow rate R_{s} is much larger than the reactor feed $F_{0\mathrm{s}}$ (see Tables 7.1 and 7.2), we have $\varepsilon \ll 1$. Also, we define the scaled inputs $u_0 = F_0/F_{0\mathrm{s}}$, $u_{\mathrm{F}} = F/F_{\mathrm{s}}$, $u_{\mathrm{R}} = R/R_{\mathrm{s}}$, and $u_{\mathrm{c}} = F_{\mathrm{c}}/F_{\mathrm{cs}}$, and the $\mathcal{O}(1)$ quantity $k_{\mathrm{F}} = F_{\mathrm{s}}/F_{0\mathrm{s}}$. The model of Equations (7.24) thus becomes

$$\begin{aligned}\dot{M} &= F_{0\mathrm{s}}(u_0 - k_{\mathrm{F}} u_{\mathrm{f}}) \\ \dot{\mathbf{C}} &= \mathbf{S}\mathbf{r} + \frac{F_{0\mathrm{s}}}{M} u_0(\mathbf{C}_0 - \mathbf{C}) \\ \dot{T} &= -\frac{1}{C_p}\,\mathbf{\Delta H}^{\mathrm{T}}\,\mathbf{r} + \frac{F_{0\mathrm{s}}}{M} u_0 (T_0 - T) + \frac{1}{\varepsilon}\frac{F_{0\mathrm{s}}}{M} u_{\mathrm{R}}(T_{\mathrm{R}} - T) \\ \dot{T}_{\mathrm{R}} &= \frac{1}{\varepsilon}\frac{F_{0\mathrm{s}}}{M_{\mathrm{R}}} u_{\mathrm{R}}(T - T_{\mathrm{R}}) - \frac{UA}{C_p M_{\mathrm{R}}}(T_{\mathrm{R}} - T_{\mathrm{c}}) \\ \dot{T}_{\mathrm{c}} &= \frac{F_{\mathrm{cs}}}{M_{\mathrm{c}}} u_{\mathrm{c}}(T_{\mathrm{c}0} - T_{\mathrm{c}}) + \frac{UA}{C_{p\mathrm{c}} M_{\mathrm{c}}}(T_{\mathrm{R}} - T_{\mathrm{c}})\end{aligned} \tag{7.26}$$

For useful energy removal, the rate of heat removal from the reactor by the recycle stream, $(RC_p(T - T_{\mathrm{R}}))_{\mathrm{s}}$, must be of the same magnitude as the rate of heat generation by the chemical reactions, $\Delta H_{\mathrm{s}} = (-\mathbf{\Delta H}^{\mathrm{T}}\,\mathbf{r}M)_{\mathrm{s}}$:

$$k_{\Delta H} = \frac{\Delta H_{\mathrm{s}}}{(RC_p(T - T_{\mathrm{R}}))_{\mathrm{s}}} = \mathcal{O}(1) \tag{7.27}$$

Equivalently,

$$\Delta H_{\mathrm{s}} = \frac{1}{\varepsilon} k_{\Delta H} F_{0\mathrm{s}} C_p (T - T_{\mathrm{R}})_{\mathrm{s}} \tag{7.28}$$

Our assumption that the heat capacities of the coolant and of the reaction mixture are of comparable magnitude, i.e.,

$$\frac{C_p}{C_{p\mathrm{c}}} = k_{\mathrm{c}p} = \mathcal{O}(1) \tag{7.29}$$

implies that the flow rate of the external cooling utility stream in the heat exchanger will be in direct relationship with the reaction mass throughput, i.e., a high recycle rate will require a high coolant flow rate. Hence, we can assume that $F_{\mathrm{cs}}/R_{\mathrm{s}} = k_{\mathrm{r}} = \mathcal{O}(1)$ and consequently $F_{0\mathrm{s}}/F_{\mathrm{cs}} = \mathcal{O}(\varepsilon)$. At steady state (assuming zero losses), the heat-transfer rate in the heat exchanger and the net rate at which heat is input to the heat exchanger by the recycle stream R are identical.

$$\frac{(UA(T_{\mathrm{R}} - T_{\mathrm{c}}))_{\mathrm{s}}}{(RC_p(T - T_{\mathrm{R}}))_{\mathrm{s}}} = 1$$

Additionally, we assume that the time constants for heat transfer and mass transport are of the same order of magnitude, i.e.,

$$\frac{UA/(C_p M_{\mathrm{R}})}{R_{\mathrm{s}}/M_{\mathrm{R}}} = k_{\mathrm{h}} = \mathcal{O}(1)$$

or, using Equation (7.25),

$$\frac{UA}{C_p} = k_{\mathrm{h}} \frac{F_{0\mathrm{s}}}{\varepsilon} \tag{7.30}$$

With the above notation, the dynamic model of the process in Figure 7.13 can be written as

$$\begin{aligned}
\dot{M} &= F_{0\mathrm{s}}(u_0 - k_{\mathrm{F}} u_{\mathrm{f}}) \\
\dot{\mathbf{C}} &= \mathbf{Sr} + \frac{F_{0\mathrm{s}}}{M} u_0 (\mathbf{C}_0 - \mathbf{C}) \\
\dot{T} &= \frac{F_{0\mathrm{s}}}{M} u_0 (T_0 - T) + \frac{1}{\varepsilon}\frac{F_{0\mathrm{s}}}{M} u_{\mathrm{R}} (T_{\mathrm{R}} - T) \\
&\quad - \frac{1}{\varepsilon}\frac{k_{\Delta H}}{\Delta H_{\mathrm{s}}} F_{0\mathrm{s}} (T - T_{\mathrm{R}})_{\mathrm{s}}\, \mathbf{\Delta H}^{\mathrm{T}}\, \mathbf{r} \\
\dot{T}_{\mathrm{R}} &= \frac{1}{\varepsilon}\frac{F_{0\mathrm{s}}}{M_{\mathrm{R}}} u_{\mathrm{R}} (T - T_{\mathrm{R}}) - \frac{1}{\varepsilon}\frac{k_{\mathrm{h}} F_{0\mathrm{s}}}{M_{\mathrm{R}}} (T_{\mathrm{R}} - T_{\mathrm{c}}) \\
\dot{T}_{\mathrm{c}} &= \frac{1}{\varepsilon}\frac{k_{\mathrm{r}} F_{0\mathrm{s}}}{M_{\mathrm{c}}} u_{\mathrm{c}} (T_{\mathrm{c}0} - T_{\mathrm{c}}) + \frac{1}{\varepsilon}\frac{k_{\mathrm{h}} k_{cp} F_{0\mathrm{s}}}{M_{\mathrm{c}}} (T_{\mathrm{R}} - T_{\mathrm{c}})
\end{aligned} \tag{7.31}$$

which is in the form of Equations (7.1). Owing to the presence of streams with flow rates of different magnitudes, and of fast heat transfer (captured by the singular perturbation parameter ε) the above model is stiff. As highlighted in the first part of this chapter, the rational control approach for this process entails the use of separate controllers for the fast dynamics, with any model-based controllers being synthesized using appropriate reduced-order models.

Let us thus proceed with the derivation of reduced-order models, starting from the fast time scale. To this end, we define the stretched, fast time scale $\tau = t/\varepsilon$ in which Equations (7.31) become

$$\begin{aligned}
\frac{dM}{d\tau} &= \varepsilon F_{0\mathrm{s}}(u_0 - k_{\mathrm{F}} u_{\mathrm{f}}) \\
\frac{d\mathbf{C}}{d\tau} &= \varepsilon \left[\mathbf{Sr} + \frac{F_{0\mathrm{s}}}{M} u_0 (\mathbf{C}_0 - \mathbf{C}) \right] \\
\frac{dT}{d\tau} &= \varepsilon \frac{F_{0\mathrm{s}}}{M} u_0 (T_0 - T) + \frac{F_{0\mathrm{s}}}{M} u_{\mathrm{R}} (T_{\mathrm{R}} - T) - \frac{k_{\Delta H}}{\Delta H_{\mathrm{s}}} F_{0\mathrm{s}} (T - T_{\mathrm{R}})_{\mathrm{s}} \, \mathbf{\Delta H}^{\mathrm{T}} \, \mathbf{r} \\
\frac{dT_{\mathrm{R}}}{d\tau} &= \frac{F_{0\mathrm{s}}}{M_{\mathrm{R}}} u_{\mathrm{R}} (T - T_{\mathrm{R}}) - \frac{k_{\mathrm{h}} F_{0\mathrm{s}}}{M_{\mathrm{R}}} (T_{\mathrm{R}} - T_{\mathrm{c}}) \\
\frac{dT_{\mathrm{c}}}{d\tau} &= \frac{k_{\mathrm{r}} F_{0\mathrm{s}}}{M_{\mathrm{c}}} u_{\mathrm{c}} (T_{\mathrm{c}0} - T_{\mathrm{c}}) + \frac{k_{\mathrm{h}} k_{cp} F_{0\mathrm{s}}}{M_{\mathrm{c}}} (T_{\mathrm{R}} - T_{\mathrm{c}})
\end{aligned} \tag{7.32}$$

Then, we consider the limit $\varepsilon \to 0$, corresponding to infinitely large recycle and cooling medium flow rates and infinitely fast heat transfer in the heat exchanger. In this limit, we obtain the following description of the fast dynamics of the process:

$$\begin{aligned}
\frac{dT}{d\tau} &= \frac{F_{0\mathrm{s}}}{M} u_{\mathrm{R}} (T_{\mathrm{R}} - T) - \frac{k_{\Delta H}}{\Delta H_{\mathrm{s}}} F_{0\mathrm{s}} (T - T_{\mathrm{R}})_{\mathrm{s}} \, \mathbf{\Delta H}^{\mathrm{T}} \, \mathbf{r} \\
\frac{dT_{\mathrm{R}}}{d\tau} &= \frac{F_{0\mathrm{s}}}{M_{\mathrm{R}}} u_{\mathrm{R}} (T - T_{\mathrm{R}}) - \frac{k_{\mathrm{h}} F_{0\mathrm{s}}}{M_{\mathrm{R}}} (T_{\mathrm{R}} - T_{\mathrm{c}}) \\
\frac{dT_{\mathrm{c}}}{d\tau} &= \frac{k_{\mathrm{r}} F_{0\mathrm{s}}}{M_{\mathrm{c}}} u_{\mathrm{c}} (T_{\mathrm{c}0} - T_{\mathrm{c}}) + \frac{k_{\mathrm{h}} k_{cp} F_{0\mathrm{s}}}{M_{\mathrm{c}}} (T_{\mathrm{R}} - T_{\mathrm{c}})
\end{aligned} \tag{7.33}$$

In order to obtain a description of the slow dynamics, we first recognize that the equations describing the energy balance can be replaced, in the slow time scale, by the corresponding quasi-steady-state constraints. These constraints are obtained by multiplying Equations (7.31) by ε and considering the limit $\varepsilon \to 0$:

$$\begin{aligned}
0 &= \frac{F_{0\mathrm{s}}}{M} u_{\mathrm{R}} (T_{\mathrm{R}} - T) - \frac{k_{\Delta H}}{\Delta H_{\mathrm{s}}} F_{0\mathrm{s}} (T - T_{\mathrm{R}})_{\mathrm{s}} \, \mathbf{\Delta H}^{\mathrm{T}} \mathbf{r} \\
0 &= \frac{F_{0\mathrm{s}}}{M_{\mathrm{R}}} u_{\mathrm{R}} (T - T_{\mathrm{R}}) - \frac{k_{\mathrm{h}} F_{0\mathrm{s}}}{M_{\mathrm{R}}} (T_{\mathrm{R}} - T_{\mathrm{c}}) \\
0 &= \frac{k_{\mathrm{r}} F_{0\mathrm{s}}}{M_{\mathrm{c}}} u_{\mathrm{c}} (T_{\mathrm{c}0} - T_{\mathrm{c}}) + \frac{k_{\mathrm{h}} k_{cp} F_{0\mathrm{s}}}{M_{\mathrm{c}}} (T_{\mathrm{R}} - T_{\mathrm{c}})
\end{aligned} \tag{7.34}$$

It is straightforward to verify that the algebraic constraints in Equations (7.34) are generically linearly independent and hence they can be solved for the quasi-steady-state values $\boldsymbol{\theta}^\star(M, \mathbf{C}) = [T^\star, T_{\mathrm{R}}^\star, T_{\mathrm{c}}^\star]$ of the variables $[T, T_{\mathrm{R}}, T_{\mathrm{c}}]$. Substituting the value for $T^\star$, we then obtain

$$\begin{aligned}
\dot{M} &= F_{\mathrm{s}}(k_{\mathrm{f}} u_0 - u_{\mathrm{F}}) \\
\dot{\mathbf{C}} &= \mathbf{Sr}(\boldsymbol{\theta}^\star) + \frac{k_{\mathrm{f}} F_{\mathrm{s}}}{M} u_0 (\mathbf{C}_0 - \mathbf{C})
\end{aligned} \tag{7.35}$$

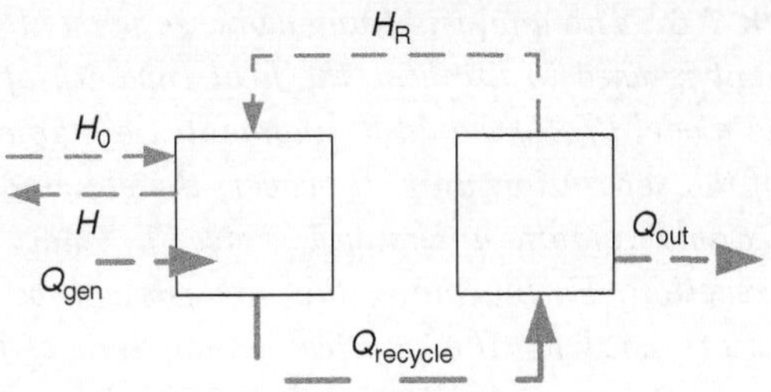

Figure 7.14 Energy flows of different magnitudes in a reactor–external-heat-exchanger system.

which represents the model of the slow dynamics of the reactor–external-heat-exchanger system.

Remark 7.4. *Figure 7.14 illustrates the material and energy flows in the reactor–heat exchanger. The rate of heat generation by the highly exothermic reactions,* Q_{gen}*, the rate of heat removal from the reactor by the large recycle stream acting as an energy carrier,* Q_{recycle}*, and the rate of heat removal from the process by the coolant,* Q_{out}*, are of comparable magnitude. These terms are much larger than the rate of heat removal by the reactor effluent (*Q_{F}*) and thus dominate the energy balance of the system.*

On the other hand, the material throughput of the process is small, owing to the small reactor feed flow rate F*. While the recycle rate* R *is much larger than the feed flow rate* F*, under the assumption that no reaction occurs outside the reactor, its composition remains constant. Therefore,* the large recycle stream has no influence on the material balance *and* the material-balance equations do not contain any large terms. *From these features, one can infer that the energy dynamics of the process, being dominated by the large terms corresponding to the generation and removal of heat through the heat exchanger, i.e., by the high energy throughput, is faster than the dynamics of the material balance, which is characterized by the small material throughput. This conclusion is consistent with the results of the analysis presented above.*

Remark 7.5. *The material- and energy-balance equations of the process are not decoupled: the rates of heat generation from the* $\mathcal{R}$ *reactions are the product of two terms,* ΔH_i *and* r_i*, corresponding to the heat of reaction and reaction rate, respectively. Consequently, a high rate of heat generation by reaction could be due to fast reactions with moderate reaction enthalpy, or to reactions that have moderate rates and a high heat of reaction. If the reaction chemistry involves both of the aforementioned reaction categories, the material balance of Equations (7.35) will itself be in a nonstandard singularly perturbed form (Baldea and Daoutidis 2005), and further reduction steps will be necessary in order to obtain non-stiff descriptions of an intermediate and a slow dynamics.*

Remark 7.6. *The analysis framework we presented is also applicable if an inert component is used to increase the heat capacity of the reaction mixture. In this case, the model (7.24) would be augmented by the equations corresponding to the model of the separation unit. However, the stoichiometric matrix* $\mathbf{S}$ *and reaction rates* $\mathbf{r}$ *would remain unchanged, since the inert component does not partake in any reaction. Furthermore, the analysis can be applied if more complex correlations are used for the physical parameters of the system (e.g., temperature dependence of heat capacities and densities), as long as the basic assumptions (7.27), (7.29), and (7.30) apply.*

7.6.3 Control implications and controller implementation

The arguments presented above indicate that the large recycle and coolant flow rates u_R and u_C are the only manipulated inputs available in the fast time scale, and should be used to control the process temperatures. Likewise, the dynamics of the material-balance variables in the slow time scale are affected only by the small feed and effluent flow rates u_0 and u_F, which are thus the manipulated inputs that must be used to tackle control objectives involving the material balance. $\boldsymbol{\theta}_{sp}$, the setpoints of the temperature controllers in the fast time scale, are also available as manipulated inputs in the slow time scale, a choice that leads to cascaded control configurations between the "energy-" and "material-balance" controllers.

In most cases, the only objective in the fast time scale is the control of the reactor temperature, for which there are two available manipulated inputs, u_R and u_C. Thus, several control system design options are available.

(i) Control the reactor temperature using the coolant flow rate u_C as a manipulated input, while fixing R at its nominal value ($u_R = 1$).
(ii) Use two controllers, one to control the reactor temperature using the recycle flow rate u_R, and the other to control the recycle stream temperature T_R with the coolant flow rate u_C.
(iii) Use a single controller to control the reactor temperature and manipulate u_C **and** u_R, keeping the two flow rates in a fixed ratio. The ratio u_C/u_R depends on the cost of circulating the reaction mass and the cost of the coolant, and constitutes a design parameter.

Notice that in the first case the energy transfer between the reactor and the heat exchanger is a limiting factor in the overall energy flow, which could lead to reactor runaway because of insufficient heat-removal capacity. In the second case one overcomes this problem at the cost of a more elaborate control structure, while the third approach combines the benefits of the preceding two, i.e., it avoids the heat-transfer limitations and relies on a simple control structure. In principle, (iii) can be regarded as using the net rate of heat removal from the reactor as a manipulated input for controlling the reactor temperature.

On reverting to the particular case considered in Section 7.6.1, and applying the model-reduction framework outlined above, we obtain the following description of the fast dynamics:

$$\begin{aligned}
\frac{dT}{d\tau} &= \frac{F_{0\mathrm{s}}}{M}u_{\mathrm{R}}(T_{\mathrm{R}} - T) - \frac{k_{\Delta H}}{\Delta H_{\mathrm{s}}}F_{0\mathrm{s}}(T - T_{\mathrm{R}})_{\mathrm{s}} \\
&\quad \times \left(\Delta H_1\, k_{10}e^{-E_{\mathrm{a1}}/(RT)}C_{\mathrm{A}} + \Delta H_2\, k_{20}e^{-E_{\mathrm{a2}}/(RT)}C_{\mathrm{B}}\right) \\
\frac{dT_{\mathrm{R}}}{d\tau} &= \frac{F_{0\mathrm{s}}}{M_{\mathrm{R}}}u_{\mathrm{R}}(T - T_{\mathrm{R}}) - \frac{k_{\mathrm{h}}F_{0\mathrm{s}}}{M_{\mathrm{R}}}(T_{\mathrm{R}} - T_{\mathrm{c}}) \\
\frac{dT_{\mathrm{c}}}{d\tau} &= \frac{k_{\mathrm{r}}F_{0\mathrm{s}}}{M_{\mathrm{c}}}u_{\mathrm{c}}(T_{\mathrm{c}0} - T_{\mathrm{c}}) + \frac{k_{\mathrm{h}}k_{cp}F_{0\mathrm{s}}}{M_{\mathrm{c}}}(T_{\mathrm{R}} - T_{\mathrm{c}})
\end{aligned} \tag{7.36}$$

According to our analysis, we address the control of the reactor temperature T in the fast time scale, keeping the ratio $u_{\mathrm{C}}/u_{\mathrm{R}}$ constant and using the proportional-integral feedback law:

$$u_{\mathrm{C}} = 1 + K_{\mathrm{C}}\left[T - T_{\mathrm{sp}} + \frac{1}{\tau_i}\int_0^t (T - T_{\mathrm{sp}})dt\right] \tag{7.37}$$

The constraints arising from the fast dynamics (7.36) can now be solved for the quasi-steady-state value, i.e., $T^\star = T_{\mathrm{sp}}$, which allows us to obtain a description of the slow dynamics:

$$\begin{aligned}
\dot{M} &= F_0 - F \\
\dot{C}_{\mathrm{A}} &= \frac{F_0}{M}(C_{\mathrm{A}0} - C_{\mathrm{A}}) - k_{10}e^{-E_{\mathrm{a1}}/(RT_{\mathrm{sp}})}C_{\mathrm{A}} \\
\dot{C}_{\mathrm{B}} &= -\frac{F_0}{M}C_{\mathrm{B}} + k_{10}e^{-E_{\mathrm{a1}}/(RT_{\mathrm{sp}})}C_{\mathrm{A}} - k_{20}e^{-E_{\mathrm{a2}}/(RT_{\mathrm{sp}})}C_{\mathrm{B}} \\
\dot{C}_{\mathrm{C}} &= -\frac{F_0}{M}C_{\mathrm{C}} + k_{20}e^{-E_{\mathrm{a2}}/(RT_{\mathrm{sp}})}C_{\mathrm{B}}
\end{aligned} \tag{7.38}$$

Since the flow rate of the feed stream F_0 is fixed (and subject to disturbances arising from changes in the upstream process conditions), it is not available as a manipulated input in the slow time scale. Therefore, we address the control of the inventory and the product purity C_{B} by employing, respectively, F and T_{sp} as manipulated inputs, the latter choice leading to a cascaded control configuration.

Prior to embarking on the design of the purity controller, let us consider a steady-state analysis: Figure 7.15 presents the steady-state concentration profiles as a function of the reactor temperature. C_{B} exhibits a maximum at $T_{\mathrm{max}} = 353.55$ K. However, operating the reactor at T_{max} is not feasible because $dC_{\mathrm{B}}/dT|_{T=T_{\mathrm{max}}} = 0$ (at $T = T_{\mathrm{max}}$ controllability is lost), and the first requirement, namely obtaining a product stream with a high concentration of B and a very low concentration of C, is fulfilled by operating the reactor at

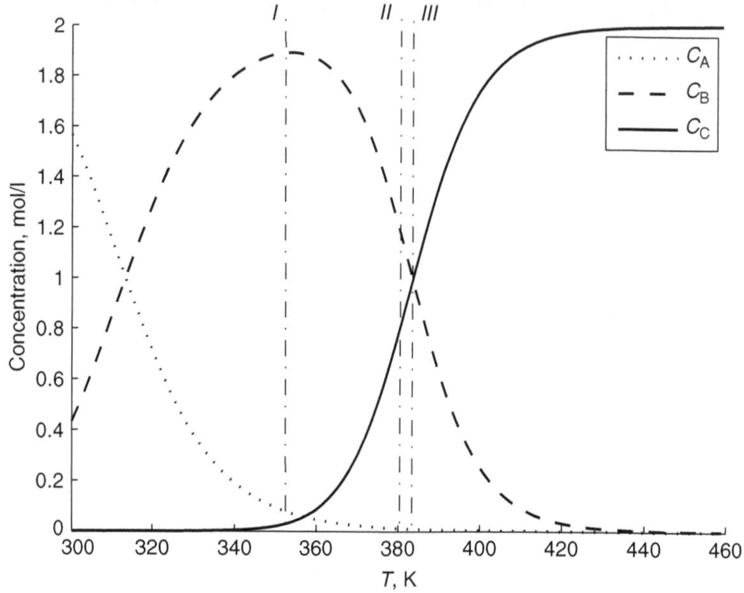

Figure 7.15 Steady-state reactor concentrations as functions of temperature.

$T_{\mathrm{sp,I}} = 353.00$ K $< T_{\max}$ (operating point I). On the other hand, the product mixtures with $C_{\mathrm{B}}/C_{\mathrm{C}} = 1.5$ and $C_{\mathrm{B}}/C_{\mathrm{C}} = 1$ can be obtained by operating the reactor at $T_{\mathrm{sp,II}} = 380.00$ K $> T_{\max}$ and at $T_{\mathrm{sp,III}} = 383.50$ K $> T_{\max}$, respectively (operating points II and III). Owing to the different signs of the steady-state gain of the process at operating point I, and at operating points II and III, any linear controller with integral action leads to instability if used *both* at operating point I *and* at operating points II and III, while a proportional controller leads to offset (Daoutidis and Kravaris 1992), a limitation that does not affect a nonlinear controller.

In what follows, we address the design of such a controller. We assume that switching between a product with $C_{\mathrm{B}}/C_{\mathrm{C}} = 1.5$ and a product with $C_{\mathrm{B}}/C_{\mathrm{C}} = 1$ is also required. Thus, in addition to good disturbance-rejection abilities at all operating points, the controller is required to exhibit good setpoint-tracking abilities between operating points II and III. Figure 7.16 shows the evolution of the product purity, initially at the steady state $C_{\mathrm{B}} = 1.2$ mol/l, $T = 380.00$ K (operating point II) in response to a 1-K increase in T_{sp}. Notice that the product purity exhibits an inverse response. This non-minimum phase behavior originates from the increased contribution of the second reaction at temperatures higher than $T_{\max}$. Namely, around operating point II, a rise in temperature leads to a rise in the rates both of the first and of the second reaction. However, the rate of the first reaction is quicker to increase than that of the second and, immediately after the temperature rise, more B is generated than is consumed. Consequently, C_{B} increases. Subsequently, the consumption of B increases because the rate

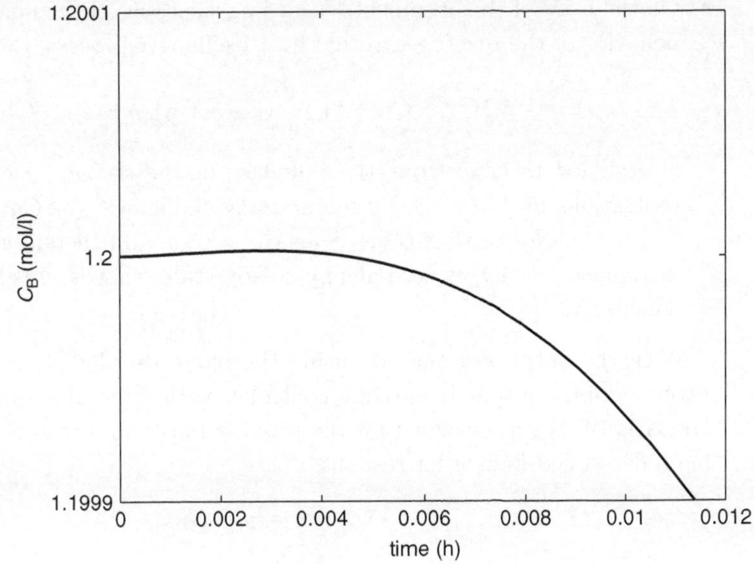

Figure 7.16 Inverse response of the product purity C_B.

of the second reaction increases, and C_B falls as expected. A similar dynamic behavior is encountered around operating point III.

Motivated by the above, we address the design of the product-purity controller for the reactor–external heat-exchanger process by following the approach of statically equivalent outputs (Kravaris *et al.* 1998) in a manner analogous to Kumar and Daoutidis (1999b). To this end, we will construct an auxiliary output $\tilde{y}$ such that (i) $\tilde{y}$ is statically equivalent to the process output C_B, i.e., $\tilde{y} = C_B$ at every steady state; and (ii) the system is minimum phase with respect to $\tilde{y}$ (and to the other output, M). Once such an output $\tilde{y}$ has been constructed, an input–output linearizing controller will yield asymptotic tracking for C_B, with closed-loop stability. We consider a statically equivalent output (notice that this choice is not unique) of the form

$$\tilde{y} = C_B + \gamma_{1,1}\frac{dC_B}{dt} + \beta\left[C_A + C_B + \gamma_{1,2}\frac{d}{dt}(C_A + C_B) - (C_{A0} - C_C)\right] \quad (7.39)$$

with $\gamma_{1,1}$, $\gamma_{1,2}$, and β being scalar parameters. The above form is motivated by the following two factors.

- The term $C_B + \gamma_{1,1}\, dC_B/dt$ is statically equivalent to C_B, and corresponds to requesting a first-order response in C_B when using a standard input–output linearizing controller. However, such a controller would lead to closed-loop instability, and the output requires a "statically equivalent" addition that would allow one to overcome this limitation.

- On the basis of the arguments regarding the cause of the non-minimum-phase behavior of the reactor–external-heat-exchanger process, the term

$$\beta\left[C_{\mathrm{A}} + C_{\mathrm{B}} + \gamma_{1,2}\frac{d}{dt}(C_{\mathrm{A}} + C_{\mathrm{B}}) - (C_{\mathrm{A0}} - C_{\mathrm{C}})\right]$$

  is designed to cancel out the influence of the second reaction on the concentrations of A and B, by maintaining the sum $C_{\mathrm{A}} + C_{\mathrm{B}}$ at its "setpoint" $C_{\mathrm{A0}} - C_{\mathrm{C}}$. Notice that $C_{\mathrm{A}} + C_{\mathrm{B}} + C_{\mathrm{C}} = C_{\mathrm{A0}}$, and therefore the second term in Equation (7.39) is zero at any steady state; that is, $\tilde{y}$ is statically equivalent to C_{B}.

With the outputs $\tilde{y}$ and M, using the reduced-order model (7.38), a multivariable input–output linearizing controller with integral action (Daoutidis and Kravaris 1994) was designed for the product purity and reactor holdup, requesting a decoupled first-order response:

$$\tilde{y} = C_{\mathrm{B,sp}} \tag{7.40}$$

$$M + \gamma_2\frac{dM}{dt} = M_{\mathrm{sp}} \tag{7.41}$$

The controller was tuned with $\gamma_{1,1} = 30$ min, $\gamma_{1,2} = 57$ min, $\beta = 0.5$, and $\gamma_2 = 20$ min, and, with the linear controller (7.37) tuned with $K_{\mathrm{C}} = 0.17\ \mathrm{K}^{-1}$ and $\tau_{\mathrm{i}} = 2.8$ min, its performance was studied through simulations.

7.6.4 Simulation results and discussion

An initial scenario considered the process to be at operating point II ($C_{\mathrm{B,sp}} = 1.2$ mol/l, corresponding to $C_{\mathrm{B}}/C_{\mathrm{C}} = 1.5$). At $t = 60$ min, we imposed a drop in the product-purity setpoint to $C_{\mathrm{B,sp}} = 0.995$ mol/l (operating point III, corresponding to $C_{\mathrm{B}}/C_{\mathrm{C}} = 1$). After 19 h of operation, the setpoint of C_{B} was raised to 1.2 mol/l, switching the operation of the process back to point II. Figures 7.17–7.19 present the evolution of C_{B}, of the reactor temperature, and of the coolant flow rate for this simulation. Observe that the proposed nonlinear controller exhibits excellent tracking performance. Figures 7.20–7.22 show the closed-loop profiles for the same scenario in the presence of unmodeled disturbances and plant–model mismatch, again evincing excellent performance.

Figures 7.23–7.27 show the closed-loop profiles for a 10% increase in the production rate at operating point I (attained by increasing F_0), and a decrease in the purity setpoint to $C_{\mathrm{B,sp}} = 1.888$ mol/l – this reduction is necessary since the nominal purity is beyond the maximum attainable purity for the increased throughput. Although controller design was carried out to account for the inverse response exhibited by the system at operating points II and III, and in spite of the plant–model parameter mismatch, the proposed control structure clearly yields good performance at operating point I as well.

Finally, Figures 7.28–7.30 present the case of operating the system at point I with an unmeasured 6.0-K rise in the coolant inlet temperature occurring at

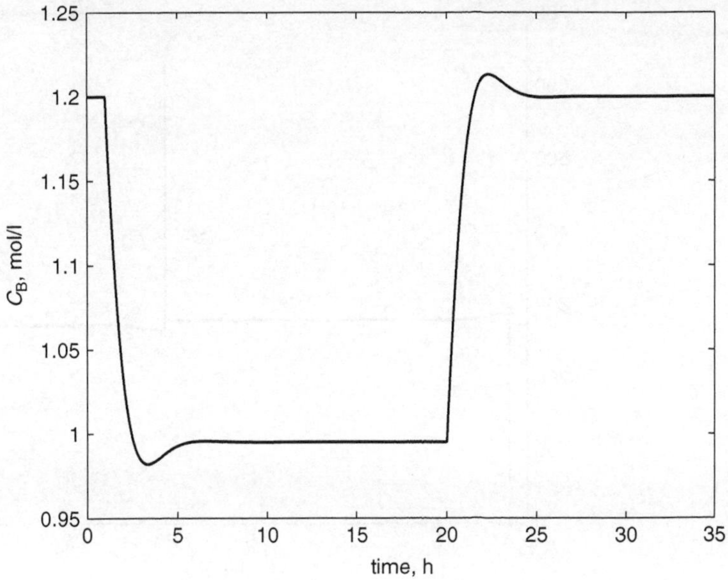

Figure 7.17 Evolution of the product purity in the case of switching between operating points II and III.

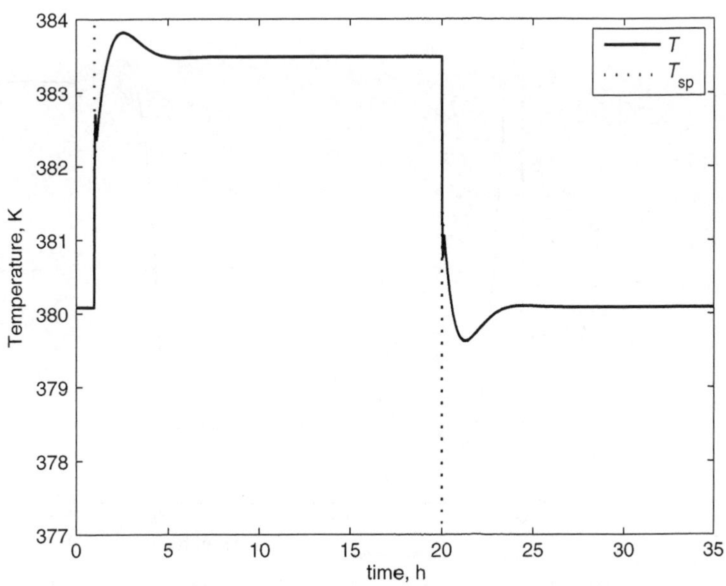

Figure 7.18 Evolution of the reactor temperature in the case of switching between operating points II and III.

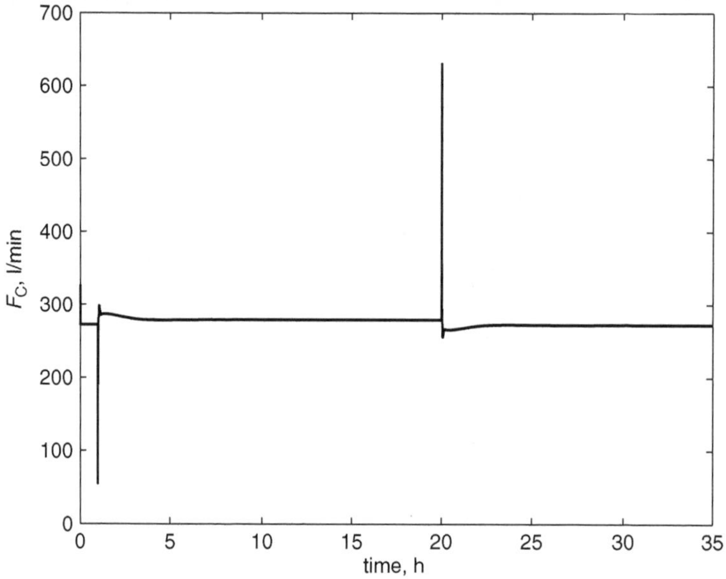

Figure 7.19 Evolution of the coolant flow rate in the case of switching between operating points II and III.

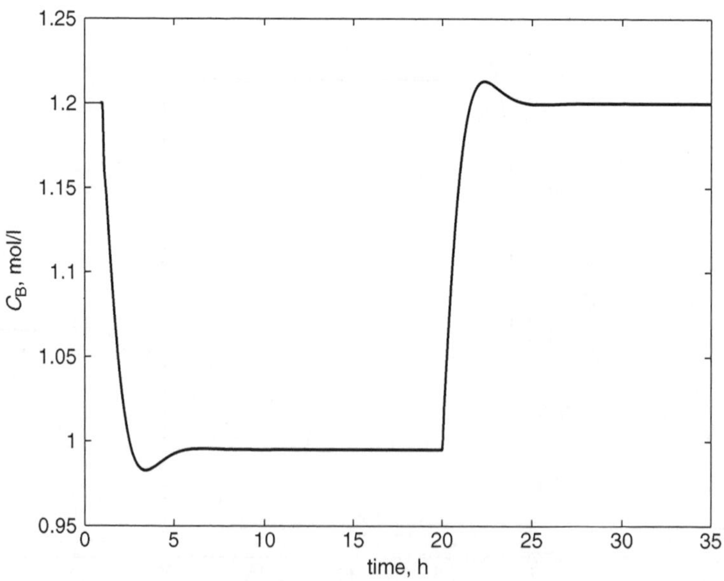

Figure 7.20 Evolution of the product purity in the case of switching between operating points II and III, in the presence of a 10-K unmeasured increase in the coolant inlet temperature and an unmodeled 20% drop in the heat transfer coefficient U. Both disturbances occur at $t = 60$ min.

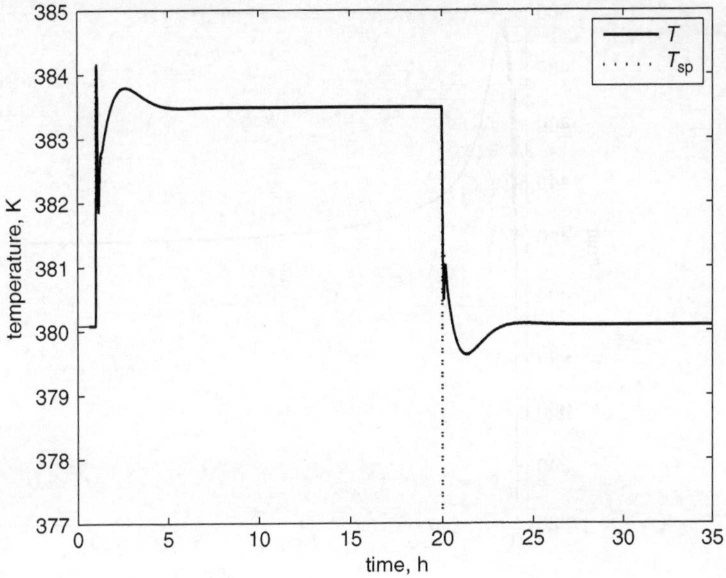

Figure 7.21 Evolution of the reactor temperature in the case of switching between operating points II and III, in the presence of a 10-K unmeasured increase in the coolant inlet temperature and an unmodeled 20% drop in the heat transfer coefficient U. Both disturbances occur at $t = 60$ min.

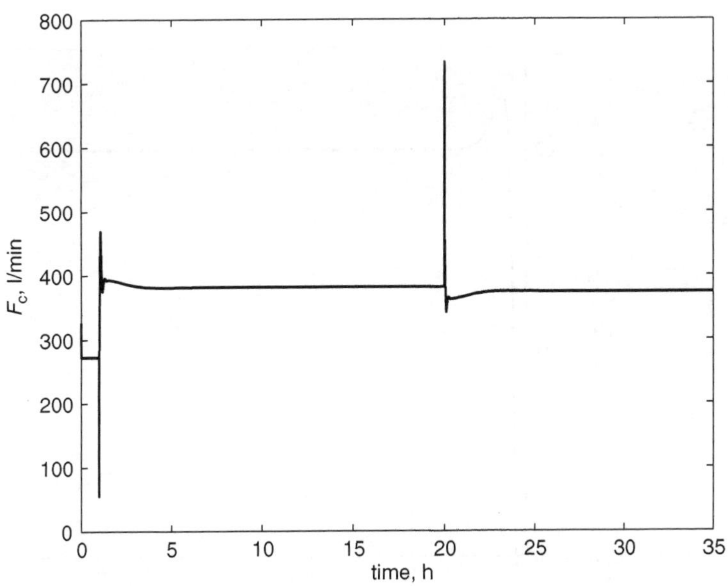

Figure 7.22 Evolution of the coolant flow rate in the case of switching between operating points II and III, in the presence of a 10-K unmeasured increase in the coolant inlet temperature and an unmodeled 20% drop in the heat transfer coefficient U. Both disturbances occur at $t = 60$ min.

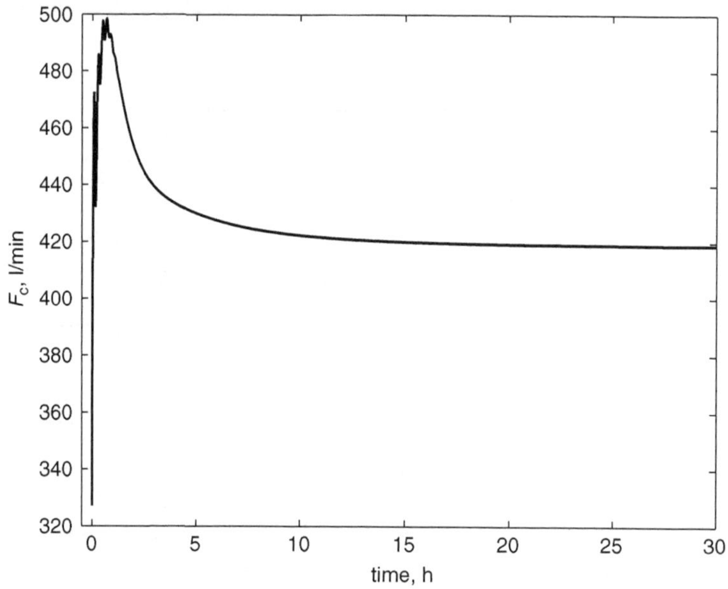

Figure 7.23 Evolution of the coolant flow rate for a 10% rise in the production rate at operating point I, under plant–model parameter mismatch. The heat transfer coefficient U in the controller model is overestimated by 10% compared with its value in the plant.

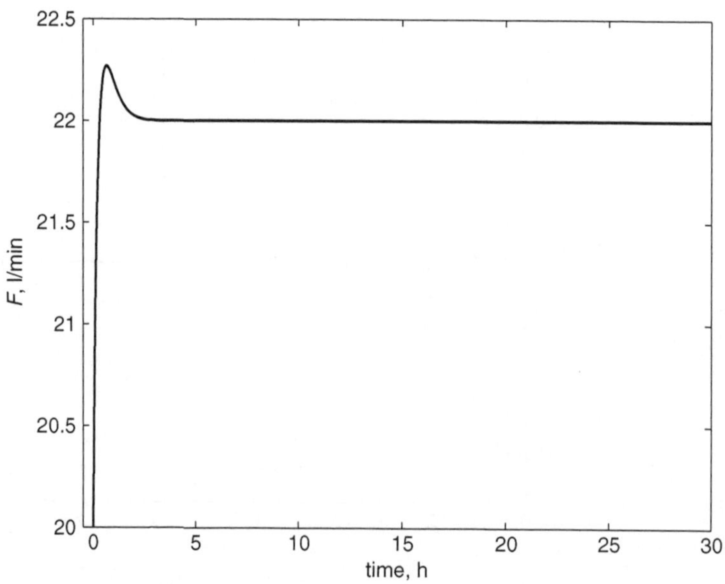

Figure 7.24 Evolution of the product flow rate for a 10% rise in the production rate at operating point I, under plant–model parameter mismatch. The heat transfer coefficient U in the controller model is overestimated by 10% compared with its value in the plant.

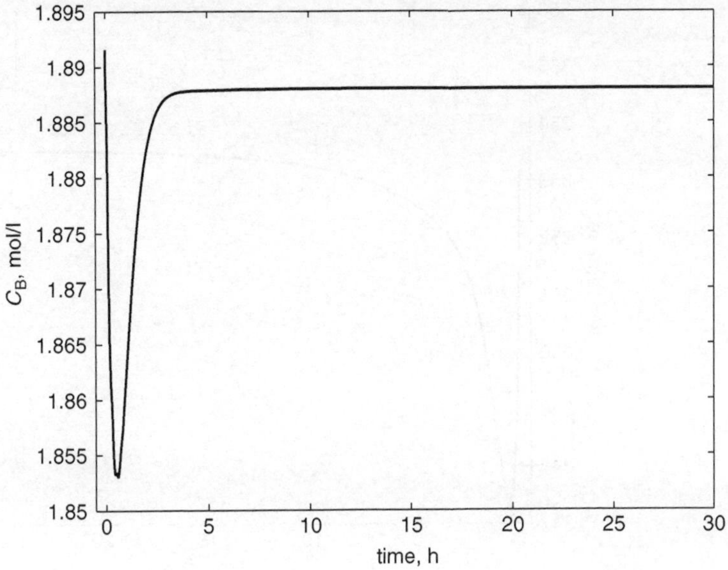

Figure 7.25 Evolution of the product purity for a 10% rise in the production rate at operating point I, under plant–model parameter mismatch. The heat transfer coefficient U in the controller model is overestimated by 10% compared with its value in the plant.

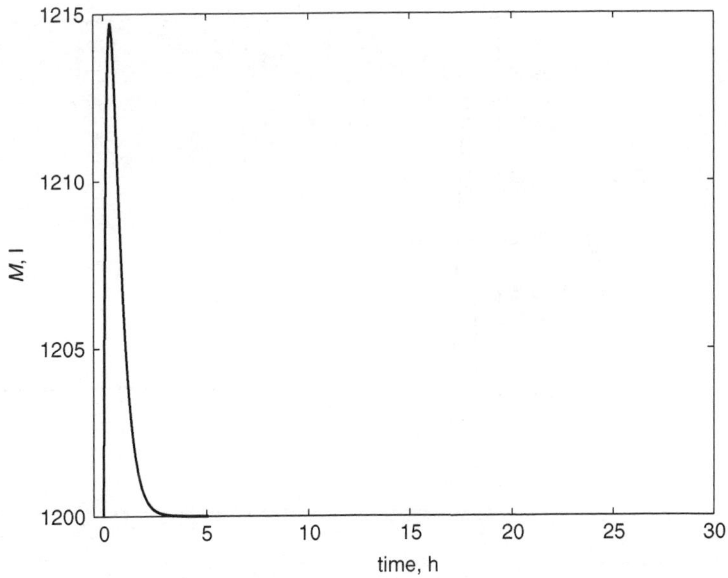

Figure 7.26 Evolution of the reactor holdup for a 10% rise in the production rate at operating point I, under plant–model parameter mismatch. The heat transfer coefficient U in the controller model is overestimated by 10% compared with its value in the plant.

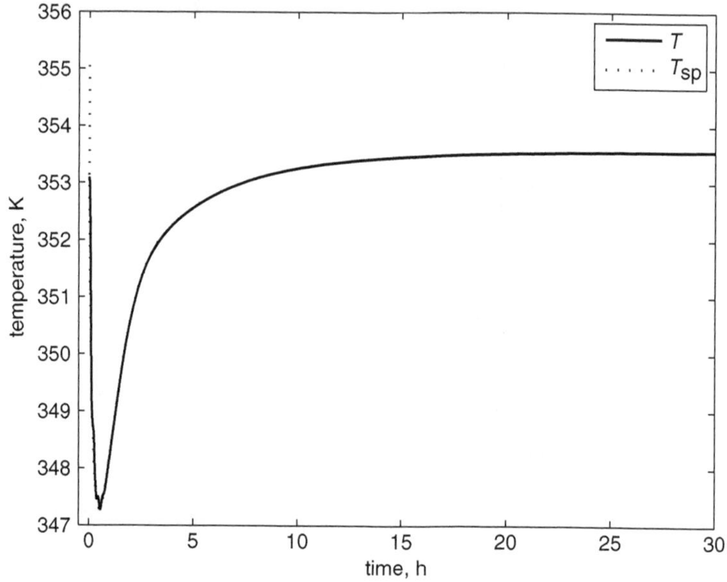

Figure 7.27 Evolution of the reactor temperature and temperature setpoint for a 10% rise in the production rate at operating point I, under plant–model parameter mismatch. The heat transfer coefficient U in the controller model is overestimated by 10% compared with its value in the plant.

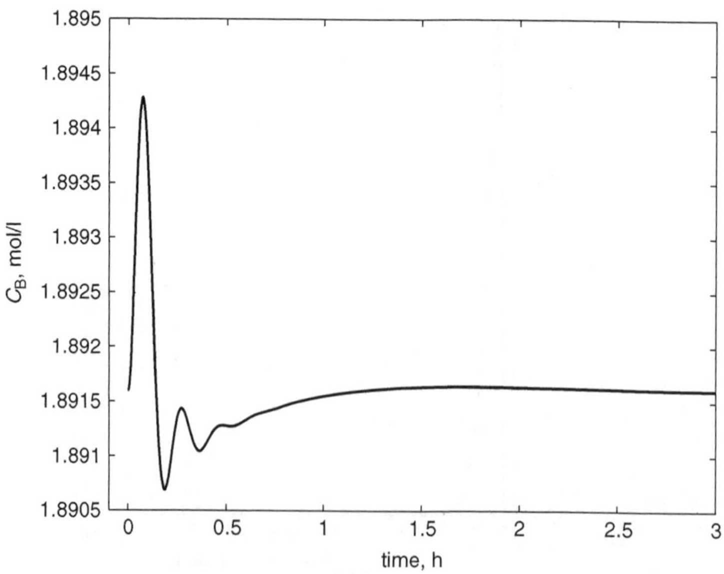

Figure 7.28 Evolution of the product purity for an unmeasured 6-K rise in the coolant inlet temperature at $t = 0$.

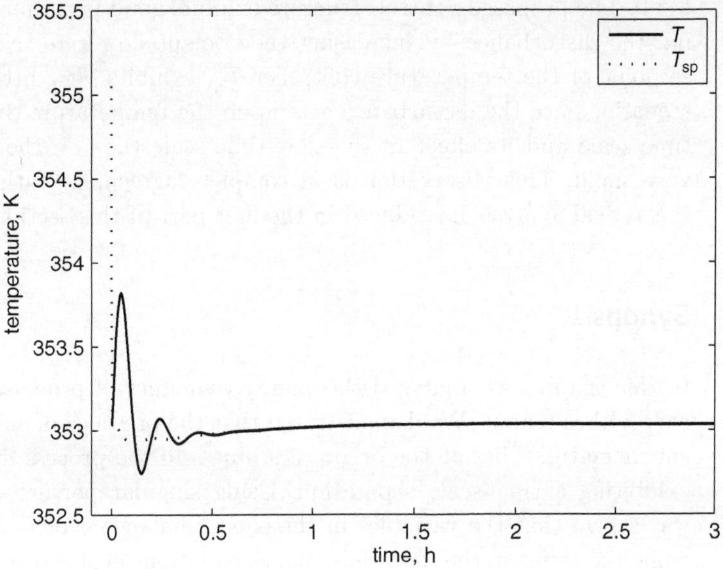

Figure 7.29 Evolution of the reactor temperature for an unmeasured 6-K rise in the coolant inlet temperature at $t = 0$.

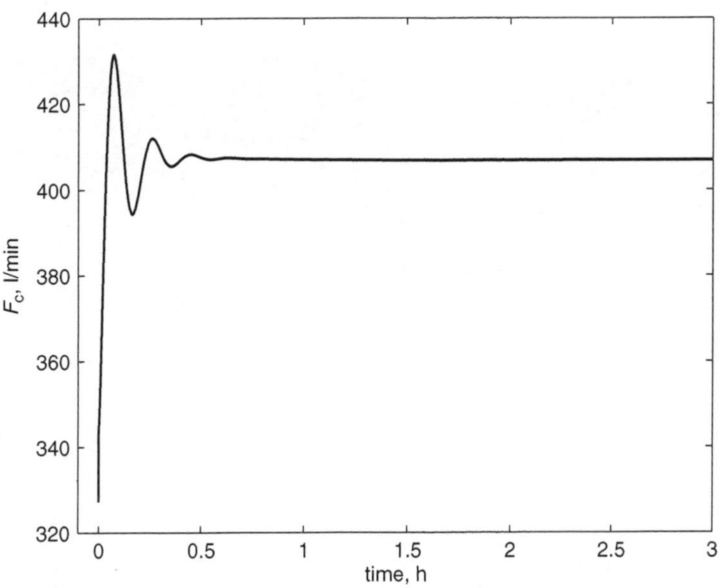

Figure 7.30 Evolution of the coolant flow rate for an unmeasured 6-K rise in the coolant inlet temperature at $t = 0$.

$t = 0$. The proposed control structure exhibits good performance, rapidly rejecting the disturbance by increasing the coolant flow rate. Notice also that the setpoint of the temperature controller T_{sp} exhibits very little variation in this scenario, since the disturbance acts upon the temperature dynamics in the fast time scale and its effect in the slow time scale (i.e., on the product purity) is very small. This observation is in complete agreement with the results of the theoretical analysis introduced in the first part of this section.

7.7 Synopsis

In this chapter we analyzed the energy dynamics of processes featuring a high energy throughput. We demonstrated that the presence of energy flows of different magnitudes lies at the origin of stiffness in the process model, its dynamics exhibiting a time-scale separation. Using singular perturbation arguments, it was shown that the variables in the energy balance evolve in a fast time scale, while the terms in the material-balance equations evolve over slower time scales. Also within the framework of singular perturbations, we derived reduced-order, non-stiff models for the dynamics in each time scale, and postulated a controller design framework predicated on the use of fast, simple, controllers for temperature control, and on addressing material-balance-related control objectives in the slower time scales.

Part IV

Appendices

Appendix A

Definitions

A.1 Lie derivatives. Involutivity

Definition A.1. *Let* $\mathbf{x} \in \mathbb{R}^n$ *be a vector,* $h(\mathbf{x}) : \mathbb{R}^n \to \mathbb{R}$ *a scalar function and* $\mathbf{g}(\mathbf{x}) : \mathbb{R}^n \to \mathbb{R}^n$ *a vector function. The Lie derivative of function* h *with respect to function* $\mathbf{g}$ *(or directional derivative of* h *along* $\mathbf{g}$*) is defined as*

$$L_{\mathbf{g}}h(\mathbf{x}) = \sum_{i=1}^{n} g_i(\mathbf{x})\frac{\partial h}{\partial x_i}(\mathbf{x}) \tag{A.1}$$

Note that $L_{\mathbf{g}}h(\mathbf{x})$ is itself a scalar function of $\mathbf{x}$ and that $L_{\mathbf{g}}h(\mathbf{x}) : \mathbb{R}^n \to \mathbb{R}$. Consequently, we can calculate its directional derivative along the vector function $\mathbf{g}$, as

$$L_{\mathbf{g}}(L_{\mathbf{g}}h(\mathbf{x})) = L_{\mathbf{g}}^2 h(\mathbf{x}) \tag{A.2}$$

or along the vector function $\mathbf{f}$, as

$$L_{\mathbf{f}}(L_{\mathbf{g}}h(\mathbf{x})) = L_{\mathbf{f}}L_{\mathbf{g}}h(\mathbf{x}) \tag{A.3}$$

with the latter representing the *mixed* Lie derivative of $L_{\mathbf{g}}h(\mathbf{x})$ with respect to the function $\mathbf{f}(\mathbf{x}) : \mathbb{R}^n \to \mathbb{R}^n$.

Similarly, higher-order Lie derivatives can be defined recursively as

$$\begin{aligned} L_{\mathbf{g}}^k h(x) &= L_{\mathbf{g}}\left(L_{\mathbf{g}}^{k-1}h(\mathbf{x})\right) \\ &\vdots \end{aligned} \tag{A.4}$$

It is also possible to define $L_{\mathbf{B}}\mathbf{f}(\mathbf{x})$ as an $(m \times m)$-dimensional matrix of Lie derivatives of the vector function $\mathbf{f}(\mathbf{x}) : \mathbb{R}^n \to \mathbb{R}^m$ along the columns $\mathbf{B}^i$, $i = 1, \ldots, m$ of the matrix function $\mathbf{B}(\mathbf{x}) : \mathbb{R}^{\mathbf{n}} \to \mathbb{R}^{\mathbf{n}} \times \mathbb{R}^{\mathbf{m}}$. $L_{\mathbf{B}}\mathbf{f}(\mathbf{x})$ is computed by multiplying the Jacobian of $\mathbf{f}(\mathbf{x})$ and $\mathbf{B}(\mathbf{x})$:

$$L_{\mathbf{B}}\mathbf{f}(\mathbf{x}) = \begin{bmatrix} L_{\mathbf{B}^1}f_1(\mathbf{x}) & \ldots & L_{\mathbf{B}^m}f_1(\mathbf{x}) \\ \vdots & \ddots & \vdots \\ L_{\mathbf{B}^1}f_m(\mathbf{x}) & \ldots & L_{\mathbf{B}^m}f_m \end{bmatrix} = \begin{bmatrix} \frac{\partial f_1(\mathbf{x})}{\partial x_1} & \cdots & \frac{\partial f_1(\mathbf{x})}{\partial x_n} \\ \vdots & \ddots & \vdots \\ \frac{\partial f_m(\mathbf{x})}{\partial x_1} & \cdots & \frac{\partial f_m(\mathbf{x})}{\partial x_n} \end{bmatrix} \mathbf{B}(\mathbf{x}) \tag{A.5}$$

Definition A.2. *Let* $\mathbf{x} \in \mathbb{R}^n$ *be a vector and* $\mathbf{f}(\mathbf{x}) : \mathbb{R}^n \to \mathbb{R}^n$ *and* $\mathbf{g}(\mathbf{x}) : \mathbb{R}^n \to \mathbb{R}^n$ *two vector functions. Then, the* Lie bracket *of the vector functions* $\mathbf{g}(\mathbf{x})$ *and* $\mathbf{h}(\mathbf{x})$ *is defined as*

$$\begin{aligned}[\mathbf{f}(\mathbf{x}), \mathbf{g}(\mathbf{x})] &= \frac{\partial \mathbf{g}}{\partial \mathbf{x}}(\mathbf{x})\mathbf{f}(\mathbf{x}) - \frac{\partial \mathbf{f}}{\partial \mathbf{x}}(\mathbf{x})\mathbf{g}(\mathbf{x}) \\ &= \begin{bmatrix} \frac{\partial g_1}{\partial x_1} & \cdots & \frac{\partial g_1}{\partial x_n} \\ \vdots & \ddots & \vdots \\ \frac{\partial g_n}{\partial x_1} & \cdots & \frac{\partial g_n}{\partial x_n} \end{bmatrix} \mathbf{f}(\mathbf{x}) - \begin{bmatrix} \frac{\partial f_1}{\partial x_1} & \cdots & \frac{\partial f_1}{\partial x_n} \\ \vdots & \ddots & \vdots \\ \frac{\partial f_n}{\partial x_1} & \cdots & \frac{\partial f_n}{\partial x_n} \end{bmatrix} \mathbf{g}(\mathbf{x})\end{aligned} \tag{A.6}$$

Since the Lie bracket is also a vector function, iterated Lie brackets can be defined using the following standard notation:

$$\begin{aligned} ad_{\mathbf{f}}^0 \mathbf{g} &= \mathbf{g} \\ ad_{\mathbf{f}}^1 \mathbf{g} &= [\mathbf{f}, \mathbf{g}] \\ ad_{\mathbf{f}}^2 \mathbf{g} &= [\mathbf{f}, [\mathbf{f}, \mathbf{g}]] \\ &\vdots \\ ad_{\mathbf{f}}^k \mathbf{g} &= \left[\mathbf{f}, ad_{\mathbf{f}}^{k-1} \mathbf{g}\right] \end{aligned} \tag{A.7}$$

The notion of *involutivity* has also been employed in the statements of Theorem A.1, and throughout the text. It is defined below.

Definition A.3. *An m-dimensional distribution*

$$\mathbf{G}(\mathbf{x}) = \text{span}\{\mathbf{g}_1(\mathbf{x}), \ldots, \mathbf{g}_m(\mathbf{x})\} \tag{A.8}$$

is involutive *if the Lie bracket* $[\mathbf{g}_i(\mathbf{x}), \mathbf{g}_j(\mathbf{x})]$, $\forall\ i, j = 1, \ldots, m$, *of any pair of vector fields belonging to* $\mathbf{G}(\mathbf{x})$ *is a vector field that belongs to* $\mathbf{G}(\mathbf{x})$.

A.2 Order of magnitude

Definition A.4. *A scalar function* $\delta(\varepsilon)$ *is said to be of* order ε, $\delta(\varepsilon) = \mathcal{O}(\varepsilon)$, *if there exist positive constants* k *and* c *such that* $|\delta(\varepsilon)| \leq k|\varepsilon|$, $\forall |\varepsilon| < c$.

A.3 Differential algebraic equations (DAEs)

Consider the nonlinear system of equations

$$\mathbf{F}(\dot{\mathbf{x}}, \mathbf{x}, \mathbf{u}(t)) = \mathbf{0} \tag{A.9}$$

where $\mathbf{x} \in \mathcal{X} \subset \mathbb{R}^n$ is the vector of state variables ($\mathcal{X}$ is an open, connected set), $\mathbf{u}(t) \in \mathbb{R}^m$ is the vector of time-dependent input (control) variables and $\mathbf{F} : \mathbb{R}^n \times \mathcal{X} \times \mathbb{R}^m \rightarrow \mathbb{R}^n$ is a smooth function. Clearly, the system in Equation (A.9) is an implicit ordinary differential equation (ODE) system if the Jacobian matrix $\partial\mathbf{F}/\partial\dot{\mathbf{x}}$ is nonsingular. However, if $\partial\mathbf{F}/\partial\dot{\mathbf{x}}$ is singular, the system (A.9) is a differential algebraic equation (DAE) system, exhibiting characteristics that differ fundamentally from those of ODE systems.

Most of the research on the analysis and numerical simulation of nonlinear DAEs has focused on systems in the fully implicit form of Equation (A.9). However, the generality of the form of the system in Equation (A.9) does not allow the development of explicit controller synthesis results. Also, the majority of chemical process applications (see examples throughout this book), as well as other engineering applications, are modeled by DAEs in a semi-explicit form, such that there is a distinct separation of the differential and algebraic equations:

$$\begin{aligned} \dot{\mathbf{x}} &= \mathbf{f}(\mathbf{x}) + \mathbf{B}(\mathbf{x})\mathbf{z} + \mathbf{G}(\mathbf{x})\mathbf{u}(t) \\ 0 &= \mathbf{k}(\mathbf{x}) + \mathbf{L}(\mathbf{x})\mathbf{z} + \mathbf{C}(\mathbf{x})\mathbf{u}(t) \end{aligned} \tag{A.10}$$

where $\mathbf{x} \in \mathcal{X} \subset \mathbb{R}^n$ is the vector of differential variables for which the explicit differential equations are available, $\mathbf{z} \in \mathcal{Z} \subset \mathbb{R}^p$ is the vector of algebraic variables that vary according to the algebraic equations, $\mathcal{X}$ and $\mathcal{Z}$ are open, connected sets, $\mathbf{u}(t) \in \mathbb{R}^m$ is the vector of input variables, $\mathbf{f}(\mathbf{x})$ and $\mathbf{k}(\mathbf{x})$ are smooth vector fields of dimensions n and p, respectively, and $\mathbf{B}(\mathbf{x})$, $\mathbf{G}(\mathbf{x})$, $\mathbf{L}(\mathbf{x})$, and $\mathbf{C}(\mathbf{x})$ are smooth matrices of appropriate dimensions. In the above description, the inputs $\mathbf{u}(\mathbf{x})$ and the algebraic variables $\mathbf{z}$ appear in the system equations in a linear fashion, which is typical of most practical applications.

The common underlying principle in the approaches for characterizing the solvability of a DAE system is to obtain, either explicitly, or implicitly, a local representation of an equivalent ODE system, for which available results on existence and uniqueness of solutions are applicable. The derivation of the underlying ODE system involves the repeated differentiation of the algebraic constraints of the DAE, and it is this differentiation process that leads to the concept of a *DAE index* that is widely used in the literature. For the semi-explicit DAE systems (A.10) that are of interest to us here, the index has the following definition.

Definition A.5. *The* index ν_{d} *of the DAE system in Equation (A.10) with specified smooth inputs* $\mathbf{u}(t)$ *is the minimum number of times the algebraic equations or their subset have to be differentiated to obtain a set of differential equations for* $\mathbf{z}$*, i.e., in order to be able to solve* $\dot{\mathbf{z}} = \mathcal{F}(\mathbf{x}, \mathbf{z}, t)$ *for* $\mathbf{z}$.

The index ν_{d} provides a measure of the "singularity" of the algebraic equations and the resulting differences from ODE systems. More specifically, consider the DAE system of Equation (A.10) in the case in which the matrix $\mathbf{L}(\mathbf{x})$

is nonsingular. Then, the algebraic equations can be solved for the algebraic variables $\mathbf{z}$:

$$\mathbf{z} = -\mathbf{L}^{-1}[\mathbf{k}(\mathbf{x}) + \mathbf{C}(\mathbf{x})\mathbf{u}(t)] \tag{A.11}$$

and one differentiation of the algebraic equations in (A.10), or equivalently of the solution for $\mathbf{z}$ (Equation (A.11)), will yield the differential equations for $\mathbf{z}$. Hence, in this case, the DAE system (A.10) has an index $\nu_{\mathrm{d}} = 1$. For such systems, a direct substitution of the solution for $\mathbf{z}$ into the differential equations for $\mathbf{x}$ will yield an equivalent ODE representation. In contrast, DAE systems with a singular algebraic equation (more exactly, with the matrix $\mathbf{L}(\mathbf{x})$ being singular) cannot be readily reduced to an ODE system, and have a high index, $\nu_{\mathrm{d}} > 1$.

In the process of deriving an equivalent ODE representation (state-space realization) for the DAE systems of Equation (A.10), repeated differentiation of the algebraic constraints may yield additional constraints that involve the manipulated inputs $\mathbf{u}(t)$. In the context of numerical simulation of a DAE system with specified inputs $\mathbf{u}(t)$, this implies that the inputs $\mathbf{u}(t)$ must vary smoothly with time. However, in the context of feedback control, u is the vector of manipulated inputs, and is not specified a priori as a function of time. It is in fact our purpose to design a control law for the inputs $\mathbf{u}$. The presence of the manipulated inputs in the algebraic constraints in $\mathbf{x}$ influences the controller design (Kumar and Daoutidis 1999a) due to the fact that the constrained state-space region for $\mathbf{x}$ (which is defined by the algebraic equations/constraints) depends on the (unknown) feedback control law for $\mathbf{u}$. The notion of regularity is used to distinguish, on the basis of the dependence of the constrained state-space region where the differential variables $\mathbf{x}$ evolve on the inputs $\mathbf{u}$, between two fundamental classes of DAE systems.

Definition A.6. *A differential algebraic equation system (A.10) is said to be regular, if*

(i) it has a finite index ν_{d} and
(ii) the state space region where the differential variables $\mathbf{x}$ are constrained to evolve is invariant under any control law for $\mathbf{u}$.

Theorem A.1 (Kumar *et al.* 1998) provides the necessary and sufficient conditions for the existence of an ε-independent coordinate change that transforms the two-time-scale system (2.36) into a standard singularly perturbed form. Also, for the systems that satisfy the conditions set forth in the theorem, it provides an explicit coordinate change.

Theorem A.1. *Consider the system of Equation (2.36), for which the slow dynamics is described by the DAE system of Equation (2.45). The system of Equation (2.36) can be transformed into a two-time-scale singularly perturbed* *system in standard form through an ε-independent nonlinear coordinate change, if and only if*

(i) *the $p \times p$ matrix $\mathbf{L_B k}$ (with the (i,j)th component given by the Lie derivative $L_{\mathbf{B}_j}\mathbf{k}_i(\mathbf{x})$, where $\mathbf{B}_j$ is the jth column of $\mathbf{B}(\mathbf{x})$ and $\mathbf{k}_i$ is the ith component of $\mathbf{k}(\mathbf{x})$) is nonsingular on $\mathcal{X}$, and*

(ii) *the p-dimensional distribution $\mathcal{B}(\mathbf{x}) = \mathrm{span}\{\mathbf{B}_1(\mathbf{x}), \ldots, \mathbf{B}_p(\mathbf{x})\}$ is involutive.*

If these conditions hold, then under the coordinate change

$$\begin{bmatrix} \boldsymbol{\zeta} \\ \boldsymbol{\eta} \end{bmatrix} = \mathbf{T}(\mathbf{x}) = \begin{bmatrix} \boldsymbol{\phi}(\mathbf{x}) \\ \mathbf{k}(\mathbf{x}) \end{bmatrix} \tag{A.12}$$

where $\boldsymbol{\zeta} \in \mathbb{R}^{n-p}$, $\boldsymbol{\eta} \in \mathbb{R}^p$, and $\boldsymbol{\phi}(\mathbf{x})$ is a vector field of dimension $n-p$ with components $\phi_i(\mathbf{x})$ such that $L_{\mathbf{B}_j}\phi_i \equiv 0$, $\forall\, i,j$, the system of Equation (2.36) takes the following standard singularly perturbed form:

$$\begin{aligned} \dot{\boldsymbol{\zeta}} &= \tilde{\mathbf{f}}(\boldsymbol{\zeta},\boldsymbol{\eta}) + \tilde{\mathbf{G}}(\boldsymbol{\zeta},\boldsymbol{\eta})\mathbf{u} \\ \varepsilon\dot{\boldsymbol{\eta}} &= \varepsilon\bar{\mathbf{f}}(\boldsymbol{\zeta},\boldsymbol{\eta}) + \varepsilon\bar{\mathbf{g}}(\boldsymbol{\zeta},\boldsymbol{\eta})\mathbf{u} + \mathbf{Q}(\boldsymbol{\zeta},\boldsymbol{\eta})\eta \\ y_i &= h_i(\boldsymbol{\zeta},\boldsymbol{\eta}),\ i = 1,\ldots,m \end{aligned} \tag{A.13}$$

where $\tilde{f}$ is the Lie derivative of $\boldsymbol{\phi}(\mathbf{x})$ along $\mathbf{f}(\mathbf{x})$, $\tilde{\mathbf{f}} = L_{\mathbf{f}}\boldsymbol{\phi}(\mathbf{x})$, $\bar{\mathbf{f}} = L_{\mathbf{f}}\mathbf{k}(\mathbf{x})$, $\tilde{\mathbf{G}} = \mathbf{L_G}\boldsymbol{\phi}(\mathbf{x})$, $\bar{\mathbf{G}} = \mathbf{L_G}\mathbf{k}(x)$, $\mathbf{Q} = \mathbf{L_B}\mathbf{k}(\mathbf{x})$, evaluated at $\mathbf{x} = \mathbf{T}^{-1}(\zeta,\eta)$, and $\mathbf{Q}(\zeta,\eta)$ is nonsingular uniformly in $\boldsymbol{\zeta}$ and $\boldsymbol{\eta}$.

Remark A.1. *Condition (i) of* Theorem A.1 *essentially means that the corresponding DAE system in Equation (2.45) has an index of two, which directly fixes the dimensions of the fast and slow variables to p and $n-p$, respectively. Condition (ii) of the theorem ensures that the $(n-p)$-dimensional slow $\boldsymbol{\zeta}$-subsystem can be made independent of the singular term $1/\varepsilon$, thereby yielding the system in Equation (A.13) in the standard singularly perturbed form. While condition (ii) is trivially satisfied for all linear systems and for nonlinear systems with $p = 1$, it is not satisfied in general for nonlinear systems with $p > 1$.*

When condition (ii) is not satisfied, a standard singularly perturbed representation for systems of the type (2.36) can still be obtained through an ε-dependent coordinate change (Kumar et al. *1998) that is singular at $\varepsilon = 0$.*

Remark A.2. *Consider the DAE system in Equation (2.45) for which condition (i) of Theorem A.1 is satisfied, i.e., the DAE has an index of two. The DAE system has exactly p constraints, and from the results on DAE systems (Kumar and Daoutidis 1999a) a minimal-order state-space realization of dimension $n-p$ can be derived, using the constraints $\mathbf{k}(\mathbf{x}) = \mathbf{0}$ as a part of a coordinate change. Condition (ii) of the theorem allows the derivation of such a state-space realization* ***without*** *evaluating a solution for the algebraic variables $\mathbf{z}$, through* *an appropriate coordinate change. In fact, the choice of the state variables* $\boldsymbol{\zeta}$ *and* $\boldsymbol{\eta}$ *in Equation (A.12) precisely comprises such a coordinate change, and the description of the slow subsystem of Equation (A.13),*

$$\begin{aligned} \dot{\boldsymbol{\zeta}} &= \tilde{\mathbf{f}}(\boldsymbol{\zeta}, \mathbf{0}) + \tilde{\mathbf{G}}(\boldsymbol{\zeta}, \mathbf{0})\mathbf{u} \\ y_i &= h_i(\mathbf{x}) \mid_{\mathbf{x}=\mathbf{T}^{-1}(\boldsymbol{\zeta}, \mathbf{0})}, \ i = 1, \ldots, m \end{aligned} \tag{A.14}$$

is the corresponding state-space realization of the index-two DAE system (2.45).

Appendix B

Systems with multiple-time-scale dynamics

Section 2.4 alluded to the possibility of expanding the methods presented in Chapter 2 to account for the presence of multiple singular perturbation parameters in a system of differential equations. This appendix is concerned with this topic, and, to this end, let us consider a multiple-time-scale (multiply perturbed) system in the standard form

$$
\begin{aligned}
\frac{d\mathbf{x}}{dt} &= f(\mathbf{x}, \mathbf{y}_1, \ldots, \mathbf{y}_M, \varepsilon),\ \mathbf{x}(0) = \mathbf{x}^0 \\
\varepsilon_j \frac{d\mathbf{y}_j}{dt} &= \mathbf{g}_j(\mathbf{x}, \mathbf{y}_1, \ldots, \mathbf{y}_M, \varepsilon),\ \mathbf{y}_j(0) = \mathbf{y}_j^0,\ j = 1, \ldots, M
\end{aligned}
\tag{B.1}
$$

where $\mathbf{x}, \mathbf{f} \in \mathbb{R}^n$, $\mathbf{y}_j$, $\mathbf{g}_j \in \mathbb{R}^{m_j}$, $j = 1, \ldots, M$, and $\boldsymbol{\varepsilon} = [\varepsilon_1, \ldots, \varepsilon_M]^{\mathrm{T}}$ is a vector of singular perturbation parameters satisfying the condition

$$
\frac{\varepsilon_{j+1}}{\varepsilon_j} \to 0 \qquad \text{as} \qquad \varepsilon_1 \to 0 \tag{B.2}
$$

In analogy with two-time-scale systems, the conditions for the regular degeneration for multiple-time-scale systems have been derived (Hoppensteadt 1971) in terms of the Jacobian matrices in each time scale. Specifically, it is required that the matrix

$$
\frac{\partial \mathbf{g}_j}{\mathbf{y}_j}(\mathbf{x}, \mathbf{y}_1, \ldots, \mathbf{y}_M, 0)
$$

be nonsingular, and that its nonsingularity be preserved for each function $\mathbf{g}_j$ with the vector $\boldsymbol{\varepsilon}$ replaced by the scalar ε_j. Under these conditions, the system will exhibit M fast time scales and one slow time scale.

Condition (B.2) implies that ε_M is the smallest singular perturbation parameter (that would yield the "fastest" fast time scale), while ε_1 is the largest singular perturbation parameter, and is responsible for the "slowest" fast time scale. Consequently, the variable $\mathbf{y}_{j+1}$ will be faster than the variable $\mathbf{y}_j$, for $j = 1, \ldots, M-1$. It is this hierarchy of fast subsystems (boundary layers) that distinguishes multiple-time-scale systems from two-time-scale systems.

The analysis of multiple-time-scale systems can, however, be carried out by extending the methods used for analyzing two-time-scale systems presented in Section 2.2. In analogy with two-time-scale systems, in the limiting case as $\boldsymbol{\varepsilon} \to \mathbf{0}$, the dimension of the state space of the system in Equations (B.1) collapses from $(n + \sum_j m_j)$ to n, and the slow subsystem is described by the differential algebraic equation system

$$\begin{aligned} \frac{d\mathbf{x}}{dt} &= \mathbf{f}(\mathbf{x}, \mathbf{y}_1, \ldots, \mathbf{y}_M, 0) \\ \mathbf{0} &= \mathbf{g}_j(\mathbf{x}, \mathbf{y}_1, \ldots, \mathbf{y}_M, 0), \; j = 1, \ldots, M \end{aligned} \tag{B.3}$$

By solving the algebraic equation system given by the constraints of Equation (B.3), the steady-state solutions $\mathbf{y}_{j,\text{ss}}$ can be obtained for the fast variables. Subsequently, $\mathbf{y}_{j,\text{ss}}$ can be used for the derivation of an equivalent n-dimensional ODE representation of the slow subsystem:

$$\frac{d\mathbf{x}}{dt} = \mathbf{f}(\mathbf{x}, \mathbf{y}_{1,\text{ss}}(\mathbf{x}), \ldots, \mathbf{y}_{M,\text{ss}}(\mathbf{x}), 0) \tag{B.4}$$

The descriptions of the fast subsystems are obtained hierarchically, starting from the "fastest" fast time scale. On introducing a "stretched" time variable $\tau_M = t/\varepsilon_M$, the system in Equation (B.1) takes the form

$$\begin{aligned} \frac{d\mathbf{x}}{d\tau_M} &= \varepsilon_M \mathbf{f}(\mathbf{x}, \mathbf{y}_1, \ldots, \mathbf{y}_M, \varepsilon) \\ \frac{d\mathbf{y}_j}{d\tau_M} &= \frac{\varepsilon_M}{\varepsilon_j} \mathbf{g}_j(\mathbf{x}, \mathbf{y}_1, \ldots, \mathbf{y}_M, \varepsilon), \; j = 1, \ldots, M-1 \\ \frac{d\mathbf{y}_M}{d\tau_M} &= \mathbf{g}_M(\mathbf{x}, \mathbf{y}_1, \ldots, \mathbf{y}_M, \varepsilon) \end{aligned} \tag{B.5}$$

In the limit as $\varepsilon \to 0$, the variables x and y_j, $j = 1, \ldots, M-1$, are in a quasi-steady state, having negligible dynamics, and we obtain a representation of the "fastest" fast dynamics in the stretched time scale τ_M:

$$\frac{d\mathbf{y}_M}{d\tau_M} = \mathbf{g}_M(\mathbf{x}, \mathbf{y}_1, \ldots, \mathbf{y}_M, 0) \tag{B.6}$$

in which the slow variables $\mathbf{x}$ and $\mathbf{y}_j$, $j = 1, \ldots, M-1$, are unchanged from their initial conditions $\mathbf{x} = \mathbf{x}^0$ and $\mathbf{y}_j = \mathbf{y}_j^0$, $j = 1, \ldots, M-1$, and are treated as constant parameters. The fast subsystem (B.6) represents the fastest boundary-layer system.

Continuing this line of reasoning, the introduction of the kth stretched time scale ($\forall k \in [1, M]$), $\tau_k = t/\varepsilon_k$, results in a description of system (B.1) of the form

$$\begin{aligned} \frac{d\mathbf{x}}{d\tau_k} &= \varepsilon_k \mathbf{f}(\mathbf{x}, \mathbf{y}_1, \ldots, \mathbf{y}_M, \varepsilon) \\ \frac{d\mathbf{y}_j}{d\tau_k} &= \frac{\varepsilon_k}{\varepsilon_j} \mathbf{g}_j(\mathbf{x}, \mathbf{y}_1, \ldots, \mathbf{y}_M, \varepsilon), \; j = 1, \ldots, k-1 \\ \frac{d\mathbf{y}_k}{d\tau_k} &= \mathbf{g}_k(\mathbf{x}, \mathbf{y}_1, \ldots, \mathbf{y}_M, \varepsilon) \\ \frac{\varepsilon_j}{\varepsilon_k} \frac{d\mathbf{y}_j}{d\tau_k} &= \mathbf{g}_j(\mathbf{x}, \mathbf{y}_1, \ldots, \mathbf{y}_M, \varepsilon), \; j = k+1, \ldots, M \end{aligned} \tag{B.7}$$

In the limit as $\varepsilon_k \to 0$, the dynamics of the slow variable $\mathbf{x}$ become negligible, and

$$\frac{d\mathbf{y}_j}{\tau_k} = 0, \ j = 1, k-1$$

since, by condition (B.2), $\varepsilon_k/\varepsilon_j \to 0$ for $j = 1, \ldots, k-1$. Hence, the variables $\mathbf{x}$ and $\mathbf{y}_j$, $j = 1, \ldots, k-1$, will be fixed at their initial values. Also, following from the same condition, $\varepsilon_j/\varepsilon_k \to 0$, $j = k+1, \ldots, M$, and therefore the differential equations for $\mathbf{y}_j$, $j = k+1, \ldots, M$, are replaced by a set of algebraic constraints:

$$\mathbf{0} = \mathbf{g}_j\left(\mathbf{x}^0, \mathbf{y}_1^0, \ldots, \mathbf{y}_{k-1}^0, \mathbf{y}_k, \ldots, \mathbf{y}_m, 0\right), \ j = k+1, \ldots, M \tag{B.8}$$

The representation of the kth boundary layer, corresponding to the fast variables $\mathbf{y}_k$, is then obtained as

$$\begin{aligned} \frac{d\mathbf{y}_k}{d\tau_k} &= \mathbf{g}_k\left(\mathbf{x}^0, \mathbf{y}_1^0, \ldots, \mathbf{y}_{k-1}^0, \mathbf{y}_k, \ldots, \mathbf{y}_M, 0\right) \\ \mathbf{0} &= \mathbf{g}_j\left(\mathbf{x}^0, \mathbf{y}_1^0, \ldots, \mathbf{y}_{k-1}^0, \mathbf{y}_k, \ldots, \mathbf{y}_m, 0\right), \ j = k+1, \ldots, M \end{aligned} \tag{B.9}$$

where the faster fast variables $\mathbf{y}_j$, $j = k+1, \ldots, M$, can be obtained by solving the algebraic equation system (B.8).

Such nested applications of single-parameter singular perturbation theory (i.e., the extension of the analysis of two-time-scale systems presented in Chapter 2 to multiple-time-scale systems) have been used for stability analysis of linear (Ladde and Siljak 1983) and nonlinear (Desoer and Shahruz 1986) systems in the standard form. However, as emphasized above (Section 2.3), the ODE models of chemical processes are most often in the nonstandard singularly perturbed form, with the general multiple-perturbation representation

$$\dot{\mathbf{x}} = \mathbf{f}(\mathbf{x}) + \mathbf{G}(\mathbf{x})\mathbf{u} + \sum_{j=1}^{M} \frac{1}{\varepsilon_j} \mathbf{B}_j(\mathbf{x})\mathbf{k}_j(\mathbf{x}) \tag{B.10}$$

where $\mathbf{x} \in \mathcal{X} \subset \mathbb{R}^n$ is the vector of state variables, $\mathbf{f}(\mathbf{x})$ and $\mathbf{k}_j(\mathbf{x})$ are smooth vector fields of dimensions n and p_j $(j = 1, \ldots, M)$, and $\mathbf{G}(\mathbf{x})$ and $\mathbf{B}_j(\mathbf{x})$ are matrices of dimensions $n \times q$ and $n \times p_j$, respectively. The matrices $\mathbf{B}_j$ and $(\partial \mathbf{k}_j(\mathbf{x})/\partial \mathbf{x})$ are assumed to have full column and row ranks, respectively. Also, in order for a system described by Equation (B.10) to exhibit a dynamic behavior with more than two time scales, the small-perturbation parameters ε_j, $j = 1, \ldots, M$, are assumed to satisfy relation (B.2). Existing results (Vora 2000) referring to the time-scale decomposition of such systems (B.10) also rely on a successive application of the methods used in the analysis of two-time-scale nonregular singularly perturbed systems presented in Section 2.3.

Specifically, a representation of the fastest dynamics is obtained by introducing the stretched fastest time scale $\tau_M = t/\varepsilon_M$. Then, in the limit as $\varepsilon_M \to 0$, we have $\varepsilon_M/\varepsilon_j=0$ and the fastest dynamics (fastest boundary-layer subsystem) is described by

$$\frac{d\mathbf{x}}{d\tau_M} = \mathbf{B}_M(\mathbf{x})\mathbf{k}_M(\mathbf{x}) \tag{B.11}$$

Similarly to the analysis of two-time-scale systems, in the slow time scale t, on multiplying Equation (B.10) by ε_M, and considering the limit as $\varepsilon_m \to 0$, we obtain a set of constraints that must be satisfied in the slow subsystem:

$$0 = k_{M_i}(\mathbf{x}),\ i = 1, \ldots, p_M \tag{B.12}$$

where $k_{M_i}(\mathbf{x})$ is the ith component of $\mathbf{k}_M(\mathbf{x})$. Since the Jacobian $[\partial\mathbf{k}_M(\mathbf{x})/\partial\mathbf{x}]$ has full rank, the above constraints are linearly independent. In the slow time scale t, the ratios k_{M_i}/ε_M become indeterminate as $\varepsilon_M \to 0$. By denoting these finite, but unknown, terms by $z_{M_i} = \lim_{\varepsilon_m \to 0} k_{M_i}/\varepsilon_M$ ($\mathbf{z}_M = [z_{M_1} \ldots z_{M_p}]^{\mathrm{T}}$), and taking the limit as $\varepsilon_M \to 0$ in Equation (B.10), we obtain

$$\begin{aligned} \dot{\mathbf{x}} &= \mathbf{f}(\mathbf{x}) + \mathbf{G}(\mathbf{x})\mathbf{u} + \sum_{j=1}^{M-1} \frac{1}{\varepsilon_j}\mathbf{B}_j(\mathbf{x})\mathbf{k}_j(\mathbf{x}) + \mathbf{B}_M(\mathbf{x})\mathbf{z}_M \\ \mathbf{0} &= \mathbf{k}_M(\mathbf{x}) \end{aligned} \tag{B.13}$$

which describes the slow dynamics after the fastest boundary layer of Equation (B.10). Note that Equation (B.13) is still a stiff system, since it contains $M-1$ parameters of different orders of magnitude. In addition, (B.13) is a DAE system of nontrivial index, since the "algebraic" variables $\mathbf{z}_M$ cannot be directly evaluated from the algebraic equations. However, for most practical cases (Vora 2000), the matrix $(L_{\mathbf{B}_M}\mathbf{k}_m(\mathbf{x}))$ is nonsingular, thus allowing us to obtain the variables $\mathbf{z}_M$ after one differentiation of the algebraic constraints:

$$\mathbf{z}_M = -(L_{\mathbf{B}_M}\mathbf{k}_m(\mathbf{x}))^{-1} \left\{ L_{\mathbf{f}}\mathbf{k}_m(\mathbf{x}) + L_{\mathbf{G}}\mathbf{k}_m(\mathbf{x})\mathbf{u} + \sum_{j=1}^{M-1} \frac{1}{\varepsilon_j} L_{\mathbf{B}_j}\mathbf{k}_j(\mathbf{x}) \right\} \tag{B.14}$$

The terms ε_j/k_j contained in the solution for the variables $\mathbf{z}_M$ (Equation (B.14)) become indeterminate as $\varepsilon_j \to 0$, and are implicitly determined by the additional constraints obtained in the jth time scale. Using the solution for $\mathbf{z}_M$ in Equation (B.14), a state-space realization of the DAE system in Equation (B.13) is obtained as

$$\begin{aligned} \dot{\mathbf{x}} &= \mathbf{f}(\mathbf{x}) + \mathbf{G}(\mathbf{x})\mathbf{u} + \sum_{j=1}^{M-1} \frac{1}{\varepsilon_j}\mathbf{B}_j(\mathbf{x})\mathbf{k}_j(\mathbf{x}) - \mathbf{B}_M(\mathbf{x})(L_{\mathbf{B}_M}\mathbf{k}_M(\mathbf{x}))^{-1} \\ &\quad \times \left\{ L_{\mathbf{f}}\mathbf{k}_M(\mathbf{x}) + L_{\mathbf{G}}\mathbf{k}_M(\mathbf{x})\mathbf{u} + \sum_{j=1}^{M-1} \frac{1}{\varepsilon_j} L_{\mathbf{B}_j}\mathbf{k}_j(\mathbf{x}) \right\} \\ \mathbf{0} &= \mathbf{k}_M(\mathbf{x}) \end{aligned} \tag{B.15}$$

or, in a slightly rearranged form that collects the terms containing the parameter ε_{M-1},

$$\begin{aligned}
\dot{\mathbf{x}} = &\left(\mathbf{f}(\mathbf{x}) - \mathbf{B}_M(\mathbf{x})(L_{\mathbf{B}_M}\mathbf{k}_M(\mathbf{x}))^{-1}L_{\mathbf{f}}\mathbf{k}_M(\mathbf{x})\right) \\
&+ \left(\mathbf{G}(\mathbf{x}) - \mathbf{B}_M(\mathbf{x})(L_{\mathbf{B}_M}\mathbf{k}_M(\mathbf{x}))^{-1}L_{\mathbf{G}}\mathbf{k}_M(\mathbf{x})\right)\mathbf{u} \\
&+ \left\{\sum_{j=1}^{M-2}\frac{1}{\varepsilon_j}\mathbf{B}_j(\mathbf{x})\mathbf{k}_j(\mathbf{x}) - \mathbf{B}_M(\mathbf{x})(L_{\mathbf{B}_M}\mathbf{k}_M(\mathbf{x}))^{-1}\right. \\
&\qquad \left. \times \sum_{j=1}^{M-2}\frac{1}{\varepsilon_j}(L_{\mathbf{B}_j}\mathbf{k}_M(\mathbf{x}))\mathbf{k}_j(\mathbf{x})\right\} \\
&+ \frac{1}{\varepsilon_{M-1}}\{\mathbf{B}_{M-1}(\mathbf{x})\mathbf{k}_{M-1}(\mathbf{x}) \\
&\qquad - \mathbf{B}_M(\mathbf{x})(L_{\mathbf{B}_M})^{-1}(L_{\mathbf{B}_{M-1}}\mathbf{k}_M(\mathbf{x}))\mathbf{k}_{M-1}(\mathbf{x})\} \\
\mathbf{0} = &\ \mathbf{k}_m(\mathbf{x})
\end{aligned} \tag{B.16}$$

Introducing the $(M-1)$st fast time scale, $\tau_{M-1} = t/\varepsilon_{M-1}$, and considering the limit $\varepsilon_{M-1} \to 0$, we obtain the following description of the $(M-1)$st fast dynamics of the system (B.10):

$$\begin{aligned}
\frac{d\mathbf{x}}{d\tau_{M-1}} &= [\mathbf{B}_{M-1}(\mathbf{x}) \quad \mathbf{B}_M(\mathbf{x})]\begin{bmatrix}\mathbf{k}_{M-1}(\mathbf{x}) \\ -(L_{\mathbf{B}_M}\mathbf{k}_M(\mathbf{x}))^{-1}(L_{\mathbf{B}_{M-1}}\mathbf{k}_M(\mathbf{x}))\mathbf{k}_{M-1}(\mathbf{x})\end{bmatrix} \\
\mathbf{0} &= \mathbf{k}_M(\mathbf{x})
\end{aligned} \tag{B.17}$$

which represents the $(M-1)$st boundary-layer subsystem. Under the assumption that the matrix $[\mathbf{B}_{M-1}(\mathbf{x}) \quad \mathbf{B}_M(\mathbf{x})]$ has full column rank, the constraints after the $(M-1)$st boundary layer, in addition to $\mathbf{k}_M(\mathbf{x}) = \mathbf{0}$, are $\mathbf{k}_{M-1}(\mathbf{x}) = \mathbf{0}$. Note that the additional constraints $\mathbf{k}_{M-1}(\mathbf{x}) = \mathbf{0}$ are the same as those that would be obtained on considering Equation (B.10) in the limit $\varepsilon_{M-1} \to 0$. On letting $\varepsilon_{M-1} \to 0$ in Equation (B.16), and denoting by $\mathbf{z}_{M-1}$ the p_{M-1}-dimensional vector of indeterminate terms $\mathbf{k}_{M-1}(\mathbf{x})/\varepsilon_{M-1}$ that arise in this limit, we obtain the following description of the slow dynamics after the $(M-1)$st boundary layer:

$$\begin{aligned}
\dot{\mathbf{x}} = &\left(\mathbf{f}(\mathbf{x}) - \mathbf{B}_M(\mathbf{x})(L_{\mathbf{B}_M}\mathbf{k}_m(\mathbf{x}))^{-1}L_{\mathbf{f}}\mathbf{k}_M(\mathbf{x})\right) \\
&+ \left(\mathbf{G}(\mathbf{x}) - \mathbf{B}_M(\mathbf{x})(L_{\mathbf{B}_M}\mathbf{k}_M(\mathbf{x}))^{-1}L_{\mathbf{G}}\mathbf{k}_M(\mathbf{x})\right)\mathbf{u} \\
&+ \left\{\sum_{j=1}^{M-2}\frac{1}{\varepsilon_j}\mathbf{B}_j(\mathbf{x})\mathbf{k}_j(\mathbf{x}) - \mathbf{B}_M(\mathbf{x})(L_{\mathbf{B}_M}\mathbf{k}_M(\mathbf{x}))^{-1}\right. \\
&\qquad \left. \times \sum_{j=1}^{M-2}\frac{1}{\varepsilon_j}(L_{\mathbf{B}_j}\mathbf{k}_M(\mathbf{x}))\mathbf{k}_j(\mathbf{x})\right\} \\
&+ [\mathbf{B}_{M-1}(\mathbf{x}) - \mathbf{B}_M(\mathbf{x})(L_{\mathbf{B}_M}\mathbf{k}_M(\mathbf{x}))^{-1}(L_{\mathbf{B}_{M-1}}\mathbf{k}_M(\mathbf{x}))]\mathbf{z}_{M-1} \\
\mathbf{0} = &\ \mathbf{k}_{M-1}(\mathbf{x}) \\
\mathbf{0} = &\ \mathbf{k}_M(\mathbf{x})
\end{aligned} \tag{B.18}$$

Proceeding in a similar fashion, in the slow time scale after the lth boundary layer, assuming that the $\left(n \times \sum_{j=l}^{M} p_j\right)$ matrix $[\mathbf{B}_l(\mathbf{x})| \cdots |\mathbf{B}_m(\mathbf{x})]$ has full column rank, the limit $\varepsilon_l \to 0$ leads to the additional constraints $\mathbf{k}_l = \mathbf{0}$. These constraints must be satisfied in the slow subsystem, together with the previously obtained constraints corresponding to the faster time scales. On defining $z_{l_i} = \lim_{\varepsilon_l \to 0} k_{l_i}(\mathbf{x})/\varepsilon_l$, and taking the limit $\varepsilon_l \to 0$ in Equation (B.10), we obtain the following description of the slow dynamics after the lth boundary layer:

$$\begin{aligned} \dot{\mathbf{x}} &= \mathbf{f}(x) + \mathbf{G}(\mathbf{x})\mathbf{u} + \sum_{j=1}^{l-1} \frac{1}{\varepsilon_j} \mathbf{B}_j(\mathbf{x})\mathbf{k}_j(\mathbf{x}) + \sum_{j=l}^{M} \mathbf{B}_j(\mathbf{x})\mathbf{z}_j \\ \mathbf{0} &= \mathbf{k}_j(\mathbf{x}), \quad j = l, \ldots, M \end{aligned} \tag{B.19}$$

The system in Equation (B.19) is "less stiff" than the original system (B.10), since it contains fewer small perturbation parameters ($(l-1) < M$). Like the other descriptions of the slow dynamics obtained so far, (B.19) is a DAE system of nontrivial index, since there are no algebraic equations that would allow the variables $\mathbf{z}_j$ to be evaluated directly. Making the assumption that the $\sum_{j=l}^{M} \times \sum_{j=l}^{M}$ matrix

$$(L_{\mathbf{B}}\mathbf{k}(\mathbf{x}))_l \triangleq \begin{bmatrix} L_{\mathbf{B}_M}\mathbf{k}_M & \ldots & L_{\mathbf{B}_M}\mathbf{k}_j & \ldots & L_{\mathbf{B}_M}\mathbf{k}_l \\ \vdots & \ddots & \vdots & \ddots & \vdots \\ L_{\mathbf{B}_j}\mathbf{k}_M & \ldots & L_{\mathbf{B}_j}\mathbf{k}_j & \ldots & L_{\mathbf{B}_j}\mathbf{k}_l \\ \vdots & \ddots & \vdots & \ddots & \vdots \\ L_{\mathbf{B}_l}\mathbf{k}_M & \ldots & L_{\mathbf{B}_l}\mathbf{k}_j & \ldots & L_{\mathbf{B}_l}\mathbf{k}_l \end{bmatrix} \tag{B.20}$$

is nonsingular, $\forall l \in [1, M]$, we fix the index of the DAE system (B.19) to two, assuring that the solution for the variables $\mathbf{z}_j$, $j = l, \ldots, M$, can be obtained after only one differentiation of the algebraic constraints. Also, the numbers of slow and fast variables are set to $n - \sum_{j=l}^{M} p_j$ and $\sum_{j=l}^{M} p_j$, respectively.

By carrying on the analysis to $l = 1$, we obtain

$$\begin{aligned} \dot{\mathbf{x}} &= \mathbf{f}(\mathbf{x}) + \mathbf{G}(\mathbf{x})\mathbf{u} + \sum_{j=1}^{M} \mathbf{B}_j(\mathbf{x})\mathbf{z}_j \\ \mathbf{0} &= \mathbf{k}_j(\mathbf{x}), \quad j = 1, \ldots, M \end{aligned} \tag{B.21}$$

which is a description of the slow dynamics of the system (B.10), and is a non-stiff system.

Note that this method ultimately leads to a set of state-space realizations for the reduced-order models for each time scale of a multiple-time-scale system, but does not identify the slow and fast variables associated with the individual time scales. They can be obtained only from a coordinate change that would yield a standard singularly perturbed form of the system (B.10). The fact that $\mathbf{k}_l(\mathbf{x})$, $l = 1, \ldots, M$, are constrained to be equal to zero in the slow systems that follow the lth boundary layer and are non-zero in the lth boundary layer indicates that $\mathbf{k}_l(\mathbf{x})$ should be used in the definition of the fast variables in a nonlinear coordinate change. Theorem B.1 (Vora 2000) states the conditions for the existence of such a coordinate transformation.

Theorem B.1. *Consider the system in Equation (B.10), and assume that*

(i) the $\left(\sum_{j=l}^{M} \times \sum_{j=l}^{M}\right)$ matrix $(L_{\mathbf{B}}\mathbf{k}(\mathbf{x}))_l$ defined in Equation (B.20) is non-singular
(ii) the $\left(\sum_{j=l}^{M} p_j \times n\right)$ matrix $[(\partial \mathbf{k}_l(\mathbf{x})/\partial \mathbf{x})^{\mathrm{T}}| \cdots |(\partial \mathbf{k}_M(\mathbf{x})/\partial \mathbf{x})^{\mathrm{T}}]^{\mathrm{T}}$ has full row rank $\forall l \in [1, M]$.

Then, there exists a coordinate change of the form

$$\begin{bmatrix} \boldsymbol{\zeta} \\ \boldsymbol{\eta}_1 \\ \vdots \\ \boldsymbol{\eta}_j \\ \vdots \\ \boldsymbol{\eta}_M \end{bmatrix} = \mathbf{T}(\mathbf{x}, \varepsilon) = \begin{bmatrix} \boldsymbol{\phi}(\mathbf{x}) \\ \mathbf{k}_1(\mathbf{x})/\varepsilon_1 \\ \vdots \\ \mathbf{k}_j(\mathbf{x})/\varepsilon_j \\ \vdots \\ \mathbf{k}_M(\mathbf{x})/\varepsilon_M \end{bmatrix} \tag{B.22}$$

where $\boldsymbol{\zeta} \in \mathbb{R}^{n-\sum_j p_j}$ and $\boldsymbol{\eta}_j \in \mathbb{R}^{p_j}$, $j = 1, \ldots, M$, under which the multiple-time-scale system of Equation (B.10) takes the following singularly perturbed form:

$$\begin{aligned} \dot{\boldsymbol{\zeta}} &= \tilde{\mathbf{f}}(\boldsymbol{\zeta}, \varepsilon\boldsymbol{\eta}) + \tilde{\mathbf{g}}(\boldsymbol{\zeta}, \varepsilon\boldsymbol{\eta})\mathbf{u} + \sum_{i=1}^{M} \left\{ \tilde{\mathbf{B}}_i(\boldsymbol{\zeta}, \varepsilon\boldsymbol{\eta})\boldsymbol{\eta}_i \right\} \\ \varepsilon_1 \dot{\boldsymbol{\eta}}_1 &= \bar{\mathbf{f}}^1(\boldsymbol{\zeta}, \varepsilon\boldsymbol{\eta}) + \bar{\mathbf{G}}^1(\boldsymbol{\zeta}, \varepsilon\boldsymbol{\eta})\mathbf{u} + \sum_{i=1}^{M} \{ \bar{\mathbf{B}}_i^1(\boldsymbol{\zeta}, \varepsilon\boldsymbol{\eta})\boldsymbol{\eta}_i \} \\ &\vdots \\ \varepsilon_j \dot{\boldsymbol{\eta}}_j &= \bar{\mathbf{f}}^j(\boldsymbol{\zeta}, \varepsilon\boldsymbol{\eta}) + \bar{\mathbf{G}}^j(\boldsymbol{\zeta}, \varepsilon\boldsymbol{\eta})\mathbf{u} + \sum_{i=1}^{M} \{ \bar{\mathbf{B}}_j^1(\boldsymbol{\zeta}, \varepsilon\boldsymbol{\eta})\boldsymbol{\eta}_i \} \\ &\vdots \\ \varepsilon_M \dot{\boldsymbol{\eta}}_M &= \bar{\mathbf{f}}^M(\boldsymbol{\zeta}, \varepsilon\boldsymbol{\eta}) + \bar{\mathbf{G}}^M(\boldsymbol{\zeta}, \varepsilon\boldsymbol{\eta})\mathbf{u} + \sum_{i=1}^{M} \{ \bar{\mathbf{B}}_M^1(\boldsymbol{\zeta}, \varepsilon\boldsymbol{\eta})\boldsymbol{\eta}_i \} \end{aligned} \tag{B.23}$$

where $\tilde{\mathbf{f}} = L_{\mathbf{f}}\phi(\mathbf{x})$, $\tilde{\mathbf{G}} = L_{\mathbf{G}}\phi(x)$, $\tilde{\mathbf{B}} = L_{\mathbf{B}}\phi(\mathbf{x})$, $\bar{\mathbf{f}}^j = L_{\mathbf{f}}\mathbf{k}_j(\mathbf{x})$, $\bar{\mathbf{G}}_j = L_{\mathbf{G}}\mathbf{k}_j(\mathbf{x})$, *and* $\tilde{\mathbf{B}}_i^j = L_{\mathbf{B}_i}k_j(\mathbf{x})$ *are evaluated at* $\mathbf{x} = \mathbf{T}^{-1}(\zeta, \varepsilon\eta)$, $\forall i, j$, *and the matrix* $\mathbf{Q}_l(\zeta, 0) = (L_{\mathbf{B}}\mathbf{k}(\mathbf{x}))_l$ *evaluated at* $\mathbf{x} = \mathbf{T}^{-1}(\zeta, 0)$ *is nonsingular uniformly in* $\zeta \in \mathbb{R}^{n-\sum_j p_j}$, $\forall l \in [1, M]$.

Notice that Theorem B.1 is an extension of Theorem A.1 to systems with multiple singular perturbation parameters. Also, note that the standard singularly perturbed form (B.23) allows one to obtain the entire hierarchy of boundary-layer subsystems and the corresponding slow subsystem for any $l \in [1, M]$.

Appendix C

Matlab code

Listing C.1. Symbolic derivation of reduced-order model of the slow dynamics, and of the input–output linearizing temperature controller for the reactor–feed effluent heat exchanger system in Section 6.6

```
% define symbolic variables
syms f
syms B B11 B12 B13  B21 B32 B43
syms Fins Tins
syms Fin Tin VH VC VF UA V DH QH alpha  F
syms kr ki kc
syms kreac cA
syms rhocp
syms Tvec TC_ TH_ TR Ti
syms coordch delta eta1 eta2 eta3

syms GammaTilde k1 k2 k3
syms z
syms dTdt_slow
syms LOG_FIXED_SS
syms a cAo k0 E Rconst

% vector field f corresponding to  Equation (6.55); cA and V are not
% included in the derivation since they are slow variables

f = [Fin / VC * Tin
     - F * (1 - alpha) / VH * TH_ - alpha * F * TR / VH
     - kreac * cA * DH/ rhocp
     QH / rhocp / VF];

% construct matrix B corresponding to  Equation (6.55); cA and V are
% not included in the derivation since they are slow variables
B11 = -1 / VC;
B12 = -1 / VC;
B13 = -1 / VC;
B21 =  1 / VH;
B32 =  1 / V;
B43 =  1 / VF;
B = [B11 B12 B13 ;
     B21 0   0   ;
     0   B32 0   ;
     0   0   B43 ];
B = Fins * Tins * B;
```

```
% build vector field GammaTilde of linearly independent constraints
% corresponding to Equation (6.57). Underwood approximation of LMTD
% is used
k1 = Fin * rhocp * TR * (1-alpha) ...
     - UA * (((TH_-Tin)^(1/3) + (TR-TC_)^(1/3))/2)^(3);

k2 = Ti - TR;
k3 = TC_ - Ti;

GammaTilde = [ k1; k2; k3];

% vector of state variables in energy balance
Tvec = [TC_ ; TH_; TR ; Ti];

% compute algebraic variables z from Equation (6.57)
z = - inv ( jacobian(GammaTilde, Tvec) * B) ...
    * jacobian(GammaTilde, Tvec) * f;


% build slow subsystem in DAE form
dTdt_slow = simplify(f + B* z);

% apply coordinate change of Equations (6.61) and (6.62)
% to obtain minimal-order state-space
% realization of the slow dynamics

A = solve('delta = TR',...
    'eta1 = Fin * rhocp * TR - UA * (((TH_-Tin)^(1/3)
          + (TR-TC_)^(1/3))/2)^(3)',...
          'eta2 = Ti - TR','eta3 = TC_ - Ti', TH_, TR, TC_, Ti);

dTdt_slow_new = subs(dTdt_slow, TH_, A.TH_(1));

dTdt_slow_new = subs(dTdt_slow_new, TC_, A.TC_(1));

dTdt_slow_new = subs(dTdt_slow_new, TR, A.TR(1));

dTdt_slow_new = subs(dTdt_slow_new, Ti, A.Ti(1));

eta1 = 0; eta2 = 0; eta3 = 0 ;

dTdt_slow_new = subs(dTdt_slow_new);

% separate the expression for the evolution of the reactor
% temperature in the slow time scale and obtain Equation (6.63)

ddeltadt = simplify(dTdt_slow_new(3));

% controller design
% The third equation in (6.63) is in the format
%
% d delta/ dt = f_slow + g_slow * u
%
% with h = delta; u = QH
%
% the corresponding input/output linearizing controller
```

```
% (Daoutidis and Kravaris, 1992) will be in the form
%
% u = (T_R_sp - T_R - beta * f_slow)/beta/g_slow
%
% with

f_slow = simplify( subs(ddeltadt, QH, 0))

g_slow = simplify(ddeltadt - f_slow);

g_slow = subs(g_slow, QH, 1)
```

Listing C.2. S-function illustrating the use of the reduced-order model of the slow dynamics in the derivation of a control system (including an input–output linearizing temperature controller) for the reactor–feed effluent heat exchanger system in Section 6.6

```
function [sys,x0,str,ts] = sf_FEHE(t,x,u,flag)
% Matlab s-function implementation of Case Study 6.6,
% control of a reactor-FEHE process

% parameters
        prm.k0          = 12267000      ; % 1/s
        prm.E           = 142870        ; % J/mol
        prm.DH          = -54.8283e3    ; % J/mol
        prm.rhocp       = 4.184E6       ; % J/m3/K
        prm.UA          = 83680         ; % W/K
        prm.VH          = 0.1           ; % m3
        prm.VC          = 0.09          ; % m3
        prm.VF          = 0.01          ; % m3
        prm.L           = 5.0           ; % m
        prm.cAo         = 1000          ; % mol/m3
        prm.nnode       = 1001          ; % nodes

        prm.Fs          = 5.7667e-4     ; % m3/s

        prm.tau_alpha = 10              ; % s
        prm.k_alpha     = 0.0018        ; % 1/K

        prm.tau_v       = 18 *60        ; % s
        prm.k_v         = - 5           ; % 1/s

        prm.NLCbeta     = 15 * 60       ; % s


switch flag,
    case 0,
        [sys,x0,str,ts] = mdlInitializeSizes(prm);
    case 1,
        sys = mdlDerivatives(t,x,u,prm);
    case 2,
        sys = mdlUpdate(t,x,u,prm);
    case 3,
        sys = mdlOutputs(t,x,u,prm);
    case 9,
        sys = mdlTerminate(t,x,u,prm);
    otherwise
        error(['Unhandled flag = ',num2str(flag)]);
```

```
end

function [sys,x0,str,ts] = mdlInitializeSizes(prm)

sizes = simsizes;

sizes.NumContStates  = (2 * prm.nnode + 4) + 7;
sizes.NumDiscStates  = 0;
sizes.NumOutputs     = (2 * prm.nnode + 4) + 11;
sizes.NumInputs      = 5;
sizes.DirFeedthrough = 1;
sizes.NumSampleTimes = 1;    % at least one sample time is needed

sys = simsizes(sizes);

x0 = load_x0(0);
str = [];
ts  = [0 0];

% end mdlInitializeSizes

function sys = mdlDerivatives(t,x,u,prm)

% inputs
     Tin          = u(1)    ; %   inlet temperature
     Fin          = u(2)    ; %   inlet flow rate
     Texitsp      = u(3)    ; %   exit temperature setpoint
     Vsp          = u(4)    ; %   reactor holdup setpoint
     TR_sp        = u(5)    ; %   reactor temperature setpoint

%==================================================
%                CONTROL SYSTEM                  %
%==================================================

%==== PROCESS OUTLET TEMPERATURE (Texit) CONTROLLER ======

% isolate process outlet temperature from the state vector
Texit = SysVariableMapping (x,prm,'Texit');

% isolate time integral of outlet temperature error
integralerror_texit = SysVariableMapping ( x, prm, ...
                                           'integralerror_texit')

% PI controller for process outlet temperature
alpha = PIcontrol(Texit, Texitsp, ...
        integralerror_texit, 0.1, prm.k_alpha, prm.tau_alpha);

% define additional state, required to compute
% integral action for outlet temperature controller
dintegralerror_texit = -Texit + Texitsp;

% first-order filter for process outlet temperature
dTexit_dt = - 1/2*(Texit - ((alpha * x(3) + (1 - alpha) * x(5))));

%==== REACTOR HOLDUP CONTROLLER ======================
```

```
% integral action for holdup controller
integralerror_v = SysVariableMapping ( x, prm, 'integralerror_v');

% PI controller for reactor holdup
F = PIcontrol (x(1), Vsp,...
    integralerror_v, prm.Fs, prm.k_v, prm.tau_v);

% define additional state, required to compute
% integral action for reactor holdup controller
dintegralerror_v = Vsp - x(1);



%==== REACTOR TEMPERATURE CONTROLLER =================

% input-output linearizing output feedback controller
% synthesized using the reduced-order model of the slow dynamics

V     = SysVariableMapping ( x, prm, 'v_red');
cA    = SysVariableMapping ( x, prm, 'cA_red');
delta = SysVariableMapping ( x, prm, 'delta');

kreac = prm.k0 * exp(-prm.E/8.314/delta);


integralerror_t = SysVariableMapping(x,prm,'integralerror_t');

[f_slow, g_slow] = ReducedOrderSlowModel (x,u,prm,F,alpha);

% input-output linearizing controller
% calculate Q heater using the slow model

% integral action for reactor temperature controller
d_integralerror_t = 1/ prm.NLCbeta * (TR_sp - x(3));

QH = ReactorTemperatureControl (x(3), TR_sp,...
                              integralerror_t, prm, f_slow, g_slow);

% state-space realization of the reduced-order model (used as
% state observer for the reactor temperature controller)

% overall reactor mass balance
dV_red_dt = Fin - F;
% mass balance for component A
dcA_red_dt = Fin * (prm.cAo - cA) / V - kreac * cA;
% reactor energy balance
ddelta_dt = f_slow + g_slow * QH;

% vector of the time derviatives of the variables in the slow model
ddt_slow_red = [dV_red_dt; dcA_red_dt; ddelta_dt];


% Full-order model
% a finite-difference scheme is used to discretize
% the partial differential
% equations used to model the heat exchanger
```

```
V               =   SysVariableMapping(x,prm,'V');
cA              =   SysVariableMapping(x,prm,'cA');
TR              =   SysVariableMapping(x,prm,'TR');
Ti              =   SysVariableMapping(x,prm,'Ti');
Tcout           =   SysVariableMapping(x,prm,'Tcout');

kreac           =   prm.k0 * exp( - prm.E / 8.314 / TR);
TH(1:prm.nnode) =   x(5: (prm.nnode + 4));
TC(1:prm.nnode) =   x((prm.nnode+5): (2*prm.nnode + 4));

% reactor balance equations
dVdt            =   Fin - F;
dcAdt           =   Fin * (prm.cAo - cA) / V - kreac * cA;
dTRdt           =   Fin * (Ti - TR) ...
                    / V  - kreac * cA * prm.DH / prm.rhocp;

% energy balance for furnace
dTidt =  Fin * (Tcout - Ti) / prm.VF + QH /prm.rhocp / prm.VF;

% discretized model of the feed-effluent heat exchanger
dTCdt = zeros(prm.nnode,1);
dTHdt = zeros(prm.nnode,1);

Ah = prm.VH / prm.L; % cross section of hot leg
vh = F*(1-alpha)/Ah;

Ac = prm.VC / prm.L; % cross section of cold leg
vc = Fin/Ac;

TH(1:prm.nnode) = x(5: (prm.nnode + 4));
TC(1:prm.nnode) = x((prm.nnode+5): (2*prm.nnode + 4));

dz = prm.L/(prm.nnode-1); % set discretization interval length

% finite difference discretization of HX PDE
dTCdt(1) = - vc * (TC(1)- Tin) / dz;

dTHdt(1) = - vh * (TH(1) - TH(2)) / dz ...
           - prm.UA/prm.rhocp * (TH(1) - TC(1)) / (prm.VC);


dTCdt(prm.nnode) = - vc * (TC(prm.nnode)- TC(prm.nnode-1))/dz ...
                   + prm.UA/prm.rhocp * (TH(prm.nnode) ...
                   - TC(prm.nnode) ) / (prm.VC);

dTHdt(prm.nnode) = - vh * (TH(prm.nnode) - TR)/dz;



for i=2:(prm.nnode-1)
    dTCdt(i) = - vc * (TC(i)- TC(i-1)) / dz ...
        + prm.UA/prm.rhocp * (TH(i) - TC(i))/ (prm.VC);
    dTHdt(i) = - vh * (TH(i) - TH(i+1))/dz ...
        - prm.UA/prm.rhocp * (TH(i) - TC(i))/(prm.VH);
end
```

```
% assemble vector of time derivatives of the states of both the
% reduced-order and full-order models

sys = [dVdt ; dcAdt; dTRdt ; dTidt; dTHdt; dTCdt; ...
    ddt_slow_red ;d_integralerror_t;dTexit_dt; ...
    dintegralerror_texit; dintegralerror_v];

% end mdlDerivatives

function sys = mdlUpdate(t,x,u,prm)

sys = [];

% end mdlUpdate

function sys = mdlOutputs(t,x,u,prm)

% inputs
    Tin          = u(1)   ; %   inlet temperature
    Fin          = u(2)   ; %   inlet flow rate
    Texitsp      = u(3)   ; %   exit temperature setpoint
    Vsp          = u(4)   ; %   reactor holdup setpoint
    TR_sp        = u(5)   ; %   reactor temperature setpoint

% recalculate controller outputs for reporting purposes
% bypass ratio
Texit = SysVariableMapping (x,prm,'Texit');

integralerror_texit = SysVariableMapping ( x, prm, '
    integralerror_texit');

alpha = PIcontrol(Texit, Texitsp, ...
        integralerror_texit, 0.1, prm.k_alpha, prm.tau_alpha);


% reactor effluent flow rate
integralerror_v = SysVariableMapping ( x, prm, 'integralerror_v');

F = PIcontrol (x(1), Vsp,...
    integralerror_v, prm.Fs, prm.k_v, prm.tau_v);


% QH
integralerror_t = SysVariableMapping(x,prm,'integralerror_t');

[f_slow, g_slow] = ReducedOrderSlowModel (x,u,prm,F,alpha);

QH = ReactorTemperatureControl (x(3), TR_sp, ...
                                integralerror_t, prm, f_slow, g_slow
                                    );

% calculate reactor conversion
conversion = (prm.cAo-x(2))/prm.cAo * 100;

% assemble output vector
```

```
sys = [x;QH;conversion;alpha; F];

% end mdlOutputs


function sys = mdlTerminate(t,x,u,prm)

sys = [];

% end mdlTerminate


function out = PIcontrol(Ym,Ysp,integr_error,bias, Kc,tau_I)

out = bias +  Kc * ((Ysp - Ym) + 1/tau_I * integr_error);

% end PIcontrol


function out = SysVariableMapping(x,prm,VariableSelector)
% isolate specified variable from state vector x
    if strcmp(variableSelector,'Textit')
        % process outlet temperature
        out = x (2 * prm.nnode + 4+5);
    elseif strcmp (VariableSelector, 'integralerror_texit')
        out = x (2 * prm.nnode + 4 + 6);
    elseif strcmp (VariableSelector, 'integralerror_v')
        % integral action for holdup controller
        out = x (2 * prm.nnode + 4 + 7);
    elseif strcmp (VariableSelector, 'V_red')
        % reactor holdup based on reduced-order model
        out = x (2 * prm.nnode + 4 + 1);
    elseif strcmp (VariableSelector, 'cA_red')
        % reactor composition based on reduced-order model
        out = x (2 * prm.nnode + 4 + 2);
    elseif strcmp (VariableSelector, 'delta')
        % reactor temperature based on reduced-order model
        out = x (2 * prm.nnode + 4 + 3);
    elseif strcmp (VariableSelector, 'integralerror_t')
        % integrated error of reactor temperature
        out = x (2 * prm.nnode + 4 + 4);
    elseif strcmp (VariableSelector, 'V')
        % reactor holdup
        out = x (1);
    elseif strcmp (VariableSelector, 'cA')
        % reactor composition
        out = x (2);
    elseif strcmp (VariableSelector, 'TR')
        % reactor temperature
        out = x (3);
    elseif strcmp (VariableSelector, 'Ti')
        % furnace temperature
        out = x (4);
    elseif strcmp (VariableSelector, 'Tcout')
        % cold leg outlet temperature
        out = x (4 + 2 * prm.nnode);
```

```
    else
        disp('Unknown string');
        out = NaN
    end;
% end SysVariableMapping


function [f_slow, g_slow] = ReducedOrderSlowModel (x,u,prm,F,alpha)
% functions f and g in the reduced-order model of the slow dynamics

% states of the reduced-order model
V       = SysVariableMapping (x , prm , 'V_red') ;
cA      = SysVariableMapping (x , prm , 'cA_red');
delta   = SysVariableMapping (x , prm , 'delta') ;

% system inputs (same as full-order model)

    Tin         = u(1)   ; %   inlet temperature
    Fin         = u(2)   ; %   inlet flow rate
    Texitsp     = u(3)   ; %   exit temperature setpoint
    Vsp         = u(4)   ; %   reactor holdup setpoint
    TR_sp       = u(5)   ; %   reactor temperature setpoint

kreac = prm.k0 * exp( -prm.E / 8.314 / delta);

DEN = ( prm.VC * prm.UA + prm.UA * prm.VF ...
            + prm.UA * V + 8 * prm.VH * Fin * prm.rhocp ...
            - 8 * prm.VH * Fin * prm.rhocp * alpha) * prm.rhocp;


f_slow = ( - kreac * cA * prm.DH * V * prm.UA ...
           + Fin * Tin * prm.rhocp * prm.UA ...
           - prm.rhocp * F * Tin * prm.UA ...
           - 8 * prm.rhocp^2 * F * Fin * delta ...
           + prm.rhocp * F * Tin * alpha * prm.UA ...
           + 8 * prm.rhocp^2 * F * alpha * Fin * delta ...
           - delta * prm.rhocp * F * alpha * prm.UA) ...
           / DEN ;


g_slow = prm.UA / DEN ;

% end SlowModel

function QH = ReactorTemperatureControl (TR, TR_sp, integralerror_t

                                            prm, f_slow, g_slow)
% nonlinear controller for reactor temperature

QH = ((integralerror_t + TR_sp - TR ...
    - prm.NLCbeta * f_slow ) / (prm.NLCbeta * g_slow ));

% end ReactorTemperatureControl
```

References

Adams, J. P., Collis, A. J., Henderson, R. K., and Sutton, P. W. (2009). Biotransformations in small-molecule pharmaceutical development. In J. Whittall and P. Sutton, eds., *Practical Methods for Biocatalysis and Biotransformations*, pp. 1–82. New York: John Wiley & Sons.

Ali, E. and Alhumaizi, K. L. (2000). Temperature control of ethylene to butene-1 dimerization reactor. *Ind. Eng. Chem. Res.*, **39**, 1320–1329.

Andrecovich, M. J. and Westerberg, A. W. (1985a). A simple synthesis method based on utility bounding for heat-integrated distillation sequences. *AIChE J.*, **31**, 363–375.

Andrecovich, M. J. and Westerberg, A. W. (1985b). An MILP formulation for heat-integrated distillation sequence synthesis. *AIChE J.*, **31**, 1461–1474.

Annakou, O. and Mizsey, P. (1996). Rigorous comparative study of energy-integrated distillation schemes. *Ind. Eng. Chem. Res.*, **35**, 1877–1885.

Antelo, L. T., Otero-Muras, I., Banga, J. R., and Alonso, A. A. (2007). A systematic approach to plant-wide control based on thermodynamics. *Comput. Chem. Eng.*, **31**, 677–691.

Antoniades, C. and Christofides, P. D. (2001). Integrating nonlinear output feedback control and optimal actuator/sensor placement for transport–reaction processes. *Chem. Eng. Sci.*, **56**, 4517–4535.

Baldea, M. and Daoutidis, P. (2005). Dynamics and control of integrated process networks with multi-rate reactions. In *Proceedings of the 16th IFAC World Congress*, Prague.

Baldea, M. and Daoutidis, P. (2006). Model reduction and control of reactor–heat exchanger networks. *J. Proc. Contr.*, **16**, 265–274.

Baldea, M. and Daoutidis, P. (2007). Control of integrated process networks – a multi-time scale perspective. *Comput. Chem. Eng.*, **31**, 426–444.

Baldea, M., Daoutidis, P., and Kumar, A. (2006). Dynamics and control of integrated networks with purge streams. *AIChE J.*, **52**, 1460–1472.

Baldea, M., Daoutidis, P., and Nagy, Z. K. (2010). Nonlinear Model Predictive Control of integrated process systems. In *Proceedings Nonlinear Control Systems (NOLCOS 2010)*.

Banerjee, A. and Arkun, Y. (1995). Control configuration design applied to the Tennessee Eastman plant-wide control problem. *Comput. Chem. Eng.*, **19**, 453–480.

Bao, J. and Lee, P. L. (2007). *Process Control: The Passive Systems Approach.* New York: Springer.

Belanger, P. W. and Luyben, W. L. (1998). Plantwide design and control of processes with inerts. 1. Light inerts. *Ind. Eng. Chem. Res.*, **37**, 516–527.

Bildea, C. S. and Dimian, A. C. (1998). Stability and multiplicity approach to the design of heat integrated PFR. *AIChE J.*, **44**, 2703–2712.

Bildea, C. S., Dimian, A. C., and Iedema, P. D. (2000). Nonlinear behavior of reactor–separator–recycle systems. *Comput. Chem. Eng.*, **24**, 209–214.

Buckley, P. S. (1964). *Techniques of Process Control.* New York: Wiley.

Chen, J. J. J. (1987). Comments on improvements on a replacement for the logarithmic mean. *Chem. Eng. Sci.*, **42**, 2488–2489.

Chen, R. and McAvoy, T. J. (2003). Plantwide control system design: methodology and application to a vinyl acetate process. *Ind. Eng. Chem. Res.*, **42**, 4753–4771.

Chen, R., McAvoy, T. J., and Zafiriou, E. (2004). Plantwide control system design: extension to multiple-forcing and multiple-steady-state operation. *Ind. Eng. Chem. Res.*, **43**, 3685–3694.

Chen, Y. H. and Yu, C. C. (2003). Design and control of heat integrated reactors. *Ind. Eng. Chem. Res.*, **42**, 2791–2808.

Chow, J. H. and Kokotović, P. V. (1976). A decomposition of near optimum regulators for systems with slow and fast modes. *IEEE Trans. Automat. Contr.*, **21**, 701–705.

Chow, J. H. and Kokotović, P. V. (1978). Two-time-scale feedback design of a class of nonlinear systems. *IEEE Trans. Automat. Contr.*, **23**, 438–443.

Christofides, P. D. and Daoutidis, P. (1996a). Feedback control of two-time-scale nonlinear systems. *Int. J. Contr.*, **63**, 965–994.

Christofides, P. D. and Daoutidis, P. (1996b). Compensation of measurable disturbances in two-time-scale nonlinear systems. *Automatica*, **32**, 1553–1573.

Christofides, P. D., Davis, J. F., El-Farra, N. H., Clark, D., Harris, K. R. D., and Gipson, J. N. (2007). Smart plant operations: vision, progress and challenges. *AIChE J.*, **53**, 2734–2741.

Contou-Carrère, M. N., Baldea, M., and Daoutidis, P. (2004). Dynamic precompensation and output feedback control of integrated process networks. *Ind. Eng. Chem. Res.*, **43**, 3528–3538.

Dadebo, S. A., Bell, M. L., McLellan, P. J., and McAuley, K. B. (1997). Temperature control of industrial gas phase polyethylene reactors. *J. Proc. Contr.*, **7**, 83–95.

Daoutidis, P. and Kravaris, C. (1992). Dynamic output feedback control of minimum-phase nonlinear processes. *Chem. Eng. Sci.*, **47**, 837–849.

Daoutidis, P. and Kravaris, C. (1994). Dynamic output feedback control of minimum-phase multivariable nonlinear processes. *Chem. Eng. Sci.*, **49**, 433–447.

Denn, M. M. and Lavie, R. (1982). Dynamics of plants with recycle. *Chem. Eng. J.*, **24**, 55–59.

Desoer, C. A. and Shahruz, S. M. (1986). Stability of nonlinear systems with three time scales. *Circ. Syst. Sig. Proc.*, **5**, 449–464.

Desoer, C. A. and Vidyasagar, M. (2009). *Feedback Systems: Input–Output Properties.* Philadelphia, PA: Society for Industrial and Applied Mathematics.

Diehl, M., Amrit, R., and Rawlings, J. B. (2011). A Lyapunov function for economic optimizing model predictive control. *IEEE Trans. Automat. Contr.*, **56**, 703–707.

Diez, E., Langston, P., Ovejero, G., and Romero, M. D. (2009). Economic feasibility of heat pumps in distillation to reduce energy use. *Appl. Therm. Eng.*, **29**, 1216–1223.

Dimian, A. C., Groenendijk, A. J., and Iedema, P. D. (2001). Recycle interaction effects on the control of impurities in a complex plant. *Ind. Eng. Chem. Res.*, **40**, 5784–5794.

Douglas, J. M. (1988). *Conceptual Design of Chemical Processes.* New York: McGraw-Hill.

Downs, J. J. and Vogel, E. F. (1993). A plant-wide industrial process control problem. *Comput. Chem. Eng.*, **17**, 245–255.

Downs, J. J. and Skogestad, S. (2009). An industrial and academic perspective on plantwide control. In *IFAC Symposium on Advanced Control of Chemical Processes*, pp. 119–130.

Edgar, T. F. (2004). Control and operations: when does controllability equal profitability? *Comput. Chem. Eng.*, **29**, 41–49.

Edgar, T. F. and Davis, J. F. (2009). Smart process manufacturing – a vision of the future. In A. A. Linninger and M. M. El-Halwagi, eds., *Design for Energy and the Environment: Proceedings of the Seventh International Conference on the Foundations of Computer-Aided Process Design*, pp. 149–165. Boca Raton, FL: CRC Press.

El-Farra, N. H., Gani, A., and Christofides, P. D. (2005). Fault-tolerant control of process systems using communication networks. *AIChE J.*, **51**, 1665–1682.

El-Halwagi, M. M. (2006). *Process Integration.* Amsterdam: Elsevier Academic Press.

Farschman, C. A., Viswanath, K. P., and Ydstie, B. E. (1998). Process systems and inventory control. *AIChE J.*, **44**, 1841–1857.

Fenichel, N. (1979). Geometric singular perturbation theory for ordinary differential equations. *J. Diff. Equat.*, **31**, 53.

Floudas, C. A. and Paules, G. E. (1988). A mixed-integer nonlinear programming formulation for the synthesis of heat-integrated distillation sequences. *Comput. Chem. Eng.*, **12**, 531–546.

Foss, A. S. (1973). Critique of chemical process control theory. *AIChE J.*, **19**, 209–214.

Georgakis, C. (1986). On the use of extensive variables in process dynamics and control. *Chem. Eng. Sci.*, **41**, 1471–1484.

Gerdtzen, Z. P., Daoutidis, P., and Hu, W. S. (2004). Non-linear reduction for kinetic models of metabolic reaction networks. *Metabolic Eng.*, **6(2)**, 140–154.

Gilliland, E. R., Gould, L. A., and Boyle, T. J. (1964). Dynamic effects of material recycle. In *Preprints of the Joint American Control Conference*, pp. 140–146.

Haberman, R. (1998). *Elementary Applied Partial Differential Equations*, third edition. Upper Saddle River, NJ: Prentice-Hall.

Henderson, L. S. and Cornejo, R. A. (1989). Temperature control of continuous, bulk styrene polymerization reactors and the influence of viscosity: an analytical study. *Ind. Eng. Chem. Res.*, **28**, 1644–1653.

Hoppensteadt, F. (1971). Properties of solutions of ordinary differential equations with small parameters. *Commun. Pure Appl. Math.*, **XXIV**, 807–840.

Illanes, A. (2008). *Enzyme Biocatalysis: Principles and Applications.* New York: Springer.

Isidori, A. (1995). *Nonlinear Control Systems.* Berlin: Springer-Verlag.

Jacobsen, E. and Berezowski, M. (1998). Chaotic dynamics in homogeneous tubular reactors with recycle. *Chem. Eng. Sci.*, **23**, 4023–4029.

Jacobsen, E. W. (1999). On the dynamics of integrated plants – non-minimum phase behavior. *J. Proc. Contr.*, **9**, 439–451.

Jillson, K. R. and Ydstie, Y. B. (2007). Process networks with decentralized inventory and flow control. *J. Proc. Contr.*, **17**, 399–413.

Jogwar, S. S. and Daoutidis, P. (2010). Energy flow patterns and control implications for integrated distillation networks. *Ind. Eng. Chem. Res.*, **49**, 8048–8061.

Jogwar, S. S., Baldea, M., and Daoutidis, P. (2009). Dynamics and control of process networks with large energy recycle. *Ind. Eng. Chem. Res.*, **48**, 6087–6097.

Jogwar, S. S., Torres, A. I., and Daoutidis, P. (2011). Networks with large solvent recycle: dynamics, hierarchical control and a biorefinery application. *AIChE J.*, DOI:10.1002/aic.12708.

Kanadibhotla, R. S. and Riggs, J. B. (1995). Nonlinear model based control of a recycle reactor process. *Comput. Chem. Eng.*, **19**, 933–948.

Kapoor, N., McAvoy, T. J., and Marlin, T. E. (1986). Effect of recycle structure on distillation tower time constants. *AIChE J.*, **32**, 411–418.

Kevorkian, J. and Cole, J. D. (1996). *Multiple Scale and Singular Perturbation Methods.* New York: Springer.

Khalil, H. K. (2002). *Nonlinear Systems*, third edition. Upper Saddle River, NJ: Prentice-Hall.

King, C. J. (1980). *Separation Processes.* New York: McGraw-Hill.

Kiss, A. A., Bildea, C. S., Dimian, A. C., and Iedema, P. D. (2002). State multiplicity in CSTR–separator–recycle polymerization systems. *Chem. Eng. Sci.*, **57**, 535–546.

Kiss, A. A., Bildea, C. S., Dimian, A. C., and Iedema, P. D. (2005). Design of recycle systems with parallel and consecutive reactions by nonlinear analysis. *Ind. Eng. Chem. Res.*, **44**, 576–587.

Kobayashi, S. (2009). Recent developments in lipase-catalyzed synthesis of polyesters. *Macromol. Rapid Commun.*, **30**, 237–266.

Kokotović, P. V., Khalil, H. K., and O'Reilly, J. (1986). *Singular Perturbations in Control: Analysis and Design.* London: Academic Press.

Kothare, M. V., Shinnar, R., Rinard, I., and Morari, M. (2000). On defining the partial control problem: concepts and examples. *AIChE J.*, **46**, 2456–2474.

Kravaris, C. and Kantor, J. C. (1990). Geometric methods for nonlinear process control, parts 1–2. *Ind. Eng. Chem. Res.*, **29**, 2295–2323.

Kravaris, C., Niemiec, M., Berber, R., and Brosilow, C. B. (1998). Nonlinear model-based control of nonminimum-phase processes. In R. Berber and C. Kravaris, eds., *Nonlinear Model Based Process Control*, pp. 115–143. Dordrecht: Kluwer Academic Publishers.

Kumar, A. and Daoutidis, P. (1996). Feedback regularization and control of nonlinear differential-algebraic-equation systems. *AIChE J.*, **42**, 2175–2198.

Kumar, A. and Daoutidis, P. (1999a). *Control of Nonlinear Differential Equation Systems.* Boca Raton, FL: Chapman & Hall/CRC Press.

Kumar, A. and Daoutidis, P. (1999b). Modeling, analysis and control of ethylene glycol reactive distillation column. *AIChE J.*, **45**, 51–68.

Kumar, A. and Daoutidis, P. (2002). Dynamics and control of process networks with recycle. *J. Proc. Contr.*, **12**, 475–484.

Kumar, A. and Daoutidis, P. (2003). Nonlinear model reduction and control for high-purity distillation columns. *Ind. Eng. Chem. Res.*, **42**, 4495–4505.

Kumar, A., Christofides, P. D., and Daoutidis, P. (1998). Singular perturbation modeling of nonlinear processes with non-explicit time-scale separation. *Chem. Eng. Sci.*, **53**, 1491–1504.

Kuster, B. F. M. and Temmink, H. M. G. (1977). The influence of pH and weak-acid anions on the dehydration of fructose. *Carbohyd. Res.*, **54**, 185–191.

Ladde, G. S. and Siljak, D. D. (1983). Multiparameter singular perturbation of linear systems with multiple time scales. *Automatica*, **19**, 385–394.

Lakshminarayanan, S., Onodera, K., and Madhukar, G. M. (2004). Recycle effect index: a measure to aid in control system design for recycle processes. *Ind. Eng. Chem. Res.*, **43**, 1499–1511.

Larsson, T. and Skogestad, S. (2000). Plantwide control – a review and a new design procedure. *Modeling, Identification Contr.*, **21**, 209–240.

Larsson, T., Hestetun, K., Hovland, E., and Skogestad, S. (2001). Self-optimizing control of a large-scale plant: the Tennessee Eastman process. *Ind. Eng. Chem. Res.*, **40(22)**, 4889–4901.

Larsson, T., Govatsmark, M. S., Skogestad, S., and Yu, C. C. (2003). Control structure selection for reactor, separator, and recycle processes. *Ind. Eng. Chem. Res.*, **42(6)**, 1225–1234.

Linnhoff, B. and Hindmarsh, E. (1983). The pinch design method for heat exchanger networks. *Chem. Eng. Sci.*, **38**, 745–763.

Linnhoff, B., Dunford, H., and Smith, R. (1983). Heat integration of distillation-columns into overall processes. *Chem. Eng. Sci.*, **38**, 1175–1188.

Liu, J., Muñoz de la Peña, D., Ohran, B. J., Christofides, P. D., and Davis, J. F. (2008). A two-tier architecture for networked process control. *Chem. Eng. Sci.*, **63**, 5394–5409.

Liu, J., Muñoz de la Peña, D., and Christofides, P. D. (2009). Distributed model predictive control of nonlinear process systems. *AIChE J.*, **55**, 1171–1184.

Luyben, M. L. and Tyreus, B. D. (1998). An industrial design/control study for the vinyl acetate monomer process. *Comput. Chem. Eng.*, **22(7–8)**, 867–877.

Luyben, M. L., Tyreus, B. D., and Luyben, W. L. (1997). Plantwide control design procedure. *AIChE J.*, **43**, 3161–3174.

Luyben, W. L. (1993a). Dynamics and control of recycle systems. Parts 1–4. *Ind. Eng. Chem. Res.*, **32**, 466–486, 1142–1162.

Luyben, W. L. (1993b). Dynamics and control of recycle systems. 3. Alternative process designs in a ternary system. *Ind. Eng. Chem. Res.*, **32**, 1142–1153.

Luyben, W. L. (1994). Snowball effects in reactor/separator processes with recycle. *Ind. Eng. Chem. Res.*, **33**, 299–305.

Luyben, W. L. (2000). Design and control of gas-phase reactor/recycle processes with reversible exothermic reactions. *Ind. Eng. Chem. Res.*, **39**, 1529–1538.

Lyman, P. R. and Georgakis, C. (1995). Plant-wide control of Tennessee Eastman problem. *Comput. Chem. Eng.*, **19**, 321–331.

Marroquin, G. and Luyben, W. L. (1973). Practical control studies of batch reactors using realistic mathematical models. *Chem. Eng. Sci.*, **28**, 993–1003.

Mayne, D. Q., Rawlings, J. B., Rao, C. V., and Scokaert, P. O. (2000). Constrained model predictive control: stability and optimality. *Automatica*, **36**, 789–814.

McAvoy, T. J. and Ye, N. (1994). Base control for the Tennessee Eastman problem. *Comput. Chem. Eng.*, **18**, 383.

McAvoy, T. J. (1999). Synthesis of plantwide control systems using optimization. *Ind. Eng. Chem. Res.*, **38**, 2984–2994.

Mészáros, I. and Fonyó, Z. (1986). A new bounding strategy for synthesizing distillation schemes with energy integration. *Comput. Chem. Eng.*, **10**, 545–550.

Mhaskar, P., Gani, A., McFall, C., Christofides, P. D., and Davis, J. F. (2007). Fault-tolerant control of nonlinear process systems subject to sensor faults. *AIChE J.*, **53**, 654–668.

Mizsey, P. and Kalmar, I. (1996). Effects of recycle on control of chemical processes. *Comput. Chem. Eng.*, **20**, S883–S888.

Mizsey, P., Hau, N. T., Benko, N., Kalmar, I., and Fonyó, Z. (1998). Process control for energy integrated distillation schemes. *Comput. Chem. Eng.*, **22**, 427–434.

Morari, M. and Faith III, D. C. (1980). The synthesis of distillation trains with heat integration. *AIChE J.*, **26**, 916–928.

Morari, M., Arkun, Y., and Stephanopoulos, G. (1980). Studies in the synthesis of control structures for chemical processes. Part 1. *AIChE J.*, **26**, 220–232.

Morud, J. and Skogestad, S. (1994). Effects of recycle on dynamics and control of chemical processing plants. *Comput. Chem. Eng.*, **18**, S529–S534.

Morud, J. and Skogestad, S. (1996). Dynamic behavior of integrated plants. *J. Proc. Contr.*, **6**, 145–156.

Morud, J. and Skogestad, S. (1998). Analysis of instability in an industrial ammonia reactor. *AIChE J.*, **44**, 888–895.

Muhrer, C. A., Collura, M. A., and Luyben, W. L. (1990). Control of vapor recompression distillation columns. *Ind. Eng. Chem. Res.*, **29**, 59–71.

Ng, C. and Stephanopoulos, G. (1996). Synthesis of control systems for chemical plants. *Comput. Chem. Eng.*, **20**, S999–S1004.

Nishida, N., Stephanopoulos, G., and Westerberg, A. W. (1981). A review of process synthesis. *AIChE J.*, **27**, 321–351.

Paterson, W. R. (1984). A replacement for the logarithmic mean. *Chem. Eng. Sci.*, **39**, 1635–1636.

Ponton, J. W. and Laing, D. M. (1993). A hierarchical approach to the design of process control systems. *Chem. Eng. Res. Des.*, **71**, 181–188.

Prett, D. M. and Garcia, C. E. (1988). *Fundamental Process Control.* London: Butterworths.

Price, R. M. and Georgakis, C. (1993). Plantwide regulatory control design procedure using a tiered framework. *Ind. Eng. Chem. Res.*, **32**, 2693–2705.

Pushpavanam, S. and Kienle, A. (2001). Nonlinear behavior of an ideal reactor separator network with mass recycle. *Chem. Eng. Sci.*, **57**, 2837–2849.

Qin, S. J. and Badgwell, T. A. (2003). A survey of industrial model predictive control technology. *Contr. Eng. Prac.*, **11**, 733–764.

Ramchandran, B., Riggs, J. B., and Heichelheim, H. R. (1992). Nonlinear plantwide control: application to a supercritical fluid extraction process. *Ind. Eng. Chem. Res.*, **31**, 290–300.

Rathore, R. N. S., Van Wormer, K. A., and Powers, G. J. (1974). Synthesis strategies for multicomponent separation systems with energy integration. *AIChE J.*, **20**, 491–502.

Rawlings, J. B. and Stewart, B. T. (2008). Coordinating multiple optimization-based controllers: new opportunities and challenges. *J. Proc. Contr.*, **18**, 839–845.

Reyes, F. and Luyben, W. L. (2000a). Steady-state and dynamic effects of design alternatives in heat-exchanger/furnace/reactor processes. *Ind. Eng. Chem. Res.*, **39**, 3335–3346.

Reyes, F. and Luyben, W. L. (2000b). Steady-state and dynamic effects of design alternatives in heat-exchanger/furnace/reactor processes. *Ind. Eng. Chem. Res.*, **39**, 3335–3346.

Ricker, N. L. (1996). Decentralized control of the Tennessee Eastman challenge process. *J. Proc. Contr.*, **6**, 205.

Ricker, N. L. and Lee, J. H. (1995). Nonlinear model predictive control of the Tennessee Eastman challenge process. *Comput. Chem. Eng.*, **19**, 961.

Rojas, O. J., Setiawan, R., Bao, J., and Lee, P. L. (2009). Dynamic operability analysis of nonlinear process networks based on dissipativity. *AIChE J.*, **55**, 963–982.

Roman-Leshkov, Y., Chheda, J. N., and Dumesic, J. A. (2006). Phase modifiers promote efficient production of hydroxymethylfurfural from fructose. *Science*, **312**, 1933–1937.

Saberi, A. and Khalil, H. (1985). Stabilization and regulation of non-linear singularly perturbed systems-composite control. *IEEE Trans. Automat. Contr.*, **30**, 739–747.

Scali, C. and Ferrari, F. (1999). Performance of control systems based on recycle compensators in integrated plants. *J. Proc. Contr.*, **9**, 425.

Scattolini, R. (2009). Architectures for distributed and hierarchical Model Predictive Control – a review. *J. Proc. Contr.*, **19**, 723–731.

Seider, W. D., Seader, J. D., and Lewin, D. R. (1999). *Process Design Principles.* New York: Wiley.

Skogestad, S. (2000). Plantwide control: the search for the self-optimizing control structure. *J. Proc. Contr.*, **10**, 487–507.

Skogestad, S. (2004). Control structure design for complete chemical plants. *Comput. Chem. Eng.*, **28**, 219–234.

Sophos, A., Stephanopoulos, G., and Morari, M. (1978). Synthesis of optimum distillation sequences with heat integration schemes. In *National AIChE Meeting*, Miami, FL.

Stephanopoulos, G. (1983). Synthesis of control systems for chemical plants – a challenge for creativity. *Comput. Chem. Eng.*, **7**, 331.

Stewart, B. T., Venkat, A. N., Rawlings, J. B., Wright, S. J., and Pannocchia, G. (2010). Cooperative distributed model predictive control. *Syst. Contr. Lett.*, **59**, 460–469.

Sun, Y. and El-Farra, N. H. (2008). Quasi-decentralized model-based networked control of process systems. *Comput. Chem. Eng.*, **32**, 2016–2029.

Sun, Y. and El-Farra, N. H. (2010). A quasi-decentralized approach for networked state estimation and control of process systems. *Ind. Eng. Chem. Res.*, **49**, 7957–7971.

Sureshkumar, M. and Lee, C. K. (2009). Biocatalytic reactions in hydrophobic ionic liquids. *J. Mol. Catal. B – Enzym.*, **60**, 1–12.

Tatara, E., Cinar, A., and Teymour, F. (2007). Control of complex distributed systems with distributed intelligent agents. *J. Proc. Contr.*, **17**, 415–427.

Tetiker, M. D., Artel, A., Teymour, F., and Cinar, A. (2008). Control of grade transitions in distributed chemical reactor networks – an agent-based approach. *Comput. Chem. Eng.*, **32**, 1984–1994.

Tian, Z. and Hoo, K. A. (2005). Multiple model-based control of the Tennessee–Eastman process. *Ind. Eng. Chem. Res.*, **44**, 3187–3202.

Tikhonov, A. N. (1948). On the dependence of the solutions of differential equations on a small parameter. *Mat. Sb.*, **22**, 193–204.

Torres, A. I., Daoutidis, P., and Tsapatsis, M. (2010). Continuous production of 5-hydroxymethylfurfural from fructose: a design case study. *Energy Environ. Sci.*, **3**, 1560–1572.

Tyreus, B. D. (1999). Dominant variables for partial control. 1. A thermodynamic method for their identification. *Ind. Eng. Chem. Res.*, **38**, 1432–1443.

Umeda, T., Kuriyama, T., and Ichikawa, A. (1978). A logical structure for process control system synthesis. In *Proceedings of the IFAC Congress*, Helsinki.

Underwood, A. J. V. (1970). Simple formula to calculate mean temperature difference. *Chem. Eng.*, **77**, 192.

van Rantwijk, F., Madeira Lau, R., and Sheldon, R. A. (2003). Biocatalytic transformations in ionic liquids. *Trends Biotechnol.*, **21**, 131–138.

Vasudevan, S. and Rangaiah, G. P. (2009). Development of guidelines for plantwide control of gas-phase industrial processes, from reactor–separator–recycle results. *Ind. Eng. Chem. Res.*, **50**, 297–337.

Vasudevan, S. and Rangaiah, G. P. (2010). Criteria for performance assessment of plantwide control systems. *Ind. Eng. Chem. Res.*, **49**, 5955–5970.

Venkat, A. N., Rawlings, J. B., and Wright, S. J. (2006). Implementable distributed model predictive control with guaranteed performance properties. In *Proceedings of the 2006 American Control Conference*, Minneapolis, MN.

Venkat, A. N., Hiskens, I. A., Rawlings, J. B., and Wright, S. J. (2008). Distributed MPC strategies with application to power system automatic generation control. *IEEE Trans. Contr. Syst. Tech.*, **16**, 1192–1206.

Verhulst, F. (2005). *Methods and Applications of Singular Perturbations: Boundary Layers and Multiple Timescale Dynamics.* New York: Springer.

Verykios, X. and Luyben, W. L. (1978). Steady-state sensitivity and dynamics of reactor/distillation column system with recycle. *ISA Trans.*, **17**, 49–55.

Vinson, D. R. (2006). Air separation control technology. *Comput. Chem. Eng.*, **30**, 1436–1446.

Vora, N. P. (2000). *Nonlinear Model Reduction and Control of Multiple Time Scale Chemical Processes: Chemical Reaction Systems and Reactive Distillation Columns.* PhD thesis, University of Minnesota – Twin Cities.

Vora, N. P. and Daoutidis, P. (2001). Dynamics and control of an ethyl acetate reactive distillation column. *Ind. Eng. Chem. Res.*, **40**, 833–849.

Vora, N. P., Contou-Carrère, M. N., and Daoutidis, P. (2006). Model reduction of multiple time scale processes in non-standard singularly perturbed form. In A. N. Gorban, N. Kazantzis, I. G. Kevrekidis, H. C. Öttinger, and K. Theodoropoulos, eds., *Coarse Graining and Model Reduction Approaches for Multiscale Phenomena*, pp. 99–116. Berlin: Springer-Verlag.

Wang, K., Qian, Y., Yuan, Y., and Yao, P. (1998). Synthesis and optimization of heat integrated distillation systems using an improved genetic algorithm. *Comput. Chem. Eng.*, **23**, 125–136.

Wang, P. and McAvoy, T. J. (2001). Synthesis of plantwide control systems using a dynamic model and optimization. *Ind. Eng. Chem. Res.*, **40**, 5732—5742.

Wang, X. H., Li, Y. G., Hu, Y. D., and Wang, Y. L. (2008). Synthesis of heat-integrated complex distillation systems via Genetic Programming. *Comput. Chem. Eng.*, **32**, 1908–1917.

Wei-Zhong, A. and Xi-Gang, Y. (2009). A simulated annealing-based approach to the optimal synthesis of heat-integrated distillation sequences. *Comput. Chem. Eng.*, **33**, 199–212.

Westerberg, A. W. (2004). A retrospective on design and process synthesis. *Comput. Chem. Eng.*, **28**, 447–458.

Wheeler, C., West, K. N., Eckert, C. A., and Liotta, C. L. (2001). Ionic liquids as catalytic green solvents for nucleophilic displacement reactions. *Chem. Commun.*, 887–888.

Wu, K. and Yu, C. (1997). Operability for processes with recycles: interaction between design and operation with application to the Tennessee Eastman challenge process. *Ind. Eng. Chem. Res.*, **36**, 2239–2251.

Yang, Z. and Pan, W. (2005). Ionic liquids: green solvents for nonaqueous biocatalysis. *Enzyme Microb. Tech.*, **37**, 19–28.

Ydstie, B. E. (2002). Passivity based control via the second law. *Comput. Chem. Eng.*, **26**, 1037–1048.

Yee, T. F., Grossmann, I. E., and Kravanja, Z. (1990). Simultaneous optimization models for heat integration – I. Area and energy targeting and modeling of multi-stream exchangers. *Comput. Chem. Eng.*, **14**, 1151–1164.

Yeomans, H. and Grossmann, I. E. (1999). Nonlinear disjunctive programming models for the synthesis of heat integrated distillation sequences. *Comput. Chem. Eng.*, **23**, 1135–1151.

Yu, H., Fang, H., Yao, P., and Yuan, Y. (2000). A combined genetic algorithm/simulated annealing algorithm for large scale system energy integration. *Comput. Chem. Eng.*, **24**, 2023–2035.

Zaks, A. and Klibanov, A. M. (1985). Enzyme-catalyzed processes in organic solvents. *Proc. Natl. Acad. Sci. USA*, **82**, 3192–3196.

Zavala, V. M. and Biegler, L. T. (2009). The advanced-step NMPC controller: optimality, stability and robustness. *Automatica*, **45**, 86–93.

Zheng, A., Mahajanam, R. V., and Douglas, J. M. (1999). Hierarchical procedure for plantwide control system synthesis. *AIChE J.*, **45**, 1255–1265.

Zhu, G. Y. and Henson, M. A. (2002). Model predictive control of interconnected linear and nonlinear processes. *Ind. Eng. Chem. Res.*, **41**, 801–816.

Zhu, G. Y., Henson, M. A., and Ogunnaike, B. A. (2000). A hybrid model predictive control strategy for nonlinear plant-wide control. *J. Proc. Contr.*, **10**, 449–458.

Index